G. Voss and G. Ramos (Editors)

Chemistry of Crop Protection

Further Reading from Wiley-VCH

Franz Müller
Agrochemicals
Composition, Production, Toxicology, Applications
1999, ISBN 3-527-29852-5

Fred Whitford
The Complete Book of Pesticide Management
Science, Regulation, Stewardship, and Communication
2002, ISBN 0-471-40728-3

Martin Scheringer
Persistence and Spatial Range of Environmental Chemicals
New Ethical and Scientific Concepts for Risk Assessment
2002, ISBN 3-527-30527-0

Chemistry of Crop Protection

Progress and Prospects in Science and Regulation

Edited by
G. Voss and G. Ramos

WILEY-VCH

Editors:
Dr. Günther Voss and Dr. Gerardo Ramos
Syngenta AG
WRO-1060.1.12
4002 Basel
Switzerland

Library of Congress Card No.: Applied for

British Library Cataloguing-in-Publication Data:
A catalogue record for this book
is available from the British Library

Bibliographic information published by Die Deutsche Bibliothek
Die Deutsche Bibliothek lists this publication in the Deutsche Nationalbibliografie; detailed bibliographic data is available in the Internet at <http://dnb.ddb.de>.
ISBN 3-527-30540-8

Printed on acid-free paper.

Printing: Strauss Offsetdruck GmbH, Mörlenbach
Bookbinding: J. Schäffer GmbH & Co. KG, Grünstadt

Printed in the Federal Republic of Germany.

Committees of the 10th IUPAC International Congress on the Chemistry of Crop Protection

Executive Committee

Dr. Jost Harr (Chairman)
Dr. Bernard Donzel
Mr. Michel Marti
Dr. Günther Voss

Organising Committee

Dr. Bernard Donzel (Chairman)
Dr. Jost Harr
Mr. Leonhard Loew
Mrs. S. Marconi
Mr. Michel Marti
Dr. Manfred Mirbach
Mrs. Saskia Riat
Dr. Günther Voss
Mrs. Josette Zwicky

Advisory Board

Prof. N. Amrhein
Mr. John R. Finney
Prof. W. Klein
Dr. J. Miyamoto

Scientific Committee

Dr. Günther Voss (Chairman)
Prof. Lars Bergström, Sweden
Prof. Peter Böger, Germany
Dr. Bernard Donzel, Switzerland
Prof. René Feyereisen, France
Prof. Ulrich Gisi, Switzerland
Dr. Hans-Walter Haesslin, Switzerland
Dr. Denis Hamilton, Australia
Dr. Jost Harr, Switzerland
Dr. Kenneth D. Racke, U.S.A.
Dr. Gerardo Ramos, Switzerland
Dr. Ulrich Schirmer, Germany
Prof. Jürg Stetter, Germany
Mr. Scott S. Tann, U.S.A.
Dr. Michael Uihlein, Germany
Prof. Isamu Yamaguchi, Japan

Preface

In agriculture, the conditions and rules have changed in an unprecedented way over recent years. Crop protection is questioned as never before. Not only have the rules changed, but alternative techniques are increasingly challenging the chemical approach.
Chemical solutions are now extremely sophisticated and continue to be major contributors to sufficient healthy food supplies. However, advances here hardly ever make the headlines, while smaller achievements in other technologies are widely applauded. We have to make the point again and again: without chemically protecting our agricultural commodities, there would not be enough to eat. Farmers would also be unable to make a living producing food. Our congress will again help people to understand chemical crop protection's vital role in securing healthy food supplies for a growing global population.

From the Welcome Address of Jost Harr, Chairman of the Congress

From August 4 to 9, 2002, the modern Convention Center Basle, Switzerland, hosted the 10th IUPAC International Congress on the Chemistry of Crop Protection (until 1998 known under the name IUPAC International Congress of Pesticide Chemistry). The congress takes place every 4 years, and is one of the leading scientific events in the crop protection diary. It offers insight into the latest results and developments of all relevant sciences, and provides an established forum for interactions among representatives of different countries, disciplines and institutions.
Of the past nine editions of this congress, only the first one (Vienna 1967) was not sponsored by the International Union of Pure and Applied Chemistry (IUPAC). The following events in Tel Aviv (1971), Helsinki (1974), Zurich (1978), Kyoto (1982), Ottawa (1986), Hamburg (1990), Washington D,.C. (1994) and London (1998) were all organised under the auspices of IUPAC. They have set high standards for the organisers of the 10th congress, both in their scientific content and their side events.
The intense and demanding scientific programme of "IUPAC 2002" was designed to address general and specific aspects of crop protection sciences, technology and regulation. More than 20 poster sessions allowed for a maximum of interactions and for a stimulating exchange of results and ideas, while invited lectures, selected research papers and workshops provided the opportunity for in-depth reviews and presentations of more comprehensive and controversial themes.

The articles of this volume are derived from almost all of the invited lectures given in the plenary and the six main topic sessions covering chemistry, biochemistry and molecular biology of (1) disease, (2) weed and (3) insect control, (4) formulation and application, (5) environmental fate and safety as well as (6) residues and consumer safety. They reflect the broad scope of the congress and exemplify progress and prospects in crop protection science and regulation.

On behalf of the Executive, Organising and Scientific Committees we express our gratitude to all persons and institutions who worked for and supported this Congress. As editors of this book we are especially indebted to our authors who submitted their manuscripts in time as a prerequiste for an early production of the book.

We dedicate this book to the memory of our colleague Dr. Cheah Uan Boh from the Strategic Resources Research Centre of the Malaysian Agricultural Research and Development Institute. When delivering his invited lecture (see page xyz) he suffered a heart-attack from which he did not recover. He quietly passed away on August 7, 2002, in the hospital of Basle City.

Günther Voss
Gerardo Ramos

Basle, Switzerland, September 2002

Contents

Environmental Fate and Safety

Residues and Consumer Safety

Prologue

Reacting to Change - Past Success Stories and Future Opportunities for the Agricultural Enterprise

John R. Abernathy

Dean and Professor, College of Agricultural Sciences and Natural Resources, Texas Tech University, Lubbock, Texas USA 79409-2123, E-mail: j.abernathy@ttu.edu

"Be sure to take change by the hand before it takes you by the throat."
Sir Winston Churchill

We must understand the drivers of change which impact our industry and which will continue to influence future opportunities. Understanding and predicting change will be very valuable as we view the future and make decisions regarding opportunities in the agricultural enterprise. In the beginning, early man's days were consumed with finding food and securing protection against the elements. Our great success stories began with the cultivation of grains and the ability to feed humans and animals. The development of controlling fire for cooking propelled the need for additional food sources. Through the scientific genius of chemists, plant and soil scientists, plant breeders, weed scientists, plant pathologists, and entomologists; synthetic fertilizers, pest management technology, and genetically adopted species were developed. These significant contributions allowed the agricultural enterprise to produce the food and fiber to feed and clothe today's 5.9 billion humans. Engineering and economic principles provided the machinery to till, plant, cultivate, irrigate, harvest, market, transport, and store this huge food supply. Most participants of this congress are products of universities with strong programs in agriculture, chemistry, biological sciences, and engineering.

The scientific genius supporting and developing the agricultural enterprise has been extremely successful. Today's society now takes for granted that a stable food supply and clothing will always be available at a modest cost. Net income spent on food ranges from 8.5% to 70% today with most of us spending less than 15% of our net income on food. This success now allows many humans to become more interested in the aesthetic and sustainable value of the landscape, the quality and safety of their food, and the systems utilized to produce the food. Today's agricultural enterprise also receives additional attention from sensationalized electronic and written media stories.

For our future successful achievements, we will need to understand and study in detail several assumptions, which will be fundamental to the future agricultural enterprise:

- Human population on earth will increase.
- Human population will also become more affluent - thus requiring additional food and fiber.
- Food and fiber will be produced on this planet somewhere by someone.
- Fewer farmers will produce the food; larger more cost efficient farms/livestock feeding facilities/dairies will evolve.
- Food production will become more global and intertwined.
- The agricultural sector will have less influence on international and domestic policies and politics.
- The agricultural enterprise will become more consumer driven.
- The consumer has limited understanding of science and risk.
- The consumer is easily influenced by the electronic and written media.
- There will always be insect, disease, and weed pests and fertility needs; new technology will be required.
- Conservation of natural resources and the environment will become more important.
- Regulations on technology use will become more stringent.

So how do we build these basic assumptions into agricultural enterprise opportunities? It is my opinion that outstanding opportunities in agriculture will continue to exist and even be greater, if we approach this enterprise in a more strategic manner. As an industry collectively or individually, we need to spend considerable time planning and strategizing about ways to understand and influence future opportunities. We will need to call on expertise that perhaps we have not utilized previously. Strategic business planning will need to include more communications, educational efforts, consumer research, coalition building, and political interaction in addition to the traditional activities. Perhaps the following areas will serve as key points for future agricultural enterprise marketing.

Understanding the Consumer and Their Needs

The influence of the end user consumer on the agricultural enterprise continues to increase across the globe. It is critical that we organize mechanisms in which we can gain knowledge of consumer needs and interest and involve their input in our decision making process. Perception by the consumer is easily influenced and in the real world, "perception equals reality." We must use educational and public relations techniques to showcase the scientific excellence of our technologies especially in the areas of environment, health and risk, and molecular biology. We should partner with the medical community and involve them in the educational process. The agricultural enterprise must become much better "spin doctors."

We need to develop a certification system for all inputs, products and processes involved in a food system. Certification of tolerance levels of pesticides in all stages of a food product is needed. Development and utilization of a sophisticated tracking system of all ingredients in a food product will also become increasingly important.

Political Influence

The global agricultural enterprise is increasingly subject to political influence. Each country has their own complex political interactions and the global political game is even more complicated. Regardless, the sectors of the agricultural enterprise must collectively or in coalitions discuss and plan ways in which greater voice and political influence can be organized to add stability and profitability to agricultural activities.

Food and Fiber Production and Security

The farmer has historically utilized the products of our technological development. The farmer has truly been the integrator of technologies on the landscape. As we look to the future, we must assess who will be the manager of the landscape. Will it be the traditional farmer or will it evolve to a management/consultant operation utilizing contract farmer operations to cultivate the landscape? These questions have great relevance to our industry as future business is conducted. The decision point for use of agricultural chemicals and seed may be much different with many fewer individuals involved in the process.

We must study and rationalize where specific crops will be grown on a global basis. Global climate change, water and land availability, economics, and political maneuvering may dramatically influence the production areas of many crops. New technology development in planting, spraying, tillage, and harvesting equipment may impact crop production systems. The use of GPS-GIS (Global Positioning System-Geographic Information System) systems has the potential of greatly enhancing our management and monitoring systems. We must also be very aware of changing cultural practices that dramatically change the use of agricultural chemicals. An example is the practice of conservation tillage, which continues to be used on increasing amounts of land on a global basis.

September 11, 2001, changed this planet in many ways. Securing a stable food and fiber supply has taken on many new dimensions. Ironically, this threat also creates many new opportunities in surveillance, rapid assay, control, and remediation of animal and plant pathogens.

In summary, the agricultural enterprise faces many challenges; however, if we strategize and plan and work together, the opportunities are greater than ever for our genius in the development of new technologies to be used in feeding and clothing the projected 9.0 billion humans on this planet by 2050.

Plenary Lectures

01 Crop Protection in the New Millenium

Michael Pragnell

Syngenta International AG, Schwarzwaldallee 215, CH-4002, Switzerland, E-mail: michael.pragnell@syngenta.com

1 Introduction

Agribusiness and the crop protection industry are undergoing change on a scale and with an impact not seen before in this industry. These changes and their implications for crop protection will be explored in this paper, beginning first with an overview of the global environment, then addressing agribusiness and finally considering the pressures and responses demanded of the crop protection chemicals enterprise. This sets the landscape within which innovation in crop protection chemistry must make its impact by harnessing the creativity of scientists exploiting exciting developments in technology.

2 Global Outlook

Global financial markets are currently under pressure following a downturn in the USA and many other major economies having been in recession for all or part of the past year. The situation is made worse by global security fears as a result of the dreadful events of September 11th 2 001, and conflicts in a variety of regions, most notably the Middle East. Such uncertainty not only dents business confidence but it can prompt nations to become short term in their policy outlook, placing domestic popularity before international cooperation, which often leads to trade frictions.

Information meanwhile flows across borders with increasing ease. The rapid and dramatic impact of the Internet extends to every facet of life and was unimaginable to most people just ten years ago. This information explosion has had a further, specific effect. By making it easier for people to access information, share experiences and organise themselves, consumers have gained significantly greater influence. This is often channelled through formal "Non-governmental Organisations" (NGOs), which have become important stakeholders for every enterprise. These changes are profound and demand new ways of doing business, new priorities and new business models.

3 Agribusiness Drivers

Let us now turn to agribusiness specifically and look at some of the most important drivers for this sector.

The demand for food is driven not only by increasing population but also by increasing prosperity. Longer life expectancy is a key driver of population growth worldwide although it is, of course, in developing countries where this will have the biggest impact, with forecast growth of 2 billion people in the next 20 years. It is also in these developing economies where the potential for faster economic growth exists. This is expected to drive greater meat consumption, which, because of the calorific inefficiency of meat production, will drive the need for even greater crop production. It has been estimated that the world calorie consumption will double over the next 25 years – a much greater increase than that of the global population in the same period.

This projection is all the more remarkable when you consider that the amount of available arable land has been and will remain essentially constant. Crop protection chemicals have made a significant contribution over the past four decades or so in feeding the world, reducing mortality and increasing the availability of fresh, healthy food. The key role played by crop protection chemicals will remain. They will continue to play a vital part in securing sustainable agriculture, as it continues to contribute to improved crop yields. This is the only solution to the declining number of farm hectares per capita and the even faster decline in the farm hectares per calorie consumed.

This underpins a fundamental need to maintain a healthy crop protection enterprise supporting sustainable agriculture. However, there are several other challenges to the industry.

Crop prices have always been volatile, driven by such unpredictable events as the weather and pest pressure as well as the global economy. At times of high crop prices, a higher grower investment in inputs is amply rewarded by returns from the higher resulting yield. However, in the late 1990s prices of the major crops reached an historic low (Figure 1) and have repeatedly defied forecasts of recovery. The picture masks a 1.3% decline in real terms in the prices of field crops since 1973, thanks to the contribution of modern agricultural practices. So we are victims of our success. The first message for research, therefore, is that we must ensure that our new discoveries provide cost-effective benefits at today's crop prices, and perhaps at lower prices still.

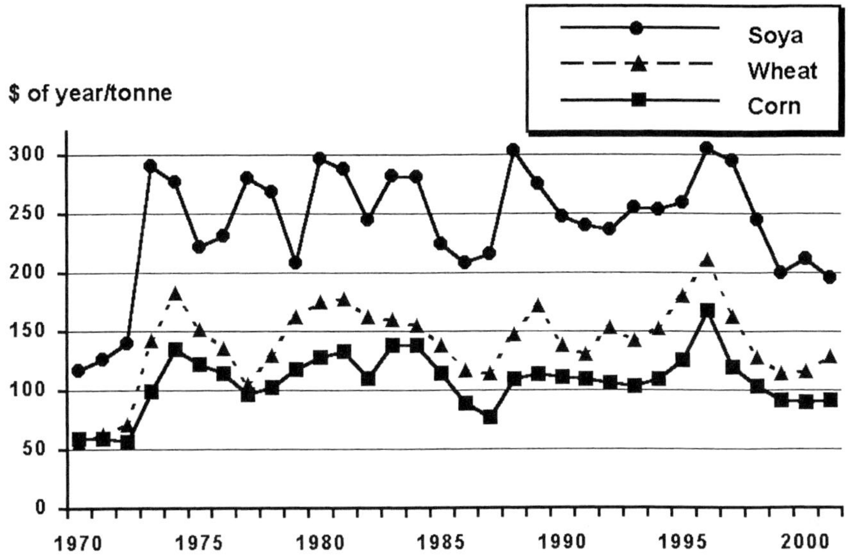

Figure 1. Commodity crop prices in dollars of the year. (Source : IMF).

Another significant factor affecting farm economics is farm subsidies. However, there are significant pressures to reform farm subsidy, evidenced by the debate over the US Farm Bill and EU CAP reform, and this will impact a huge proportion of the agribusiness market. The arguments favouring reform include sustainability and the benefits of free trade, although these arguments are being resisted by an equally strong lobby to preserve the farm economy, not least in recognition of the essential role that agriculture plays in managing and sustaining the landscape.

Regulatory legislation also has a direct impact on agribusiness and the requirements of regulators are becoming ever more stringent as today's empowered consumer demands ever higher standards. The critical importance of ensuring consumer and food safety through effective regulatory processes has also been heightened by the food scares in Europe, which have created consumer distrust in the regulatory framework.

Care of the environment has always been important in agriculture, and the regulatory framework to encourage sustainable agriculture is taking shape. This is overlaid with regulatory initiatives to remove some older, less environmentally friendly compounds, from sale. We are currently faced with the prospect of increasing tension between the strong push for free trade and ever-growing consumer concerns. The net result of these trends to more stringent regulation to protect the consumer is a greater challenge for research and an escalating cost for agriculture and the food chain.

Nowhere is the impact of regulation more evident in agriculture today than in biotechnology. Those promoting the benefits of biotechnology have faced objections on food safety, environmental and even ethical grounds. The drama continues to unfold essentially on a regional basis, with the Americas leading the way. The US already plants some 90m biotech acres per year and in Latin America, Argentina has been enthusiastically adopting GM soya. A decision with respect to the growing of GM crops is awaited in Brazil. Progress is slower in the rest of the world, with significant consumer resistance in Europe that has led to a regulatory log-jam and, recently, restrictive labelling requirements. The situation is similar in Japan, although there are some brighter prospects in other parts of Asia, notably China and India.

The influence of consumers in today's information intensive society has already been mentioned. Their influence is not only with governments and regulators but also throughout the food value chain. Retailers are quick to respond to informed consumer demand for fresher food, healthier food, residue-free fruit and vegetables, non-GM produce, paper from sustainable forests, free-range chickens and so on. The retailers pass these needs through to their suppliers and onwards up the value chain. As a result, growers are increasingly required to observe particular crop management protocols. The demands on the agribusiness industry for appropriate products and information about those products continues to increase.

While this appears to be another layer of regulation over and above that required by the regulatory authorities, we cannot ignore such demands from the end-consumers of the food, feed and fibre that the crop protection industry aims to protect. Consumer acceptance is critical.

All of these factors combine so forcefully in today's rapidly changing environment that we have to embrace a new industry structure. This transformation began over 10 years ago.

4. Crop Protection Industry Developments

Let us now look at how the crop protection industry has developed to the shape it is today.

In each significant crop protection market sector, the number of products has at least doubled in the twenty years from 1975 to 1995. The marketing of mixture products in certain crops, such as wheat, further increases market crowding. The major crop protection chemical companies have products in nearly every major sector and the intensification of competition not only puts pressure on pricing but also means that so-called "me-too" products cannot justify their development costs – leading to the clear conclusion that step-changes in effect are required to achieve substantial market penetration.

The net result has been a slow-down in market growth over the last thirty years (see Figure 2). The last five years have seen a slight contraction. Industry analysts are forecasting a mild recovery over the next five years, with a compound growth of about +2.5% per annum forecast through to 2006.

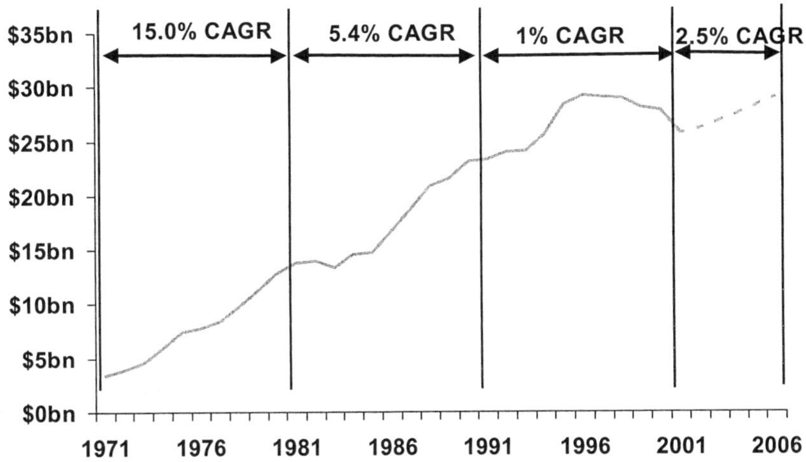

Figure 2. Slowing market growth: Crop Protection Chemical Market (Source: Philips McDougall).

The level of growth in crop protection chemicals is unlikely to support the overall investment in R&D made by the industry in the past decade. We have estimated that for each dollar invested in R&D by the industry as a whole in 1971, just over seven dollars was returned fifteen years later. In contrast, just four dollars was returned in 1995 from a dollar investment made in 1980. This is only a proxy measure of R&D returns and far from accurate, but the trend is clear.

Fortunately not every company has an average performance and the leading companies are achieving acceptable returns on R&D today, but the downward pressure remains. The winners will have to work hard to beat the trend, but there are exciting opportunities for new technology to accelerate innovation. However, these new technologies also incur further cost, perhaps most notably biotechnology.

Despite the consumer resistance in Europe and despite the complex, high volume global trade in crops, biotechnology has made an enormous impact on agriculture very quickly. It is the fastest agricultural revolution ever. The major significance of this rapid penetration in the context of crop protection chemicals is the ability of biotechnology to substitute genes for chemicals. There has been a decline in sales of selective herbicides in some sectors and an impact on insecticides in some crops. More is likely to come.

The commercialisation of biotechnology has brought a substantial investment by the industry in seeds, as delivery vehicle for these new traits. The investment brings the promise of growth and increased returns in a traditionally low margin industry.

The crowded market, the slowing market growth, the declining returns on R&D and the impact of biotechnology have all contributed to drive the industry to consolidate. In just ten years, today's six major companies have been created through the merger or acquisition of sixteen companies that existed in 1992. The savings made in R&D through this consolidation have considerably reduced industry R&D expenditure.

5. Innovation Opportunities

Having explained the current environment, the remainder of this paper will focus on viewing the market in the light of the two revolutions now underway : the need to harness the explosion in technology and the need to respond as never before to the demanding needs of consumers.

The technology explosion has been gathering momentum for several years – driven partly by genomics and partly by the pharmaceutical industry seeking technology to solve research challenges similar to crop protection. Many of these technologies require considerable investment and so the skill is to harness technology opportunity selectively to deliver the promise that exists, and to do so cost-effectively.

Arguably the most exciting breakthrough is genomics. It has applications in several fields of interest to agribusiness – so investment here brings many options. This is critical, because possession of a complete genomics capability is not inexpensive. However, genomics does bring the promise of a step-change in understanding and innovative capability. Not surprisingly, industry and academic research quickly realised the potential impact. Genomics has huge potential to be mined for new products: it is good to see a session at this conference on this topic.

With respect to the second "revolution", the research community must respond to the greater focus on consumer needs. In this context, perhaps the most tangible impact is the increasingly stringent technical hurdles that must be cleared by new products. This extends well beyond simple product efficacy, to embrace parameters such as suitability for use in integrated pest and crop management programmes and applicability within resistance management programmes. Retailers, as indicated earlier, are specifying crop management protocols to their suppliers, reflecting their customer's demand for high quality food. Such a market will therefore only embrace new active ingredients or products if they fill gaps in the market, or are better than existing products. This last point requires that new inventions be constantly compared with established products, and substantial benefits be shown.

Excellence in regulatory science allows agribusiness to respond to requirements for effective products that are safe to the environment, to users and to consumers. Regulatory science is the combination of health assessment and toxicology with environmental safety. It is these fields of science that enable us to meet those stringent technical requirements outlined above: it is increasingly important to be thinking about the regulatory end-points earlier and earlier in the invention process.

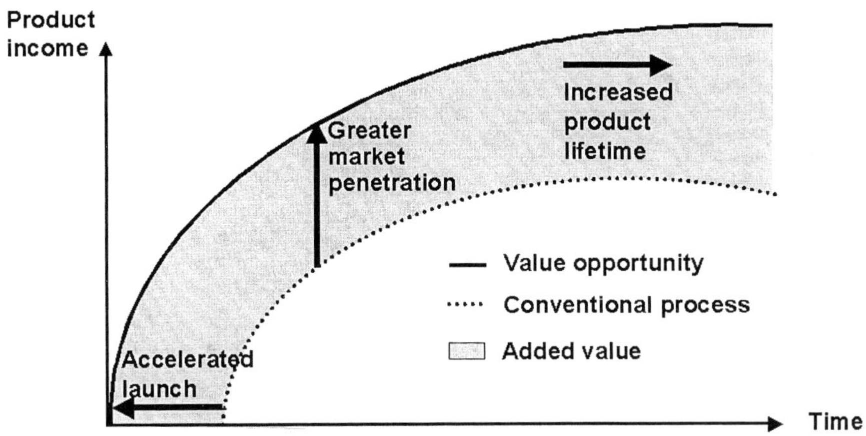

Figure 3. Schematic showing the added value of reduced-risk products.

Not surprisingly, such customer focus is rewarded in the market-place (see Figure 3). High quality products mean "reduced risk" registrations and earlier sales, because of the faster registration process. Their broad applicability and lack of use restrictions mean more market penetration. The product also enjoys a longer lifetime, because it is harder to displace from the market.

Chemicals will remain a vital component of crop protection into the foreseeable future. Although the overall market may have recently declined and is quite crowded, new technical demands continue to arise because obsolete products are withdrawn, because of resistance development, or because of a need to complement a biotechnology effect. Successful companies meeting these challenges can win market share and increase margins with the right products. There are many strategies to meeting these challenges, but there are compelling arguments pointing to the attractiveness of new modes of action.

Why are new modes of action so important? Historically, they have introduced those differentiated effects demanded by the market-place, met the stringent technical challenges required and displaced older products. Each new mode of action often brings families of products, tailored to particular needs, which means substantial reward for the innovator. Figure 4 illustrates this using data for the growth of sales of product classes that share a common mode of action. Note the rapid increase in sales in nearly every case, as the new biological effect associated with each new mode of action is appreciated by the grower.

Most importantly, the chart indicates how two classes with new modes of action, the neonicotinoids and strobilurins, both introduced in the last decade, continue to stimulate the market, suggesting that a focus on new modes of action remains an excellent recipe for success.

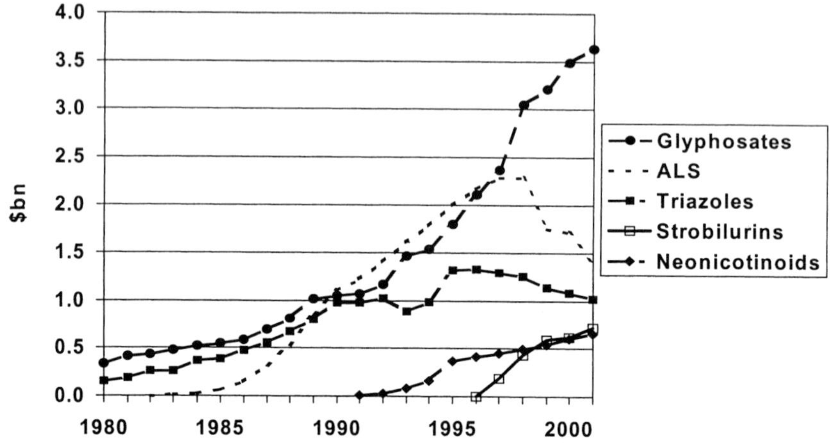

Figure 4. Impact of new modes of action: Global sales in $ of the year of selected chemical classes sharing a common mode of action. (Source : Philips McDougall)

New modes of action are not enough to ensure successful products. Properties must be fine-tuned in a wide range of scientific fields requiring a broad range of skills and technologies. Informatics, for example, is a revolutionary technology of critical importance. Progress here enables scientists to assimilate and draw conclusions from very large complex datasets in a way unheard of a decade ago. Each of these technologies requires investment of money and skilled human resources. To be successful in this millennium, crop protection businesses will require the full set, together with the skills and experience for wise decision-making as these technologies are harnessed and integrated.

Managing the life-cycle of existing active ingredients will grow in importance as it becomes harder to find new active ingredients and as the need to squeeze the last drop of value from our inventions becomes an imperative. Success in product life-cycle management means we can add as much value as a new active ingredient through the generation of new products. In its most simple form, research might be directed to finding ways of improving performance, which can be in terms of raw activity, or broadening the spectrum, or perhaps by using combination products with other active ingredients. It can also mean finding ways to increase sustainability by reducing the impact on non-target organisms or reducing the amount and mobility of material which escapes into the environment. Whilst biology is critical, we must also be looking for ways to drive down manufacturing costs and use our science to hone the cutting edge of technical marketing and to support global registrations.

A whole range of formulation and presentation technologies are now available and continue to deliver new effects. Seed treatments are an outstanding example of exploiting the potential of an active ingredient. They bring benefits beyond excellent biological effects, in terms of reduced environmental load.

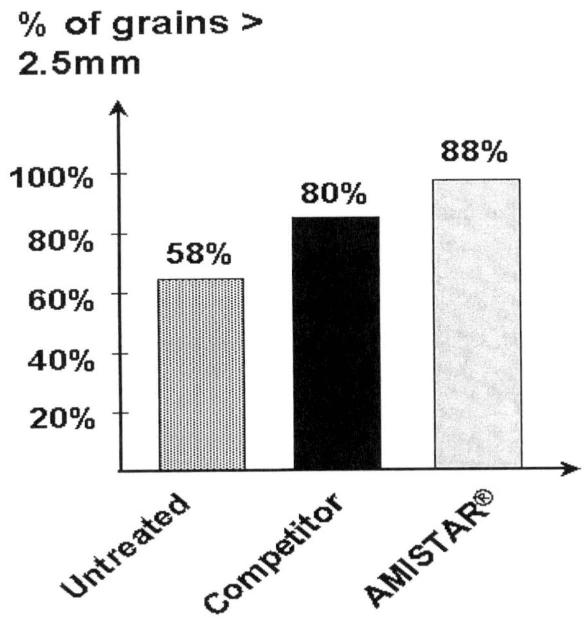

Figure 5. Percentage of barley showing a grain size of >2.5mm in a trial with different fungicide treatments.

Finally, changing customer needs offers opportunities for tailored solutions for new and existing customers. Those involved in fungicide invention might think that customers were most interested in fungal control and the benefit this brings in terms of yield and grain weights – as such measurements have always been done by our industry. However, in France, the maltsters (who make the malt for beer production) are mainly interested in grain size and grain germination. Figure 5 shows the effect of good fungal control on barley grain size. In trials, we see an amazing 30% increase in the numbers of grains that are >2.5mm in diameter, which is the critical threshold for quality malting grain. The industry can share in the value of this benefit with the maltsters

This means that the industry has to consider new customers and think beyond just the control of the pest, to capture value for effects that might not have been measured before. Syngenta has many examples of adding value through crop protection chemicals for downstream customers, who might not previously have been regarded as customers. With millers and bakers, we have shared in the effects of our products on flour yield per grain and protein content, which directly influences the bread making quality of the flour. Animal feed is another field where the industry can add value through the effects of crop protection on grain starch content.

In conclusion, products and solutions can today be assembled that are targeted to the particular needs of agriculture. The banana plantation manager in Honduras has different needs from the corn grower in Illinois, who incidentally needs a different product offer compared to the corn grower just a short distance north - west in South Dakota. This requires new marketing thinking and new technical support from the science community.

6 Conclusion: The Innovation Challenge

Notwithstanding the current pace of change, chemicals will remain essential to agriculture. There remain unmet needs, open to the judicious application of new technologies. Regulatory and food chain pressures add to technical demands. As recent successful launches of compounds with new modes of action have demonstrated, the market remains receptive to new efforts. In addition, maximising the value from existing products through "product life-cycle management" is an imperative.

We have to think differently from the past. The new environment demands it. The technology explosion demands it. And above all the consumer - the fully engaged consumer - demands it.

The answer - and hence the challenge – is value driven innovation: innovation driven by the changing environment, technological opportunity, and the consumer. Technology push will no longer suffice: Innovation must be driven by consumer pull.

This is a unique time for crop protection research. We must deliver a step change in our innovation capability. With that, I wish the participants a stimulating, productive and most of all an innovative conference.

Acknowledgements

This paper was prepared in collaboration with Dr. Martin Battersby with the assistance of Glen Slade and Michael Stopford. Their contribution was invaluable. We warmly thank colleagues throughout Syngenta who provided extensive material.

02 Homologs of Amino Acids and Explorations into the Worlds of β- and γ-Peptides

Dieter Seebach

Laboratorium für Organische Chemie der Eidgenösischen Technischen Hochschule,
ETH Hönggerberg, Wolfgang-Pauli-Strasse 10, HCI, CH-8093 Zürich, Switzerland,
E-mail: seebach@org.chem.ethz.ch

1 Introduction - From PHB to β-Peptides

In 1982 we have first become aware of the existence of the biopolymer PHB, consisting of (*R*)-3-hydroxybutanoates (HB), which we used as a source of readily available chiral building block for syntheses of natural products, such as elaiophylidin [1]. We then learned that short-chain PHB (*ca.* 150 HB units) is found in small amounts in all living cells and tissues, where it has been looked for (Figure 1), including plants, for instance

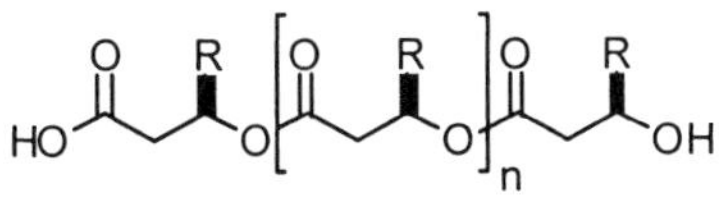

Microbial storage material (*s*-PHB), R = Me, n *ca.* 10^4
BIOPOL®, R = Me/Et, n *ca.* 10^4
PHB in genetically modified plants , R = Me, n *ca.* 10^4
cPHB (*CaPPi complex; Ca-channel)*, R = Me, n *ca.* 150

E. Coli: inner cell membrane when genetically competent (caused by Ca^{2+})

Eukaryotic organisms: highest concentration in *mitochondria*
Ca^{2+} concentration mitochondrion/cytosol/extracellular
4 : 1 : 10^3

Human blood serum: 5-15 μg/ml, mainly bound to *albumin* which is **the** transport system for lipids and which is also binding *ca.* 40% of the serum Ca^{2+} content

Figure 1. The ubiquitous high- and low-molecular-weight biopolymer PHB is a microbial storage material (carbon and reductase equivalents, *cf.* Figure 2) and is found as part of ion-transporting systems in procaryotic and eucaryotic organisms, respectively [2].

spinach or gorse. Another biopolymer with the same backbone structure is polymalic acid (PMA, up to 500 units long), found in molds and fungi in certain development stages of the organism and in certain compartments of the cells, in concentrations as high as 150g/l. PHB occurs in two varieties, as microbial storage material (sPHB, molecular weight up to 10^6 Da) and as short-chain, so-called complexing PHB (cPHB, chain lengths *ca.* 20-150 HB units). The metabolism (biosynthesis and degradation) of sPHB has been fully elucidated (Figure 2), while the biosynthetic origin of cPHB is unknown.

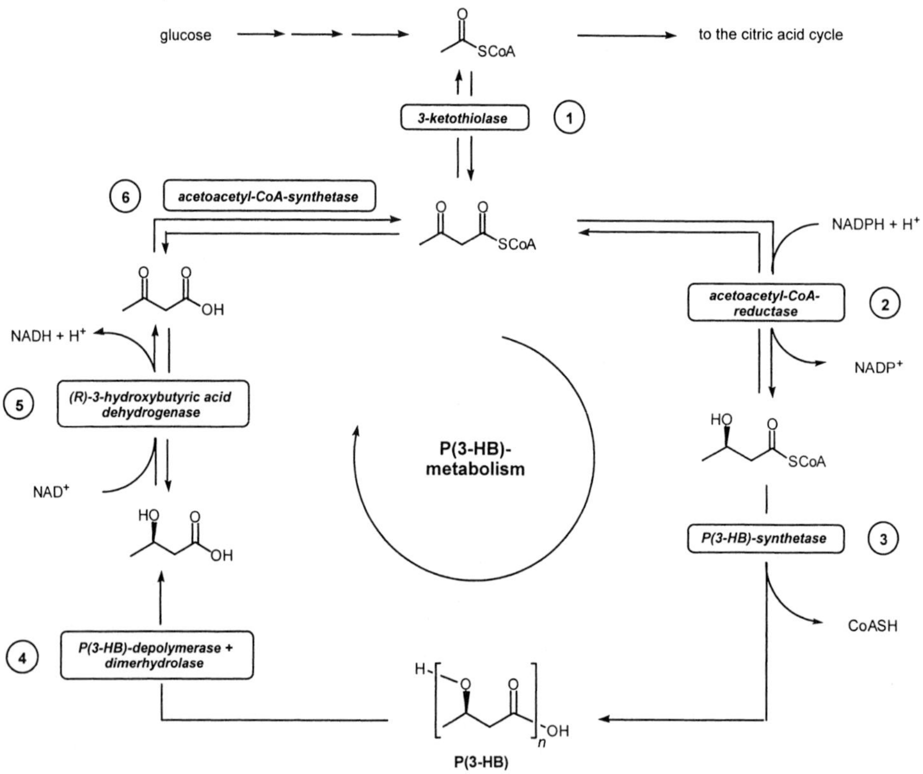

Figure 2. Biosynthesis and biodegradation of high-molecular-weight storage PHB (sPHB) by the most common mechanism. Microorganisms store PHB (acetate and NADH precursor!) as a survival measure, when glucose or other acetate precursors are available in the environment and when, at the same time, essential conditions for growth and multiplication are not met (*cf.* limitation of nitrogen, oxygen, phosphorous, trace element(s)) [2].

The function of PMA in DNA-polymerase regulation of the slime mold *Physarum polycephalum* has also been assigned (Figure 3). As chemists, we have been engaged in analyzing, synthesizing and studying structural and chemical properties of malic-acid and HB oligomers (OHBs) to help understand the various functions of these simple biopolymers with the (O-CHR-CH_2-CO) backbone. Many of the results obtained are collected in four review articles and a most recent paper [2] in which the seminal contributions by polymer chemists, biochemists, biologists, environmental scientists and medical experts are referred to (a few names are *Dawes, Doi, Holler, Lemoign, Lenz, Marchessault, Reusch, Schlegel, Sinkey, Steinbüchel, Witholt*).

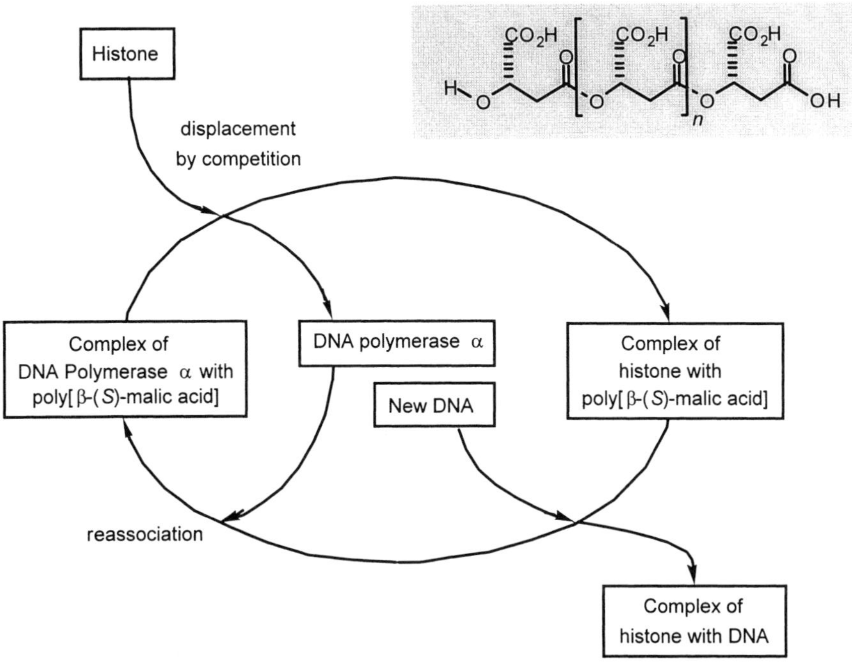

Figure 3. Proposed function of polymalate (PMA) in the reproduction of certain molds and fungi according to *E. Holler* (see references cited in [2]). Like DNA, PMA is an *anionic* polyelectrolyte, while the histones and DNA polymerase are proteins with high content of amino-acid residues carrying positively changed, *cationic* side chains (histidine, lysine, arginine).

In order to find out whether there is a preferred backbone conformation ("secondary structure") of OHBs and PHB in homogeneous solution or in phospholipid bilayers ("two-dimensional solvent") we have recently prepared derivatives labeled with various side chains, with isotopes, and with fluorescent groups for NMR and FRET measurements, to find that the polyester backbone is extremely flexible [3]; only in the crystal structures of cyclic HB derivatives (oligolides) did we see distinct folding patterns of the oligoester backbone [4] (Figure 4). One of these patterns resembled the 2_1 helix which had been identified in stretched fibers of sPHB, the other one was a 3_1 helix, the surface of which is covered with methyl groups. In the 3_1 helix all carbonyl groups are arranged parallel to the helix axis, with close proximity of the carbonyl oxygens in residue i and the chain-bound oxygens in residue (i+2) (Figure 4). Thus, it occurred to us that replacement of the latter by an NH should lead to hydrogen-bond formation and stabilization of the helix, so that one might to be able to observe it in solution. This was indeed the case, and our subsequent investigation of β-peptides has led to almost 70 papers since 1996, some of which are included in the list of references (with titles) [5-49].

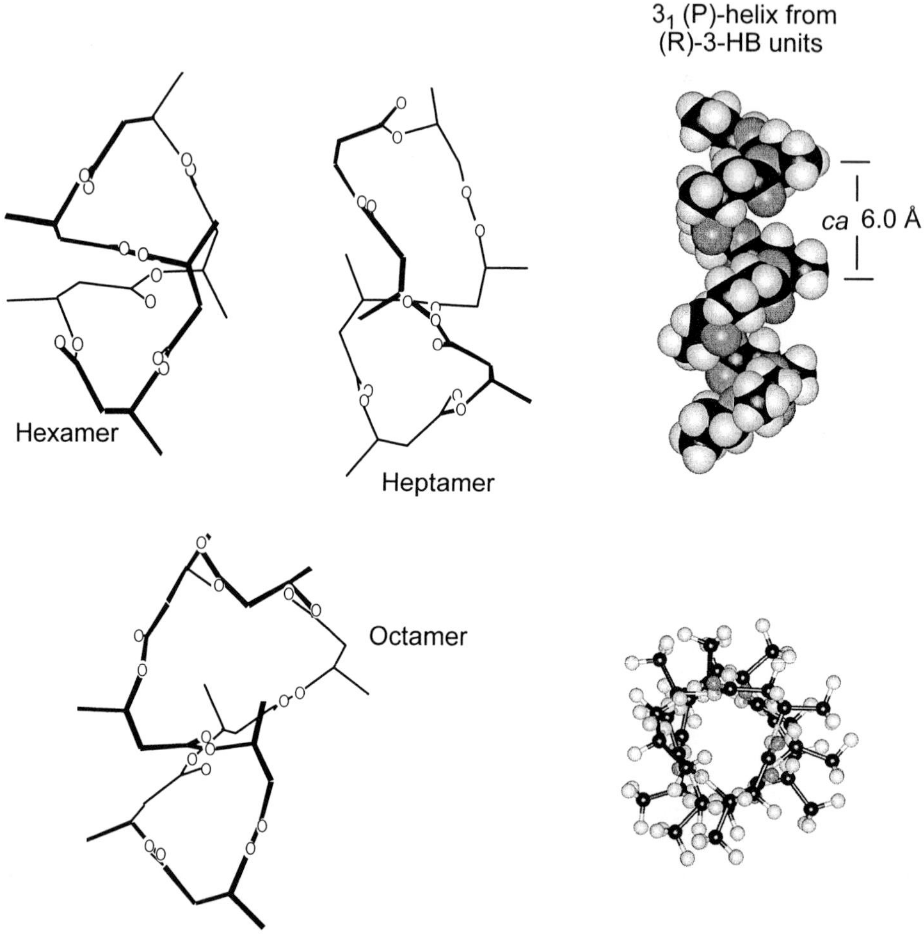

Figure 4. Folded and twisted conformations of (*R*)3-hydroxybutanoate oligolides (left) containing one (heptamer, octamer) or two (hexamer) single turns of a right-handed 3_1-helix, a model of which is shown on the right side in views from the side and along the helix axis. The helix is covered with methyl groups and has a dipole moment resulting from the unidirectional arrangement of the C=O bonds parallel to the helix axis [4].

2 β-and γ-Peptides – a Different World

General *formulae* of natural peptides (α) and of the unnatural β- and γ-peptides consisting of homologated α-amino acids are shown in Figure 5, in which an outline of the preparation of β- and γ- amino acids is also given. As can be seen, they are available by classical organic name reactions. The β^3- [50] and the γ^4-amino acid [51] derivatives are prepared from the natural α-amino acid precursors, while for the β^2-analogs enantioselective routes to substituted succinic acid esters [34,52] or enantioselective *Mannich* reactions [12] have to be employed.

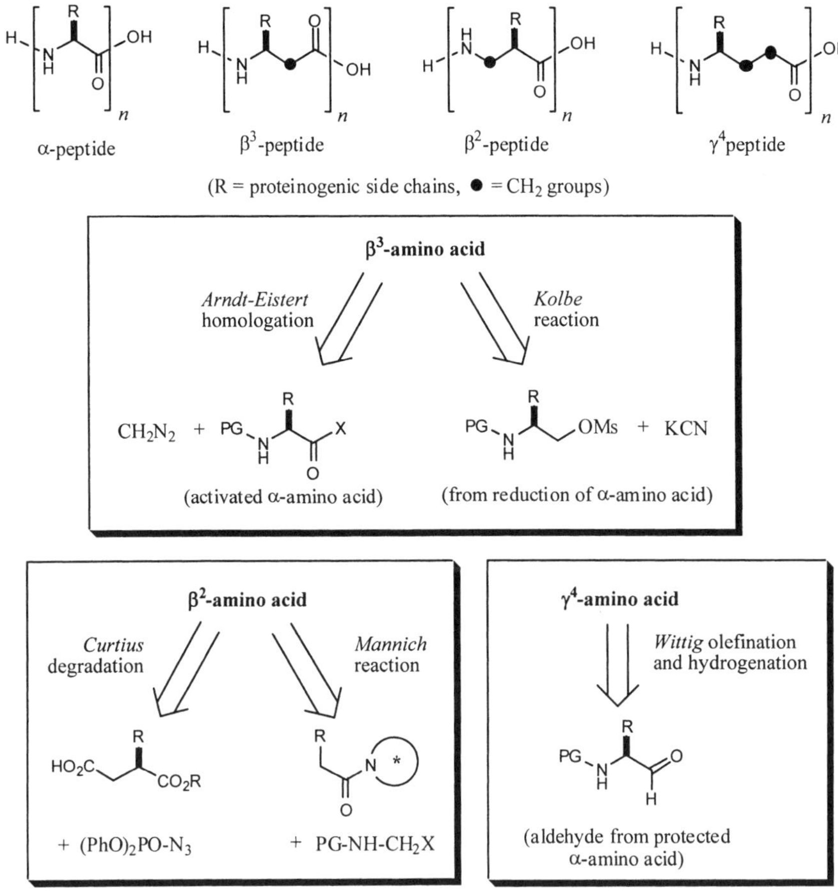

Figure 5. General structural *formulae* of the α-, β^3-, β^2- and γ^4-peptides derived from proteinogenic L-amino acids (top). Retrosynthetic presentation for the preparation of enantiopure building blocks required for the synthesis of β- and γ-peptides (bottom). – The oligomers (with n > 10 for α-, n = 6 for β- and n = 4 for γ-peptides) form helices in methanol solution; when we go from α- to β^3-to β^2- to γ^4-peptides the helicity is *P* (right-handed), *M* (left-handed), *P*, and *P*, respectively [12].

To our surprise, no oligomers of β-amino acids with the proteinogenic side chains had been synthesized up until 1995. Only polymers of high molecular weight (poly-β-lactams, mainly by industrial groups) and β-peptidic derivatives of the one and only proteinogenic β-amino acids, aspartic acid and asparagine (by a Spanish group [53]) had been described, and the effect of incorporation of single β-amino acid residues into α-peptidic chains had been studied (by medicinal chemists).

The assembly of the β- and γ-amino-acid building blocks to peptidic chains was achieved by simply using the established methods of peptide synthesis - in solution [6], on solid phase [11], or in a synthesizer machine [39]; also, the so-called "native ligation" can be applied with β-peptides [54]. Furthermore, the methods of analyzing and studying the structures of α-peptides and natural proteins can mostly be applied to β-peptides as well (the same is true for γ-peptides [51,55-60]). These methods are CD [35,37] and NMR [6, 49] spectroscopy, mass spectrometry [27,35], X-ray analysis [6,21,24,25,36], molecular dynamics (MD) calculations [9,13,18,31,38] and biological investigations [6, 15,20,26,30,41-43,45,46,48]. All of this sounds like routine, but the results are rather spectacular.

In the six years of research on peptides consisting of homologated proteinogenic amino acids we [5-52,54-59] and others [60-64] have embarked for a trip into an entirely new world, in which almost everything we know about α-peptides had to be disregarded. The homologous peptides form secondary structures, such as helices and turns, with as few as two to six residues [10,12,21,42,58], whereas α-peptides require more than ten residues for helix and turn formation under the same conditions (MeOH solution, NMR detection); parallel [6,21,64] and antiparallel [21,42,58] sheets and stacks (of cyclic β-peptides [8,22,36,45]) are found in solution and in the solid state (Figure 6). All of these secondary structures can be designed by choosing the "right" substitution pattern (constitution) and (relative and absolute) configuration of the residues in the β-and γ-peptidic chains, and MD calculations (GROMOS96 program, including solvent) furnish all the experimentally determined structures within a couple of nanoseconds ("*in silico*") [18,31,38]. Unlike natural α-peptide chains, which fold and unfold in a cooperative way, β-peptide folding is non-cooperative [19,49]. The shape, the handedness, the resulting dipole moments of sheets and helices are all different [65] as we go from α- to β- to γ-peptides, with increasing stability of the secondary structures in this order [44].

May be even more surprising than the structural properties of β-peptides are the results of their biological investigations [40,44]. (*i*) All the different types of β-and γ-peptides are absolutely stable to the peptide-cleaning enzymes [6,15,41] (Figure 7). (*ii*) One particular ^{14}C-labelled β-peptide was even metabolically stable in rats: samples collected for 96 hours from urine and feces, after *i.v.* administration, contained essentially no other radioactive compound but the β-peptide originally injected [66] (Figure 8). (*iii*) Even microorganisms in soil or in a sewer-water-treatment plant have difficulties growing on a simple β-tripeptide as sole carbon and nitrogen source [48]. (*iv*) Yet, β-and γ-peptidic hair-pin turns can be designed which are geometrically very similar to α-peptidic so-called "β-turns", often responsible for peptide-protein or protein interactions; thus, enzymatically stable β-peptidic peptidomimetics have been identified, for instance somatostatin analogs, consisting of as few as two β-amino-acid moieties and having sub-

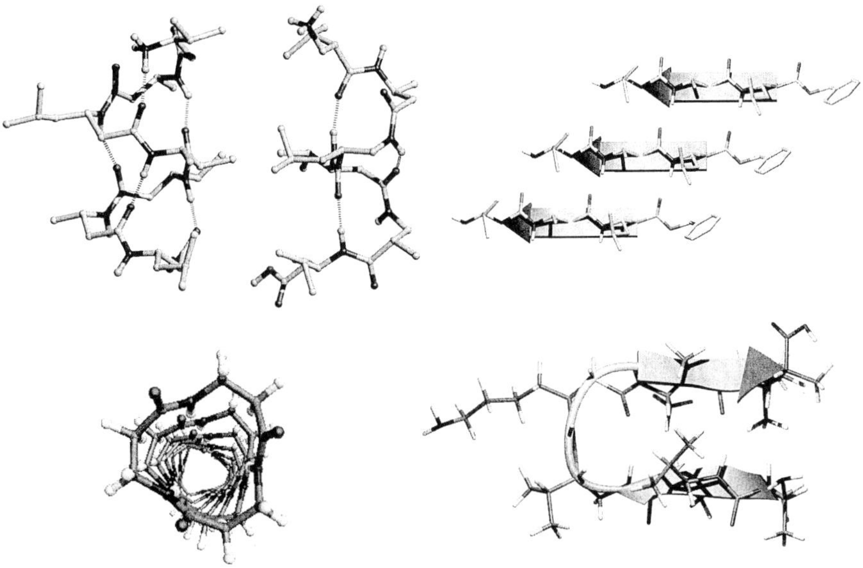

Figure 6. A 3_{14} and a 12/10 helix, a parallel and an antiparallel sheet (with hair-pin turn) and a stack (clockwise from top left) formed by β-peptides consisting exclusively of simple open-chain homologated α-amino-acid residues (Ala, Val, Leu, Lys side chains) and of α-Methyl-β-homo-Ala and Leu residues [12,21,36].

Types of β- and γ-peptides tested

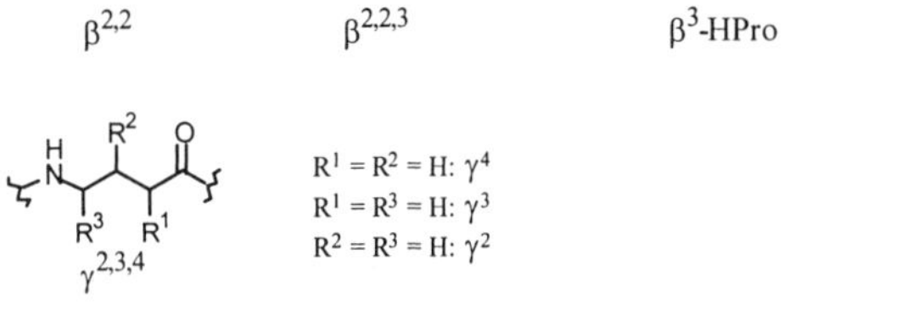

Peptidases

pepsin (*pankreas*)
trypsin (*pankreas*)
carboxypeptidase A (*pankreas*)
elastase (*pankreas*)
chymotrypsin (*pankreas*)
leucine-aminopeptidase (*kidney*)
20s proteasome (human erythrocytes)
proteinase K (*T. album*)
pronase (*S. griseus*)
penicillin amidase (*E. coli*)
amidase (*P. aeruginosa*)
β-lactamase (*E. cloacae*)

Figure 7. Complete proteolytic stability of all types of β-and γ-peptides towards a variety of peptidases. The β-peptides ranged in size from dimer to 15mer. The enzymes include all common types of peptidases (endo/exo, metallo, serine, threonine, and aspartyl proteases). After 40 hours there was no observable cleavage of any of the homologated peptides and no inhibition of the enzymes [41].

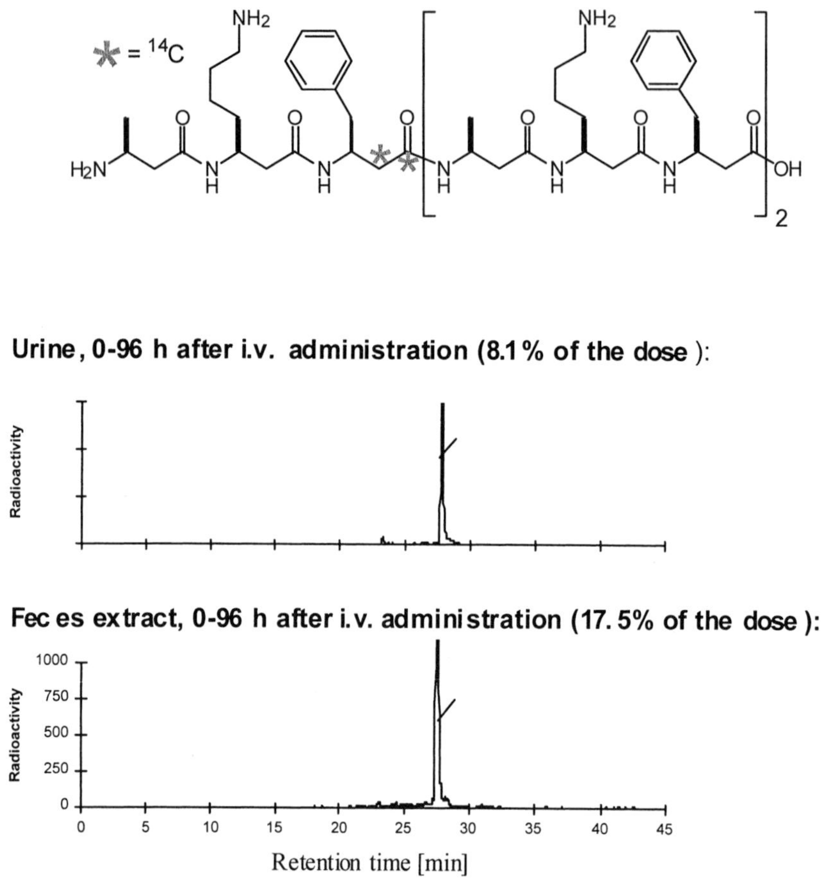

Figure 8. A β^3-nonapetide, which has been shown to be capable of mimicking an amphiphilic α-peptidic helical structure in a pepide-protein interaction [26], was ^{14}C-labelled and injected into rats. After 24 hours (in serum, not shown) and after 4 days (in urine and in feces the minor and major secretion pathways, respectively), there was hardly any metabolism, see the HPLC chromatograms with radiodetection (DSP01 is the compound, the formula of which is shown above) [68].

micromolar or even 10-20 nanomolar affinities to one of the five human receptors [20,30,42,46]. (*v*) Amphipathic helices of α-peptides, another motif for protein-protein binding, can be mimicked by β-peptidic helices, having polar and non-polar side chains positioned in the right manner [26]. (*vi*) The resistance to peptidases of β-peptides can be used to carry cargoes into cells; thus a fluorescence-labelled β-oligoarginine (a positive polyelectrolyte) is taken up by mammalian cells and migrates right to the cell nucleus docking to the *nucleoli* with their exposed DNA (a negative polyelectrolyte) [47,63]. (*vii*) Finally, antimicrobial [43,62], antiproliferative [45], but also hemolytic and cytotoxic [43,62] activities of the peptides consisting of homologated proteinogenic amino acids have been observed.

3 Conclusion

The simple idea at the outset of this research project, namely to replace an oxygen atom in a polyester chain by an NH group, has led to an almost explosive development of a new field, which is full of unexpected and promising results. For more details we refer to the list of references and to some review articles by us [10,40,44,50,65,67] and others [68-70], the latter ones also covering work on β-peptides consisting of or containing cyclic β-amino-acid residues, peptide analogs with N-N, N-O bonds or urea moieties incorporated in the backbone, or non-peptidic chains folding and turning.

4 References

[1] M. A. Sutter, D. Seebach, Liebigs Ann. Chem. 1983, 939 – 949; D. Seebach, H.-F. Chow, R. F. W. Jackson, K. Lawson, M. A. Sutter, S. Thaisrivongs, J. Zimmermann, J. Am. Chem. Soc. 1985, 107, 5292 - 5293; R. F. W. Jackson, M. A. Sutter, D. Seebach, Liebigs Ann. Chem. 1985, 2313 – 2327; D.Seebach, H.-F. Chow, R. F. W. Jackson, M. A. Sutter, S. Thaisrivongs, J. Zimmermann, Liebigs Ann. Chem. 1986, 1281 – 1308.

[2] H.-M. Müller, D. Seebach, 'Poly(hydroxyalkanoates): a Fifth Class of Physiologically Important Organic Biopolymers?', Angew. Chem. 1993, 105, 483 – 509; Angew. Chem. Int. Ed. Engl. 1993, 32, 477 – 502; D. Seebach, A. Brunner, B. M. Bachmann, T. Hoffmann, F. N.M. Kühnle, U. D. Lengweiler, 'Biopolymers and -oligomers of (*R*)-3-Hydroxyalkanoic Acids - Contributions of Synthetic Organic Chemists', Ernst Schering Research Foundation, 1995, 28, 7 – 98; D. Seebach, M. G. Fritz, 'Detection, synthesis, structure, and function of oligo(3-hydroxyalkanoates): contributions by synthetic organic chemists', Int. J. Biol. Macromol. 1999, 25, 217 – 236; R. N. Reusch, 'Transmembrane Ion Transport by Polyphosphate/Poly-(*R*)-3-hydroxybutyrate Complexes', Biochemistry (Moscow) 2000, 65, 280 – 295; S. Das, D. Seebach, R. N. Reusch, Biochemistry 2002, 41, 5307-5312.

[3] M. Rueping, A. Dietrich, V. Buschmann, M. G. Fritz, M. Sauer, D. Seebach, Macromolecules , 2001, 34, 7042 – 7048; P. Waser, M. Rueping, D. Seebach, E. Duchardt, H. Schwalbe, Helv. Chim. Acta 2001, 84, 1821-1845; P. J. Gee, F. A. Hamprecht, L. D. Schuler, W. F. van Gunsteren, E. Duchardt, H. Schwalbe, M. Albert, D. Seebach, Helv. Chem. Acta 2002, 85, 618 – 632; M. Albert, D. Seebach, E. Duchardt, H. Schwalbe, Helv. Chem. Acta 2002, 85, 633 – 658.

[4] D. A. Plattner, A. Brunner, M. Dobler, H.-M. Müller, W. Petter, P. Zbinden, D. Seebach, Helv. Chim. Acta 1993, 76, 2004 – 2033; D. Seebach, T. Hoffmann, F. N. M. Kühnle, U. D. Lengweiler, Helv. Chim. Acta 1994, 77, 2007 – 2034.

[5] J. Podlech, D. Seebach, 'The *Arndt-Eistert*-Reaction in Peptide Chemistry: A Facile Access to Homopeptides', Angew. Chem. 1995, 107, 507 – 509; Angew. Chem. Int. Ed. Engl. 1995, 34, 471 – 472.

[6] D. Seebach, M. Overhand, F. N. M. Kühnle, B. Martinoni, L. Oberer, U. Hommel, H. Widmer, 'β-Peptides: Synthesis by *Arndt-Eistert* Homologation with Concomitant Peptide Coupling. Structure Determination by NMR and CD Spectroscopy and by X-Ray Crystallography. Helical Secondary Structure of a β-Hexapeptide in Solution and its Stability towards Pepsin', Helv. Chim. Acta 1996,79, 913 – 941.

[7] D. Seebach, P. E. Ciceri, M. Overhand, B. Jaun, D. Rigo, L. Oberer, U. Hommel, R. Amstutz, H. Widmer, 'Probing the Helical Secondary Structure of Short-Chain β-Peptides', Helv. Chim. Acta, 1996, 79, 2043-2066.

[8] D.Seebach, J. L. Matthews, A. Meden, T. Wessels, C. Baerlocher, L. B. McCusker, 'Cyclo-β-peptides: Structure and Tubular Stacking of Cyclic Tetramers of 3-Aminobutanoic Acid as Determined from Powder Diffraction Data', Helv. Chim. Acta 1997, 80, 173 – 182.

[9] X. Daura, W. F. van Gunsteren, D. Rigo, B. Jaun, D. Seebach 'Studying the Stability of a Helical β-Heptapeptide by Molecular Dynamics Simulations', Chem. Eur. J. 1997, 3, 1410 – 1417.

[10] D. Seebach, J. L. Matthews, 'β-Peptides: a surprise at every turn', Chem. Commun., 1997, 2015 – 2022.

[11] G. Guichard, S. Abele, D. Seebach, 'Preparation of *N*-Fmoc-Protected β^2- and β^3-Amino Acids and Their Use as Building Blocks for the Solid-Phase Synthesis of *β* -Peptides', Helv. Chim. Acta 1998, 81, 187 – 206.

[12] D. Seebach, S. Abele, K. Gademann, G. Guichard, T. Hintermann, B. Jaun, J. L. Matthews, J. V. Schreiber, L. Oberer, U. Hommel, H. Widmer, 'β^2- and β^3-Peptides with Proteinaceous Side-Chains - Synthesis and Solution Structures of Constitutional Isomers, a Novel Helical Secondary Structure and the Role of Hydrophobic Interactions on Folding', Helv. Chim. Acta 1998, 81, 932 – 982.

[13] X. Daura, B. Jaun, D. Seebach, W. F. van Gunsteren and A. E. Mark, 'Reversible Peptide Folding in Solution by Molecular Dynamics Simulation', J. Molecular Biology 1998, 280, 925-932.

[14] J. L. Matthews, K. Gademann, B. Jaun, D. Seebach, 'Linear and Cyclic β^3-Oligopeptides with Functionalized Side-Chains (-CH_2OBn, -CO_2Bn, $CH_2CH_2CO_2Bn$) Derived from Serine and from Aspartic and Glutamic Acid', J. Chem. Soc Perkin Trans 1, 1998, 3331-3340.

[15] D. Seebach, S. Abele, J. V. Schreiber, B. Martinoni, A. K. Nussbaum, H. Schild, H. Schulz, H. Hennecke, R. Woessner, F. Bitsch, 'Biological and Pharmacokinetic Studies with β-Peptides', Chimia 1998, 52, 734 – 739.

[16] S. Abele, G. Guichard, D. Seebach, '(*S*)-β^3-Homolysine- and (*S*)-β^3-Homoserine-Containing β-Peptides: CD Spectra in Aqueous Solution', Helv. Chim. Acta 1998,81, 2141 – 2156.

[17] D. Seebach, S. Abele, T. Sifferlen, M. Hänggi, S. Gruner, P. Seiler, 'Preparation and Structure of β-Peptides Consisting of Geminally Disubstituted $\beta^{2,2}$- and $\beta^{3,3}$-Amino Acids: A Turn Motif for β-Peptides', Helv. Chim. Acta 1998, 81, 2218 – 2243.

[18] X. Daura, K. Gademann, B. Jaun, D. Seebach, W. F. van Gunsteren, A. E. Mark, 'Peptide Folding - When Simulation meets Experiment', Angew. Chem, 1999, 111, 249-253; Angew. Chem Int. Ed. Engl. 1999, 38, 236-240.

[19] K. Gademann, B. Jaun, D. Seebach, R. Perozzo, L. Scapozza, G. Folkers, 'Temperature-Dependant NMR and CD Spectra of β-Peptides. On the Thermal Stability of β-Peptide Helices – Is the Folding Process of β-Peptides non-cooperative?', Helv. Chim. Acta, 1999, 82, 1 – 11.

[20] K. Gademann, M. Ernst, D. Hoyer, D. Seebach, 'Synthesis and Biological Evaluation of a Cyclo-β-tetrapeptide as a Somatostatin Analogue', Angew. Chem. 1999, 111, 1302 - 1304; Angew. Chem. Int. Ed. Engl. 1999, 38, 1223 – 1226.

[21] D. Seebach, S. Abele, K. Gademann, B. Jaun, 'Pleated Sheets and Turns of ß-Peptides with Proteinogenic Side Chains', Angew. Chem. 1999, 111, 1700 - 1703; Angew. Chem. Int. Ed. Engl. 1999, 38, 1595 – 1597.

[22] K. Gademann, D. Seebach, 'Preparation and NMR Structure of the Cyclo-*β*-tripeptide [β^3-HGlu]$_3$ in Aqueous Solution: A New Class of Enterobactin-Type C_3-Symmetrical Ligands?', Helv. Chim. Acta 1999, 82, 957 – 962.
[23] A. Jacobi, D. Seebach, 'How to Stabilize or Break β-Peptidic Helices by Disulfide Bridges: Synthesis and CD Investigation of β-Peptides with Cysteine and Homocysteine Side Chains' Helv. Chim. Acta 1999, 82, 1150 – 1172.
[24] S. Abele, K. Vögtli, D. Seebach, 'Oligomers of β^2- and of β^3-Homoproline: What are the Secondary Structures of β-Peptides Lacking H-Bonds?', Helv. Chim. Acta 1999, 82, 1539 – 1558.
[25] S. Abele, P. Seiler, D. Seebach, 'Synthesis, Crystal Structures, and Modelling of β-Oligopeptides Consisting of 1-(Aminomethyl)cyclopropanecarboxylic Acid: Ribbon-Type Arrangement of Eight-Membered H-Bonded Rings', Helv. Chim. Acta 1999, 82, 1559 – 1571.
[26] M. Werder, H. Hauser, S. Abele, D. Seebach, 'β-Peptides as Inhibitors of Small-Intestinal Cholesterol and Fat Absorption', Helv. Chim. Acta 1999, 82, 1774 – 1783.
[27] J. V. Schreiber, M. Quadroni, D. Seebach, 'Sequencing of β-Peptides by Mass Spectrometry', Chimia 1999, 53, 621 – 626.
[28] T. Sifferlen, M. Rueping, K. Gademann, B. Jaun, D. Seebach, 'β-Thiopeptides: Synthesis, NMR Solution Structure, CD Spectra, and Photochemistry', Helv. Chim. Acta 1999, 82, 2067 – 2093.
[29] S. Abele, D. Seebach, 'Preparation of Achiral and of Enantiopure Geminally Disubstituted β-Amino Acids for β-Peptide Synthesis', Eur. J. Org. Chem. 2000, 1 – 15.
[30] K. Gademann, M. Ernst, D. Seebach, D. Hoyer, 'The Cyclo-β-Tetrapeptide (β-HPhe-β-HThr-β-HLys-β-HTrp): Synthesis, NMR Structure in Methanol Solution, and Affinity for Human Somatostatin Receptors', Helv. Chim. Acta 2000, 83, 16 – 33.
[31] D. Seebach, J. V. Schreiber, S. Abele, X. Daura, W. F. van Gunsteren, 'Structure and Conformation of β-Oligopeptide Derivatives with Simple Proteinogenic Side Chains: Circular Dichroism and Molecular Dynamics Investigations', Helv. Chim. Acta (2000, 83, 34 – 57.
[32] D. Seebach, A. Jacobi, M. Rueping, K. Gademann, M. Ernst, B. Jaun, 'Synthesis of β-Hexa- and β-Heptapeptides Containing Novel $\beta^{2,3}$-Amino Acids with Two Serine or Two Cysteine Side Chains – CD- and NMR-Spectroscopic Evidence for 3_{14}-Helical Secondary Structures in Water', Helv. Chim. Acta 2000, 83, 2115 - 2140 and in 'Hominatio – An International Tribute to Albert Eschenmoser", (Ed.: M.V. Kisakürek), Wiley-VCH, Weinheim, 2001.
[33] M. Rueping, B. Jaun, D. Seebach, 'NMR Structure in methanol of a β-hexapeptide with a disulfide clamp', Chem. Commun. 2000, 2267 – 2268.
[34] D. Seebach, T. Sifferlen, P. A. Mathieu, A. M. Häne, C. M. Krell, D. J. Bierbaum, S. Abele, 'CD Spectra in Methanol of β-Oligopeptides Consisting of β-Amino Acids with Functionalized Side Chains, with Alternating Configuration, and with Geminal Backbone Substituents – Fingerprints of New Secondary Structures?', Helv. Chim. Acta 2000, 83, 2849 –2864.
[35] J. V. Schreiber, D. Seebach, 'Solid-Phase Synthesis of a β-Dodecapeptide with Seven Functionalized Side Chains and CD-Spectroscopic Evidence for a Dramatic Structural Switch When Going from Water to Methanol Solution', Helv. Chim. Acta 2000, 83, 3139 – 3152.
[36] H. C. Le, T. Hintermann, T. Wessels, Z. Gan, D. Seebach, R. R. Ernst, 'Determination of the Amide Plane Orientations in a Cyclo-β-Peptide by Magic-Angle-Spinning Deuterium Correlation Spectroscopy, and Comparison with the Powder X-Ray Structure', Helv. Chim. Acta 2001, 84, 187 – 207.
[37] D. Seebach, J. V. Schreiber, P. I. Arvidsson, J. Frackenpohl, 'The Miraculous CD Spectra (and Secondary Structures ?) of β-Peptides as They Grow Longer', Helv. Chim. Acta Vol. 2001, 84, 271 – 279.

[38] X. Daura, K. Gademann, H. Schäfer, B. Jaun, D. Seebach, W. F. van Gunsteren, 'The β-Peptide Hairpin in Solution: Conformational Study of a β-Hexapeptide in Methanol by NMR Spectroscopy and MD Simulation', J. Am. Chem. Soc. 2001, 123, 2393-2404.

[39] P. I. Arvidsson, M. Rueping, D. Seebach, 'Design, machine synthesis, and NMR-solution structure of a β-heptapeptide froming a salt-bridge stabilised 3_{14}-helix in methanol and in water', Chem. Commun. 2001, 649 – 650.

[40] D. Seebach, M. Albert, P. I. Arvidsson, M. Rueping, J. V. Schreiber, 'From the Biopolymer PHB to Biological Investigations of Unnatural β- and γ-Peptides', Chimia 2001, 55, 345-353.

[41] J. Frackenpohl, P. I. Arvidsson, J. V. Schreiber, D. Seebach, 'The Outstanding Biological Stability of β- and γ-Peptides toward Proteolytic Enzymes: An in Vitro Investigation with Fifteen Peptidases', ChemBioChem 2001, 2, 445 – 455.

[42] K. Gademann, T. Kimmerlin, D. Hoyer, D. Seebach, 'Peptide Folding Induces High and Selectibe Affinity of a Linear and Samll β-Peptide to the Human Somatostatin Receptor 4', J. Med. Chem. 2001, 44, 2460-2468.

[43] P. I. Arvidsson, J. Frackenpohl, N. S. Ryder, B. Liechty, F. Petersen, H. Zimmermann, G. P. Camenisch, R. Woessner, D. Seebach, 'On the Antimicrobial and Hemolytic Activities of Amphiphilic β-Peptides', ChemBioChem 2001,2, 771 – 773.

[44] D. Seebach, A. K. Beck, M. Brenner, C. Gaul, A. Heckel, 'From Synthetic Methods to γ-Peptides – From Chemistry to Biology', Chimia 2001, 55, 831-838.

[45] K. Gademann, D. Seebach, 'Synthesis of Cyclo-β-tripeptides and Their Biological *in vitro* Evaluation as Antiproliferatives against the Growth of Human Cancer Cell Lines', Helv. Chem. Acta 2001, 84, 2924 – 2937.

[46] D. Seebach, M. Rueping, P. I. Arvidsson, T. Kimmerlin, P. Micuch, C. Noti, D. Langenegger, D. Hoyer, 'Linear, Peptidase-Resistant β^2/β^3-Di-and α/β^3-Tetrapeptide Derivatives with Nanomolar Affinities to a Human Somatostatin Receptor', Helv. Chem. Acta 2001, 84, 3503-3510.

[47] M. Rueping, Y. Mahajan, M. Sauer, D. Seebach, 'Cellular Uptake Studies with β-Peptides', ChemBioChem 2002,3, 257-259.

[48] J. V. Schreiber, J. Frackenpohl, F. Moser, T. Fleischmann, H.-P. Kohler, D. Seebach, 'On the Biodegradation of β-Peptides', ChemBioChem 2002, 3, 424-432.

[49] T. Etezady-Esfarjani, C. Hilty, K. Wüthrich, M. Rueping, J. Schreiber, D. Seebach, 'NMR-Structural Investigations of a β^3-Dodecapeptide with Proteinogenic Side Chains in Methanol and in Aqueous Solutions', Helv. Chem. Acta 2002, 85, 1197-1209.

[50] J. L. Matthews, C. Braun, C. Guibourdenche, M. Overhand, D. Seebach, 'Preparation of Enantiopure β-Amino Acids from α-Amino Acids Using the *Arndt-Eistert* Homologation', in 'Enantioselective Synthesis of β-Amino Acids', Chapter 5, (Ed. E. Juaristi), Wiley-VCH, New York, 1997, 105 – 126.

[51] T. Hintermann, K. Gademann, B. Jaun, D. Seebach, Helv. Chim. Acta 1998, 81, 983 – 1002.

[52] P. Micuch, D. Seebach, Helv. Chim. Acta 2002, 85, 1567 - 1577.

[53] A. Martínez de Ilarduya, C. Alemán, M. García-Alvarez, F. López-Carrasquero, S. Muñoz-Guerra, Macromolecules 1999, 32, 3257 – 3263.

[54] T. Kimmerlin, D. Seebach, D. Hilvert, Helv. Chim. Acta 2002, 85, 1812 - 1826.

[55] D. Seebach, M. Brenner, M. Rueping, B. Schweizer, B. Jaun, Chem. Commun. 2001, 207 – 208.

[56] M. Brenner, M. Rueping, D. Seebach, 'Synthesis and Structural Characterization of γ-Peptides' in Peptides 2000: Proceedings of the Twenty-Sixth European Peptide Symposium, Montpellier – France, (Eds. J. Martinez, J.-A. Fehrentz) EDK, Paris, 2001, 455-456.

[57] M. Brenner, D. Seebach, Helv. Chim. Acta 2001, 84, 1181-1189.

[58] M. Brenner, D. Seebach, Helv. Chem. Acta 2001, 84, 2155-2166.

[59] D. Seebach, M. Brenner, M. Rueping, B. Jaun, Chem. Eur. J. 2002, 8, 573-584.

[60] S. Hanessian, X. Luo, R. Schaum, S. Michnick, J. Am. Chem. Soc. 1998, 120, 8569 – 8570.

[61] B. W. Gung, D. Zou, A. M. Stalcup, C. E. Cottrell, J. Org. Chem. 1999, 64, 2176 – 2177.
[62] D. H. Lui, W. F. DeGrado, J. Am. Chem. Soc. 2001, 123, 7553 - 7559; R. P. Cheng, W. F. DeGrado, J. Am. Chem. Soc. 2001, 123, 5162 - 5163.
[63] N. Umezawa, M. A. Gelman, M. C. Haigis, R. T. Raines, S. H. Gellman, J. Am. Chem. Soc. 2002, 124, 368 - 369.
[64] S. Brenner L. B. McCusker, C. Baerlocher, J Appl. Cryst. 2002, 35, 243 – 252.
[65] D. Seebach, A. K. Beck, M. Rueping, J. V. Schreiber, H. Sellner, 'Excursion of Synthetic Organic Chemists to the World of Oligomers and Polymers', Chimia 2001, 55, 98-103.
[66] H. Wiegand, B. Wirz, A. Schweitzer, G. P. Camenisch, M. I. Rodriguez Perez, G. Gross R. Woessner, R. Voges, P. I. Arvidsson, J. Frackenpohl, D. Seebach, Biopharm. Drug Dispos. 2002, 23, in print.
[67] K. Gademann, T. Hintermann, J. V. Schreiber, 'β-Peptides: Twisting and Turning', Curr. Med. Chem. 1999, 6, 905 - 925.
[68] S. H. Gellman, 'Foldamers: A Manifesto', Acc. Chem. Res. 1998, 173 – 180.
[69] R. P. Cheng, S. H. Gellman, W. F. DeGrado, 'β-Peptides: From Structure to Function', Chem. Rev. 2001, 101, 3219 – 3232.
[70] D. J. Hill, M. J. Mio, R. B. Prince, T. S. Hughes, J. S. Moore, 'A Field Guide to Foldamers', Chem. Rev. 2001, 101, 3893 – 4011.

03 Cost Effective Production of Agrochemicals: How to Satisfy the Needs of Farmers, Industry, and the Environment

William R. Shiang and Richard L. Swafford

The Dow Chemical Company, 1710 building, Midland, MI 48674 U.S.A., E-mails: bshiang@dow.com, rlswafford@dow.com

1 Introduction

An agricultural company's primary goal is to earn a profit for its owners or shareholders. To achieve that objective, it must operate under constraints that include the economics of farming, environmental issues, industrial competition, government regulations, and public opinion. The investment value of a company is based on not only on the company's current products, but also the forecasted potential it has in its pipeline for the future. A successful company must be able to discover, develop and provide products that are aligned with today's agricultural needs while accommodating these various constraints. The current trend is such that hectares of agricultural land are decreasing while the world population continues to increase. This conflicting trend will escalate the demand for food and fiber, thus necessitating greater yields per hectare from the shrinking farmable land. The successful agricultural company must find a way to help the agricultural industry to obtain more and better products at less cost while at the same time providing a return to the investors.

In the traditional agricultural chemical business (not including seeds and biotech) according to the 2000 Wood Mackenzie report, the market was approximately $29 billion with a negative nominal growth of -3.5%. This negative growth is a good indicator that resources are limited and competition is fierce in this area. At this point it is unclear how far and how quickly biotech and genetically modified seeds will affect the traditional agricultural chemical industry, but it is speculated that there will be an effect. Biotechnology and consolidation of companies also pushed the level of competition for the limited business to greater levels in the traditional agricultural chemical company.

The 2001 Phillips McDougall annual report states that since 1950 there have been only 512 new active ingredients introduced globally, with very few introduced in the recent years. In the early years, molecules were less active, thus requiring a high application rate. In addition, the manufacturing processes were simple and inexpensive. Lack of government regulation and understanding of environmental and health risk contributed to the „use what ever amount of active it takes" mentality. The mindset for agricultural control was basically to use as much it takes to eradicate the pest without damaging the desired crop. Large volume also helps to lower the cost of manufacturing. In the past twenty years public opinion, health and environmental issues led to greater governmental control. This in turn contributed to the evolution of the agricultural active ingredients to higher activity and manufacturing complexity. Higher activity means lower application rates, leading to lower production volume. This, unfortunately, also contributes to a higher cost of manufacturing. The higher activity affords a positive impact on our environment allowing a higher tolerance for higher costs from the active. Nonetheless, the investment to find an active molecule has also escalated. These costs are associated with biology, regulatory, market research, and manufacturing R&D. A successful company must balance and evaluate several criteria to determine whether a molecule will contribute value to its portfolio.

2 Economics

Growers will buy a pesticide only if it increases their profits. They know what a given level of pest control means to their crop yields and how that translates into profit. Therefore the economics of the agricultural chemicals industry are driven by what growers can afford to pay for the desired benefit. This, in turn, sets the target for manufacturing costs. For example, a wheat farmer will pay up to \$30 per hectare to control European blackgrass. If a new herbicide controls blackgrass at 100 grams per hectare, then the net unit return per kilogram of active ingredient is \$300. If a company wants manufacturing costs to be, say, 30% of the net unit return, then the manufacturing cost per kilogram needs to be \$90 or less.

It is important to recognize that manufacturing cost is not the only manufacturing related factor that affects a project's economics. In addition to manufacturing cost, manufacturing economics also are affected by the corporate tax rate, overhead, direct fixed capital, and changes in working capital. Each of these five factors, in turn, is affected by many other variables. Two of the most important variables are production volume (kilograms per year) and process complexity.

Production volume has an inverse relation to manufacturing cost per kilogram. If production volume increases, the cost per kilogram usually decreases. If production volume decreases, the cost per kilogram typically increases. At smaller volumes a change in volume has a larger effect than at larger volumes. Process complexity, on the other hand, generally has a direct relationship to manufacturing cost per kilogram. The more complex the process, the higher the manufacturing costs. In addition to the individual effects of production volume and process complexity, they have an interactive effect. At lower volumes increasing process complexity raises costs more than it does at higher volumes.

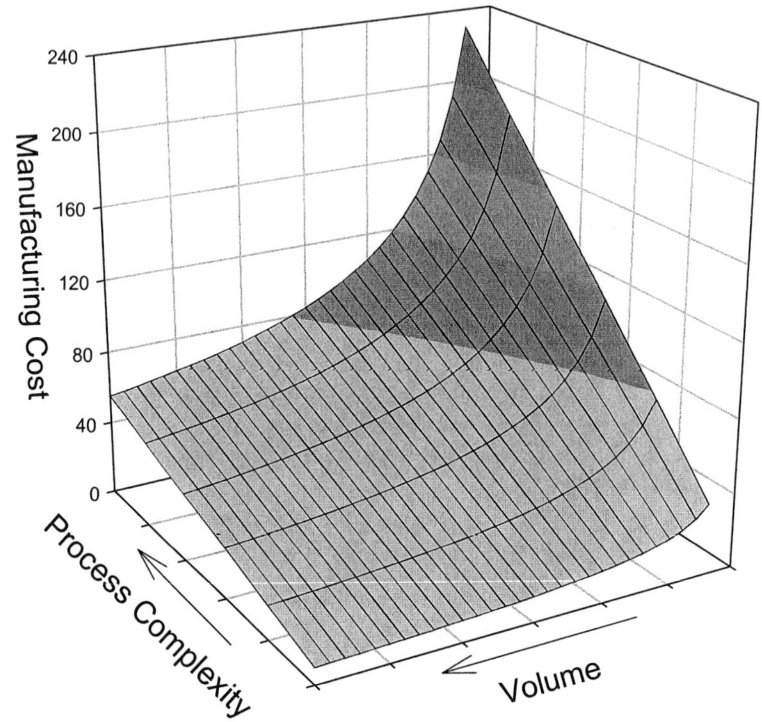

Figure 1. Effect of Production Volume and Process Complexity on Manufacturing Cost

Overall project economics are generally assessed using a cash flow model. Plotting cumulative cash flow from all activities over time allows the project team to visualize and assess some key aspects of a project's economics. First, the team can understand how deep a financial hole the project digs before it begins to make money. Second, the team can understand how long it will take the project to break even. At the break-even point, the project has paid back all of the development and other related expenses. After that, the project begins to generate net value for the company. Third, the team can assess the total value generated over the life of the project. The team can use these three pieces of information to assess the project's risks and potential return on investment.

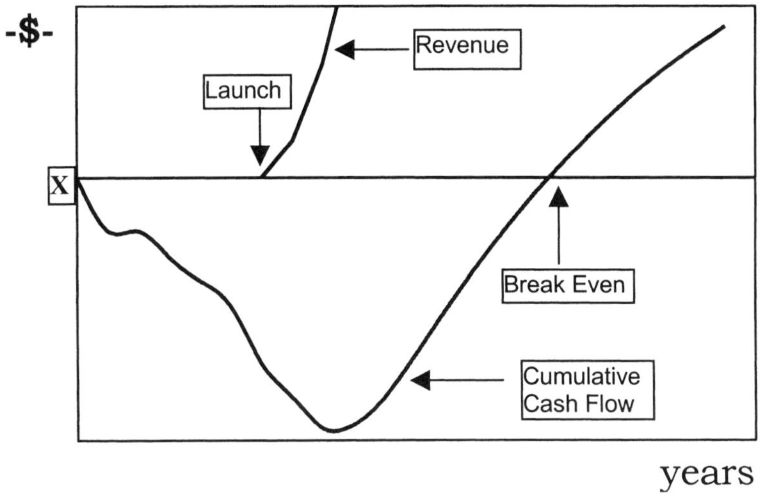

Figure 2. Cumulative Project Cash Flow

3 Project Development

There are four critical partners in a company that must function with synergy for the success of developing a new product. Those four partners are commercial, regulatory, field biology, and supply chain functions. The commercial team must provide good market analysis and prediction on the possible sales from the new product. The regulatory team ensures that all heath and environmental issues are addressed and develop the strategy for global registration of the product. The field biologists determine and evaluate the efficacy of the product. The supply chain team ensures that all activities associated to the delivery of the final product are carried out with best practice at the lowest cost.

The journey to the commercialization of new products begins with discovery of the new actives by the biology team. Once a possible candidate is identified and shows promise, it is advanced into the pre-development stage where environmental and toxicology effects are the major concern. If the molecule continues to show promise and has no toxicology or environmental issues, it is advanced to the developmental stage. At the developmental stage focus will be on the registration of the product and creation of the supply strategy of the products for launch through full commercialization. Timely introduction of the new products by the commercial group is also critical at this stage. Typically it will require several years to advance a new product from discovery to launch.

Of the four critical areas, which can a firm impact the most? The commercial team's sales prediction significantly affects the business case, but those predictions are dictated by product efficacy and market forces. Even though the forecast is variable, in most cases sales and volume forecast changes are small. The expenditure incurred by regulatory teams is dependent on the property of the active and registration requirements for the particular countries into which the product will be sold. For the biology team, the cost for determining usage rates and efficacy of formulations is linked to the type of product and is relatively consistent. In general, the cost associated to regulatory and biology can be viewed as relatively fixed and uncontrollable. However, the greatest controllable variable lies within the control of the supply team and a large segment of it resides with the process research and development segment of supply chain.

4 Process Improvement

The starting point for the process research and development group is the advancement of an active ingredient out of early stage discovery. At that point, a 3-15 kg sample of the active is requested for early stage acute toxicology and field studies. If the early biology and toxicology data is promising and the molecule is advanced to pre-development the following year, then 50-100 kg is required for sub-chronic toxicology studies and an expanded field trial. If the performance continues as positive with good economics support, then it will pass the second critical pre-development gate where 200-600 kg of active will be necessary for chronic toxicology and field studies. The amount of sample requested for pre-development studies is mostly dependent on the early toxicology data. Low toxicity compounds will require larger quantities for testing.

A target price for the manufacturing of the active can be set based on the sales projection provided by the commercial team, typically in the range of \$50-\$300/kg. As stated earlier, the cost of manufacturing an active is very much based on the application rate therefore the upper limits can range even higher. The cost of manufacturing can also vary if commercial or biology teams adjust their input. In some cases the variation can be severe enough to terminate a project.

As an example, an experimental rice herbicide (ERH-1), 2,2-dimethyl-1-(4mthylthio-5-pyrimidinyl)indane, was advanced to phase 2 along with a request for a 3 kg sample. Typically expeditious delivery of the sample was required to meet a rapid development strategy, therefore the synthetic route (Scheme 1) developed by the discovery group would be utilized to fulfill the sample requirement. The synthetic detail of the chemistry was published in the journal Organic Process Research & Development[1], therefore the chemistry will not be discussed in this transcript.

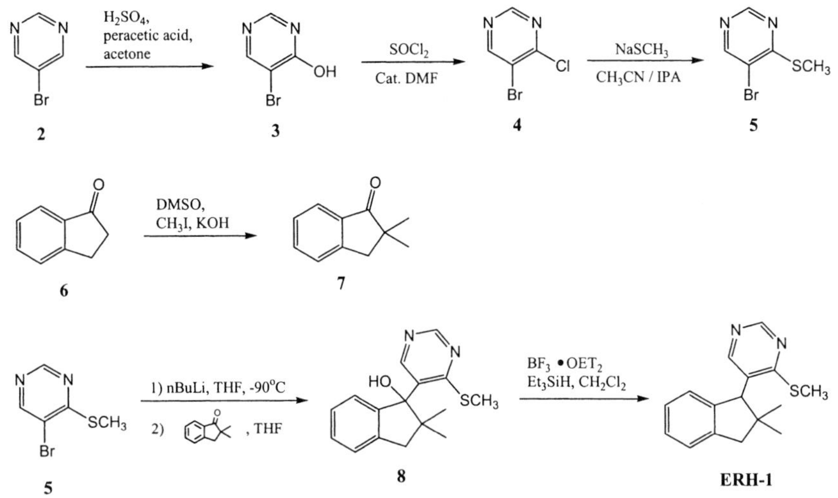

Scheme 1

There are many issues with the above synthetic route to ERH-1: (1) High raw material cost for the intermediates and reagents such as 5-bromopyrimidine, 2,2-dimethyl-1-indane, and triethylsilane. (2) The use of DMSO in the plant is not preferred due to reactive chemistry and toxicity properties. (3) Although the use of powdered KOH and cryogenic temperature requirements are easily carried out in a laboratory, they are not practical on a plant scale without incurring large capital input. (4) Due to the proximity of the plant to the community, the possibility of an odorous release is not tolerated. (5) In the laboratory it was discovered that significant amounts of explosive acetone peroxide was formed when acetone was used as solvent in the conversion of **2** to **3**. (6) Most impor(tantly, the economic evaluation of the process resulted in a cost of manufacturing that was four times the desired target.

Pending the advancement approval to the next phase, the process group has limited time to develop an improved and cost effective synthetic route to manufacture ERH-1, while simultaneously assembling a strategy for the larger sample request. Scheme 2 illustrates the synthetic process that was used to deliver 93 kg of active used for sub-chronic toxicology and field experiments. Illustrated in scheme 2, highlighted in bold, are the improvements incorporated in the synthetic process for the 93 kg sample campaign. The changes utilized in the 93 kg sample campaign provided economic and safety advantages. The substitution of acetone with water as solvent eliminated the possibility of forming the hazardous acetone peroxide in the conversion of **2** to **3** without significant effects on yield or purity. The odor issue of thiomethoxide in the conversion of **4** to **5** was circumvented by an alternate preparation of **5** using odorless thiourea as the source of sulfur[2] revealed from a literature search. In the conversion of **6** to **7**, the substitution

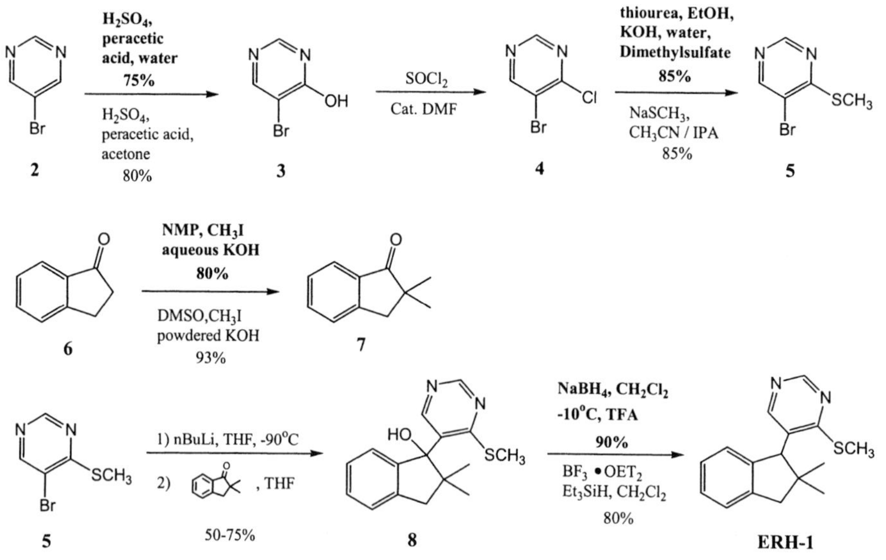

Scheme 2

of powdered KOH and DMSO with 45% aqueous KOH and N-methylpyrrolidinone were found to be exceptional in regards for ease of use and safety in the plant. The process team was able to find excellent cost effective replacements for most of the expensive reagents without loss in yield or purity. Although many improvements were incorporated in the 93 kg sample campaign, the economic analysis of the process at best was still two times that of the target manufacturing cost for the active.

In most cases convergent types of syntheses are the most cost effective and preferred for various reasons. However, that was not the case with ERH-1 as it will show in the following examples. In the convergent approach (see Scheme 3) it was assumed that barbituric acid would be a cheap source (~$4 per kg) for the pyrimidine portion since it was prepared[3] from inexpensive materials such as ethyl malonate and urea with sodium ethoxide as a condensing agent. Numerous reports [4] have shown that barbituric acid is easily converted to 2,4,6-trichloropyrimidine by heating with phosphorus oxychloride and a trialkylamine. Our intent was to couple 2,4,6-trichloropyrimidine 10 to 2,2-dimethyl-1-indanone followed by various reactions that would lead to ERH-1[5] as shown in Scheme 3.

Although the cost of barbituric acid was tenfold greater than originally assumed, it was not the only issue with the synthetic approach in scheme 3. Note that the yield of the coupling reaction to form **11** was extremely low while the rest of the reactions were reasonably high yield. A major problem with the coupling reaction was discovered where an equilibrium existed that favored the reactants. Depending on the work up procedure the yield can vary from 0-20%. At 20°C, the pyrimidine anion of **10** readily de-

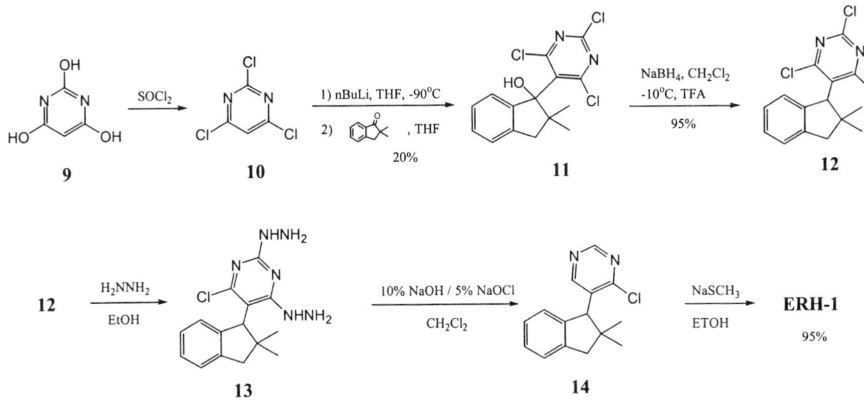

Scheme 3

composes, thus in the warmer work up, greater than 95% 2,2-dimethyl-1-indanone was recovered while little or no pyrimidine **10** was found. Economic evaluation of the barbituric acid route based on yield of the coupling reaction versus cost of manufacturing at 0.5, 1.0, and 1.5 of the projected volumes was calculated and illustrated in Figure 3. It reveals that the cost differences between 80-100% were very small and that the desired cost of manufacture cannot be met even if the coupling reaction were at 100% yield. The economic evaluation also illustrates that at 1.5 times the projected volume and at 100% yield for the coupling reaction, the cost of manufacturing falls short of the target.

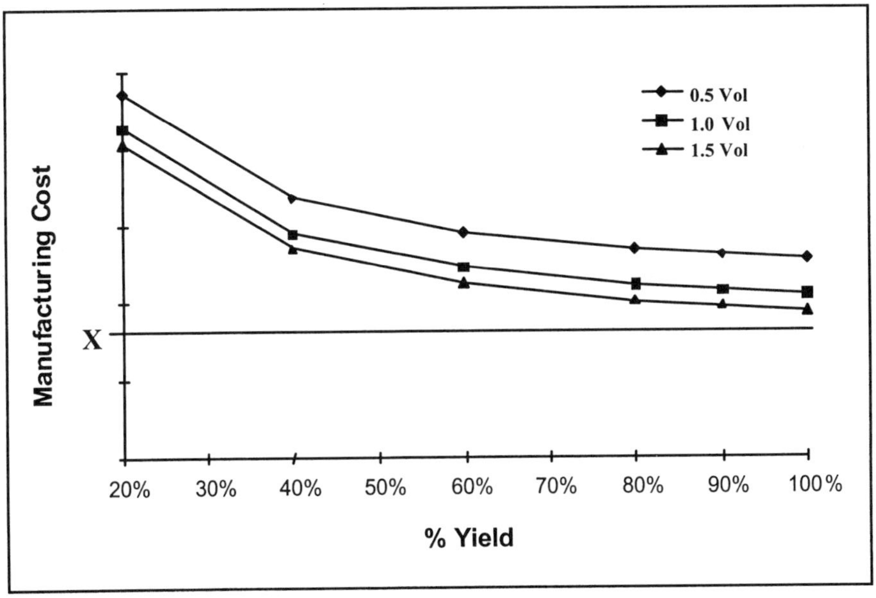

Figure 3. Effect of metallation yield on cost, trihydroxypyrimidine route

Our process team was able to develop a linear approach that met the target cost of manufacturing for ERH-1 (Scheme 4). The new manufacturing route also afforded a possibility for further improvement that may result in a further reduction in cost by ~10%. The details of the chemistry of the linear route are discussed in the journal publication [1].

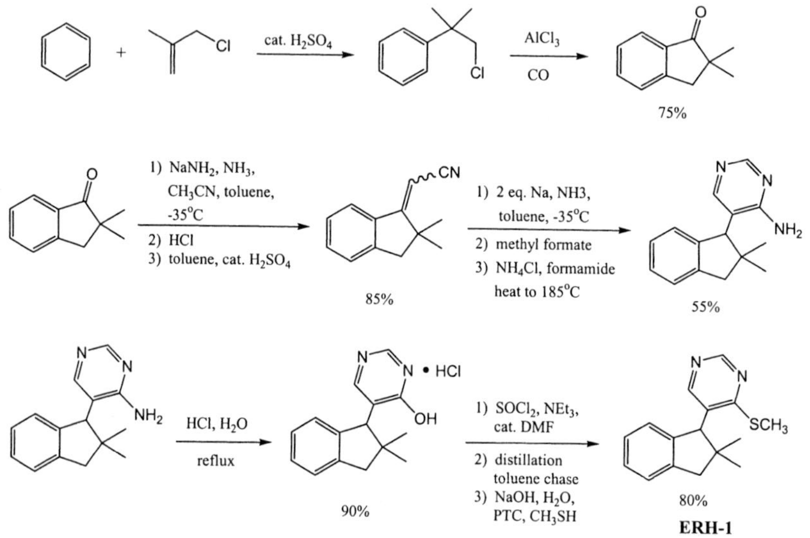

Scheme 4

The new route began with benzene and neophyl chloride to afford a low cost source for 2,2-dimethyl-1-indanone and utilized inexpensive raw materials and reagents. This new route involves carbon-carbon bond formation reaction of two nitrile-stabilized carbanions, which are generated, from inexpensive sodium metal and ammonia. A one pot dissolving metal reduction/formylation/Bredreck cyclization is employed and delivers a complex aminopyrimidine intermediate in a minimal number of unit operations. The aminopyrimidine intermediate is converted to ERH-1 using standard methods.

Unfortunately after developing this process it was discovered that the biological performance did not meet the desired objectives, thus the project was terminated. Had the molecule continued to advance and reach commercialization, our process R&D group would have been able to fulfill its responsibility to the supply chain.

What has been discussed thus far is only a small segment of the process research and development group's capability to affect cost reduction. Our production motto is that " if you make it, figure out how to sell it, use it, recycle it, treat it, or don't make it in the first place!" Safety for the employees and our community are always our first concern and will dictate many of the decisions for a project. With safety in mind we can focus on the areas with economic impact to the project such as yields, reduction of unit operations, resources, solvent selections, and choice of reagents. This is an area of production that can offer many opportunities for cost savings; however, this subject can encompass a presentation topic alone and will only be addressed briefly.

As with most chemical transformations the waste generated are many folds greater than the amount of material produced. Our waste treatment philosophy is to "eliminate, reduce, recycle, or treat" in that order of preferences. Eliminate, reduce, and recycle are basic process alternatives that are routinely part of the process strategy and are the most cost effective choices. Waste treatment is usually the last alternative and the most expensive option and is necessary in order for a chemical company to maintain its obligations to the environment and community. However, it also offers the most and diverse methods that could be tailored to a specific waste problem. Examples of methods to treat waste are incineration, wet air oxidation, chemolysis, ozonolysis, chlorinolysis, UV treatment, activated carbon, biotreatment, and supercritical oxidation.

There are many segments a successful agricultural chemical company must address from the discovery stage to commercialization. Each segment contains its own variability and constraints. The journey to launch a cost effective agrochemical product that can satisfy the needs of farmers, industry and our environment is only successful if it results in profit for the shareholders. However, it is comforting to know that most of the companies like ours are focused on a win-win outcome for all.

5 References

[1] T. J. Dietsche, D. B. Gorman, J. A. Orvik, G A. Roth, W. R. Shiang, Organic Process Research & Development 2000, 4, 275-285.

[2] D. J. Brown, R. V. Foster, Aust. J. Chem. 1966, 19, 2321-2330.

[3] D. J. Brown, "Heterocyclic Compounds; The Pyrimdines", Chap. 6, p.330, John Wiley & Sons, Inc., 1994.

[4] W. R. Shiang, An Internal Report 1996.

[5] J. W. Hull, W. R. Shiang, An Internal Report 1996.

04 Obsolete Pesticide Stocks in Developing Countries: Strategies, Policies, and Practical Steps for Their Disposal

Wolfgang A. Schimpf

Pilot Project Chemical Safety (GTZ), Tulpenfeld 2, 53113 Bonn, Germany, gtzwaschimpf@aol.com

1 Background

Studies conducted by the Food and Agriculture Organisation (FAO) of the United Nations show that there are around 100 000 tonnes of obsolete pesticide stocks in developing countries. Africa alone probably accounts for up to 48 000 tonnes. Obsolete pesticides cannot only be found in developing countries in Africa, the Near East, Asia and Latin America, they are also an environmental problem in many Central and Eastern European countries. The main hazard, however, is the acute toxicity of these pesticides. They present an acute and potential risk both for the community and the environment. These obsolete pesticides are not only a barrier to sustainable development in emerging countries, they are also a hazard for the global environment. The countries concerned are not in a position to solve the problems on their own. They have neither the technical expertise, the facilities nor the necessary financial means. For this reason, it is essential that the OECD countries provide support.

In the last decade, a number of activities have been carried out by national and international donor organisations to solve the problem and a few thousand tonnes of obsolete pesticides have been disposed of in an environmentally sound manner – some with financial support from the pesticide industry. Due to pressure from international non-governmental organisations (NGOs), such as Greenpeace and the World Wide Fund for Nature (WWF), politicians have been forced to act. Taking its lead from the United Nations, the international community has put this issue on the agenda, aiming to solve the problem through a number of international conventions, like the Montreal Protocol, and the Basel, Rotterdam (PIC) and Stockholm (POPs) Conventions. The objective of the African Stockpile Program (ASP) – a joint initiative of the World Bank and the FAO – is to eliminate this burden on both the environment and the inhabitants of developing countries.

To address this situation, the Deutsche Gesellschaft für Technische Zusammenarbeit (GTZ) GmbH launched a pilot project in 1990 geared to the disposal of obsolete pesticides. Since then, environmentally sound disposal operations have evolved into a range of activities within the field of technical co-operation. The GTZ project disposed of more than 1800 tonnes of obsolete pesticide stocks in 8 developing countries in Africa and Asia. It also carried out several large-scale analytical surveys and prevention measures.

2 Obsolete Pesticide Stockpiles

During the last 50 years pesticides have made a significant contribution to the development of these countries, securing the production of food as well as human and animal health. Due to the potential hazards, the production, trade, use and handling of pesticides is strictly regulated, managed and controlled in industrialised countries. In developing countries, however, the regulation, management and handling of agrochemicals often fails to meet international standards. There is a lack of information about the potential risks and hazards of the chemicals used, and about appropriate pesticide regulations and management. Hazardous stockpiles of obsolete pesticides present an acute and potential risk for both the community and the environment.

2.1 Occurrence and State of Obsolete Stockpiles

Obsolete pesticide stocks are present in the majority of developing countries. Since the 90s several inventories have been compiled to evaluate the situation in the individual stores. The FAO estimated stockpiles of obsolete pesticides according to inventories completed by 53 African countries [1] and identified a total of over 48 000 tonnes of pesticide waste. They have been identified in every African country, ranging from 20 tonnes in Gambia up to 4500 tonnes in Ethiopia. In Asia only a few inventories have been carried out. The GTZ carried out a detailed inventory in Pakistan. During this survey and risk assessment a joint team of the GTZ and collaborators of the Department of Agriculture identified around 1250 tonnes of obsolete pesticides in more then 190 stores in the Province of Punjab. Only estimates are available for the other provinces. According to these estimates there are an additional 2400 tonnes in some 380 stores. These figures do not include contaminated soil and other associated waste.

2.2 Classification of Stockpiles

Analytical surveys need to be carried out in developing countries in order to provide detailed information about the dimensions of the stockpiles. As a general rule, the storehouses are mostly inappropriate. It is also difficult to estimate the exact quantities of the individual stockpiles. In Pakistan, for example, several hundred different pesticide formulations with 50 different active ingredients were found in the stores, mostly insecticides. The active ingredients belong to all types of pesticides ranging from "extremely hazardous" down to "slightly hazardous" compounds. A large group belongs to the group of "banned pesticides", better known as the "Dirty Dozen", and to the group of Persistent Organic Pollutants (POPs). Environmentally these pesticides cause additional concern because of the persistence of the products and their metabolites in the environment, especially the organochlorine pesticides.

2.3 The Situation in Developing Countries

Here are examples from three different countries:

2.3.1 Pakistan

Right in the middle of the highly populated area of Malir in Karachi is one of the most dangerous pesticide stores in the world, with an estimated 200 tonnes of obsolete pesticides – Greenpeace called it "Hot Spot". There are 300 unlabelled metal drums, 200 litres each, with unknown liquids. The drums are standing outside, on the ground, without protection. Chemical analyses showed that most of the drums are filled with heptachlor. Due to corrosion a large number of these drums are already empty; the pesticides have leaked into the earth. Next to the drums there are 12 tonnes of carbofuran, lying stored in rotten jute sacks - exposed to the elements. According to the WHO classification [2], carbofuran is a Class Ib pesticide, and highly hazardous. On the same premises there are heaps with an estimated 50 tonnes of a mixture of granular substances and powders. The mostly paper and jute sacks are torn. According to the chemical analysis these heaps contain a mixture of heptachlor, endrin and diazinon. These heaps, too, are exposed to the elements. Strong wind could distribute the pesticide formulations to the residential areas opposite the store. Rainwater could leach out the active ingredient, contaminate the soil and infiltrate into the groundwater. The district court is on the same premises with hundreds of visitors every day, who are also exposed to the odour of pesticides.

2.3.2 Mali

The Plant Protection Service of the Ministry of Agriculture maintained a large pesticide store in the middle of the Sahara, 500 km east of Gao. This location, by the name of Tin-Essako, is situated in the breeding ground of the desert locust. The store was destroyed at the beginning of the 90s in the so-called Tuareg rebellion. The Tuaregs moved all the drums from the store, emptied a large number of drums and poured the insecticide in the sand. They contaminated the area heavily, including the nearby water well, the only one in the entire area. The well-water is not useable any more. A large quantity of pesticides is still lying in Tin-Essako and has not yet been safeguarded.

2.3.3 Morocco

A number of pesticide stores in the southern part of Morocco were evaluated in a joint survey carried out by the US Agency for International Development (US-AID) and the GTZ in 1991. On three areas of land, each larger than a football field, there are thousands of metal drums lying around filled with technical BHC - altogether more then 3000 tonnes. This is a typical case, where strategic stores of pesticides were being saved up for sudden calamities such as swarms of locusts. The metal drums might still be there, but because they have been stored on the ground for decades, unprotected from the moisture of the soil, they could be corroded by now and the pesticide may be leaking into the soil.

2.4 Causes of Accumulation

The previous examples are no exceptions. There are similar situations in nearly all developing countries [3]. No type or chemical group of pesticides is excluded from obsolete pesticide stockpiles. Some stocks have been accumulated over a period of 50 years; new products are being added to the stockpiles continuously. The key factors [4] that contribute to the accumulation of obsolete stockpiles in developing countries are these:

- Inadequate storage and stock management
- Inappropriate handling during transport
- Overstocking of products
- Strategic stockpiles for expected pest incidence
- Banning of products
- Purchase of inappropriate products and formulations
- Poor quality of pesticides and lack of analytical facilities
- Inappropriate packaging size and labelling
- Excessive donations and poor co-ordination among aid agencies
- Change of national policy

The factors which contribute to the accumulation of obsolete pesticides in developing countries are generally related to inadequate storage and stock management. Many products have been banned due to environmental or health risks, others are outdated or have deteriorated over time and cannot be destroyed in an appropriate manner. These pesticides are sometimes of poor quality or were delivered in inappropriate containers. By supplying excessive quantities of pesticides to developing countries and through a lack of donor-agency co-ordination, industrialised nations have also contributed to the increase in stockpiles in many African countries. The change in national policies has led to the accumulation of thousands of tonnes of obsolete stocks. The most frequent case is that pesticides were saved up in strategic stores for sudden calamities like swarms of locusts which never occurred. Often the wrong pesticides were delivered, or the formulations prevented optimum application of the agents. Unsuitable containers and inadequate labelling made the users unsure of themselves, so they left the pesticides where they found them. Developing countries often receive substandard pesticides due to the

lack of quality control. Civil wars, which tend to cause drastic downturns in production, also have negative effects on pesticide consumption. Consequently large stocks of pesticides are left behind and forgotten at abandoned storage sites.
Dieldrin was the standard insecticide for use against locusts, particularly in African countries, together with other organochlorine compounds such as aldrin, endrin, DDT and BHC. Once the range of less persistent, and less toxic alternative pesticides became available, US-AID made all further technical assistance contingent upon the use of "drins". Other donors followed, and the pesticides in question became obsolete. What those responsible for the ban had neglected to consider was that large quantities of these agents, dieldrin in particular, were still in stores all over the world, and no transition period was allowed. Without the efforts of numerous donors, there would never have been such a massive accumulation of old stocks. The producers and suppliers of pesticides have been accused of contributing towards the pointless accumulation of pesticides through their excessively aggressive product marketing activities. Donors, governments and industry have contributed to the dilemma.

2.4.1 The Stores

In most developing countries there are only a few pesticides stores available which meet an acceptable standard. They mostly belong to the private sector, and are mainly the property of international pesticide manufacturers. The government stores are very often in a bad condition and many are in an appalling state. The government authorities, mostly the Department of Agriculture, often do not have sufficient storage capacity to store the pesticides safely. Many stores are not constructed for the storage of pesticides, with insufficient ventilation and no concrete floors. Most of the storehouses are damaged, roofs are broken and the floors are not sealed. Over the years the floors and the walls of the stores have become heavily contaminated, due to leakages from the corroded drums and rotten containers. The vapours have probably diffused into the walls, made of brick or loam. The stores can normally be located by their smell. Only a few of the stores in developing countries are constructed and managed according to the basic requirements for the safe storage of toxic chemicals, recommended by the FAO and UNEP [5].

2.4.2 The Stockpiles

In most cases the stockpiles are not well managed. The recommended "first in - first out" principle is rarely common practice. Most storekeepers are not familiar with the basic rules of good stock management. Stock records may not be regularly updated. Computerised store records are hard to find. Many stores are in total disorder. Due to the rough handling of the pesticides, the floors and the surfaces are often contaminated with a thick layer of a black, evil-smelling mass of pesticides. Leaking pesticides contaminate the floors and the soil around the stores - visible by black spots on the soil. In some stores there are unprotected piles of pesticide dust or granules. The paper sacks are damaged or even rotten. Sometimes these pesticides are exposed to the elements. In some cases large quantities of highly dangerous pesticides such as sodium cyanide lie outside the stores and are not protected from human contact, especially children. Because of a lack of space in the stores pesticides are not properly stored, thereby reducing access to the products and making it difficult to monitor the condition of containers. At many locations pesticides are even stored in the open for prolonged periods of time – directly exposed to the sun. Poor storage conditions for pesticides can lead to faster deterioration due to temperature fluctuations, and to physical damage of the packaging materials. The storekeepers are not able to clean leakages or spills immediately because staff have not been trained in how to handle them. They do not have the necessary equipment and very seldom the basic protective gear for themselves.

3 Management of Obsolete Pesticide Stocks

The handling and storage of pesticides, even when the products are in good condition, present significant hazards to the public and the environment. Obsolete pesticides stocks are even more dangerous. The effects of obsolete pesticides on health and the environment range from acute toxicity to intergenerational endocrine disruptions. They are bioactive, and hence once in the environment cannot be removed. The POPs chemicals, in particular, can be transported by climatic and environmental processes over long distances. They tend to move from warmer climate regions, where most of the obsolete pesticides stockpiles are held, to colder climates, even as far as the poles, where they are accumulating in the fatty tissues of wildlife and the top of the food chain. Persistent organochlorine compounds make up a considerable part of the total historical pesticide burdens. Other pesticide classes are more dangerous because of their acute toxicity than their persistence. Pesticides can decompose and form new products with high toxic potentials. Some of the decomposition products are more toxic than the original product. Once this process of decomposition begins, the pesticide formulation is effectively unidentified and needs to be handled as such. In this process, pesticides can also change their physical state: liquids can solidify or solids can liquefy. Similarly, many pesticides form gaseous compounds when they decompose. This can lead to high pressure being generated in the containers, sometimes causing the containers to explode, or metal drums to balloon out.

When pesticides are stored in the open, the obsolete pesticides impair the quality of life of employees working in the stores and people living in their vicinity. The air in the surroundings of the stores is often heavily fouled with the odour of pesticides, and the inhabitants complain about the strong odour, especially during the hot season. Very often they also claim to suffer adverse symptoms of exposure and acute or chronic health effects. Leaking pesticides contaminate the soil and infiltrate into the ground water with the result that highly valuable drinking water could also be contaminated. This contamination could then enter the food chain.

Obsolete pesticides are also a barrier to sustainable development in emerging countries. The disposal of obsolete pesticides is an expensive process. Developing countries have neither the financial nor the technical resources to carry out disposal operations, and therefore they have to depend on external funding. The money they need for disposal operations cannot be used for the development of the country, such as improvement of the infrastructure, the health sector or education projects.

Obsolete stockpiles also have a negative effect on the global environment. Large quantities of pesticides are contaminating soil, rivers and the oceans, where they are distributed around the world. The most dangerous are the persistent organochlorine compounds. Another critical pesticide is methyl bromide, which is used as a soil fumigant and insecticide and is well known as an ozone depleting agent.

4 Technical Solution

Obsolete pesticides and associated waste are classed as hazardous waste. Extensive training and sophisticated equipment are needed to deal with obsolete pesticides safely and appropriately. In industrialised countries entire facilities are available for the removal, transport and environmentally sound disposal of hazardous waste. In developing countries there is little expertise and no appropriate facilities for the management of hazardous waste. The problem becomes more difficult as a consequence of inadequate infrastructure, the wide dispersion of obsolete stocks and their deteriorated condition. The options available for the management or environmentally sound destruction of obsolete pesticides are extremely limited.

4.1 Disposal Methods

The various techniques for disposing of hazardous chemicals can be assigned to three different categories: thermal, physicochemical and biological methods. These encompass a small number of technologies:

- High-temperature incineration
- Incineration in cement plants
- Chemical treatment
- Solidification/stabilisation
- Engineered landfill
- Biological methods of soil decontamination

4.2 Choice of Method

The biological, chemical and physical techniques tend not to have what it takes for complete and ultimate disposal. The potential advantages would include, for example, reduction of the waste volume and/or of the environmental risks. While various research institutions in industrial countries are investigating new methods of disposal, each of them tends to focus on certain specific problems, usually on a small scale. Such methods and technologies are of only limited value for developing countries with real problems, where large quantities of obsolete pesticides and baseline pollution are in need of disposal. Often the waste in question does not consist of a pure substance but of a mixture made up of several substances that are no longer individually identifiable. In almost all cases there are also contaminated containers, canisters, pallets and soil to be dealt with. For these reasons, no single technique can be recommended. Each new case requires new deliberations and new decisions.

The question of when to use which technique always calls for a case-by-case decision, but the methods must fulfil the basic requirements, they must be environmentally sound, safe and economical. While all other methods touched upon here are feasible in theory, the most appropriate is incineration in special high temperature incinerators (HTI). Suitable waste disposal facilities are only found in small number in industrialised countries.

4.2.1 High-Temperature Incineration

The incineration of obsolete pesticide stocks in a dedicated HTI at temperatures above 1000°C with efficient gas cleaning systems is the most common and practical solution. This option also includes the use of dedicated hazardous waste in cement kilns. Apart from incineration in HTIs, one example is known where a large quantity of obsolete pesticides have been disposed of in a cement plant in a developing country. In 1996 the GTZ Pesticide Disposal Project disposed of 57 tonnes of DNOC (dinitro-o-cresol) in a cement kiln in Tanzania. By modifying cement kilns, for example by a waste introduction system, cement kilns can be rendered suitable for use in the burning of hazardous waste. To be on the safe side, the proven method is applicable to the incineration of non-halogenated pesticides in cement kilns. This is a realistic option, as long as certain technical and political prerequisites are met. From a political standpoint, through, it only makes sense as long as accompanying control mechanisms effectively preclude misappropriation of the incineration facilities by unscrupulous profiteers.

4.2.2 The "Return-to-Sender" Philosophy

Up to now there are no suitable disposal facilities in any developing country; in fact, the only solution is to take the pesticides back to an OECD country for destruction in a dedicated HTI, according to the "Return-to-Sender" philosophy. This method has been used over the last 12 years. In a number of disposal operations several thousand tonnes of pesticides have been safeguarded, packed and shipped to the United Kingdom, Holland, Finland and Germany for final disposal in HTIs. This approach also entails environmental risks and is expensive. It involves huge input, beginning with the tendering procedure, for there are very few disposal enterprises that are really focused on the special situation of developing countries. Usually the scope of administrative tasks is underestimated. The legal basis for the transboundary movement of waste, in the case of repatriation of the waste back to the country of origin, is provided by the Basel Convention [6] and EU Directive 259/93 [7], as long as the country in question has no alternative.

The industrialised countries also bear a special degree of responsibility in connection with pesticides or other hazardous chemicals, for it was they who in one way or another were involved in the supply of most of the pesticides in the first place.

5 Framework Required

Whenever the GTZ has conducted disposal operations the following general requirements have been applied: disposal operations could not be a one-party operation, and the partner organisation, the owner and/or generator of the waste, must play the most active role during the operation. The GTZ prefers joint disposal operations in which as many stakeholders as possible are jointly involved in the execution of the entire operation and likes to act as a mediator between the different stakeholders.

5.1 Stakeholders

The term stakeholder is often used in disposal and other aid operations. Everybody can be regarded as a stakeholder – the only variable is the level of responsibility. Generally speaking the main stakeholders are the developing country, the former manufacturer of the pesticides and the commercial disposal company, as well as the donor organisation. The duty of each partner organisation is specified in a document such as a Memorandum of Implementation, for example.

The government of the developing country must nominate a local agency as the partner organisation – this is mostly the owner of the waste – which has to play an active role in the disposal operation. The partner organisation has to take full responsibility for the project and has to execute all the administrative steps required to meet the requirements of the Basel Convention pertaining to the transport of waste to an OECD country, including application for the import licence to the country in which the waste is to be disposed of. The partner organisation is also fully responsible for all the administrative steps within its own country, such as customs clearance and writing off the waste, and has to contribute to the operation, for example with local staff and transport. The success of the project mainly depends on the support, commitment and competence of the local government agency and their staff. The commercial disposal company has to support the partner organisation in the process of applying for the import licence. Usually these companies carry out the field work and organise the shipment as well as the final disposal in the HTI. A number of pesticide manufacturers are taking the principle of product stewardship very seriously and are supporting the disposal operations by providing technical and financial assistance. Since 1995 the former GCPF member companies have given an official commitment to the FAO that they are providing assistance for stocks that they manufactured or supplied. The donor's role is merely to offer the owners of the waste the financial, administrative, logistical and technical support they need during implementation of the disposal operations. An important factor is supervision of the work of the commercial disposal company.

5.2 Costs and Funding

The removal and destruction of obsolete pesticides is an expensive process. The cost of toxic-waste disposal, and the procedures to be employed, depend on a number of parameters, such as the quantity and type of waste, the type of packaging, the location where the waste is stored and the infrastructure, as well as the method of disposal. After all, it costs between US$ 2,500 and 4,500 per tonne to return waste to an industrialised country. The above figures cover the cost of everything from stocktaking to final incineration.

There are different ways of financing the entire disposal operation. The bilateral approach is the normal case, where the developing country sends an official request directed to the aid organisation. The second possibility is to seek technical assistance directly from an international agency, such as the FAO. In this case the FAO requests the funding for the disposal operation from third countries.

5.3 The Practical Approach

Every disposal operation contains a number of steps, starting from the identification of the problem. Sometimes this is the most difficult step, because the owners or responsible agency in the development country seldom accept that they made a mistake. Without any information on the ongoing situation on site and the quantity of obsolete pesticide stocks, disposal operations are virtually impossible to plan – therefore an analytical survey is an imperative part of any disposal operation. Surveys of this kind should also include a risk assessment of the actual situation in and around the storage site, as well as the dangers for the participants, the public at large and the environment. This information provides the basis for the safeguarding and disposal operations, as well as financing and the subsequent implementation of operations. Before the first drum of pesticides can be moved, countless administrative hurdles have to be taken. The question of financing has to be solved first. All parties involved have to sign a Memorandum of Implementation which specifies the respective duties and responsibilities. The export of waste to an OECD country requires a number of difficult and lengthy diplomatic activities, especially if the country is not a member of the Basel Convention. Obtaining the licence for importing the waste into an OECD country is very complicated. In comparison, the technical and logistical questions are easy to tackle. If the pesticide store is close to a harbour, the logistics of the project are certainly simpler than a project with stores thousands of kilometres away from the nearest port.

5.4 Disposal Operation and Prevention

The necessary field work is planned and executed on the basis of the data from the analytical survey and the plan of operations. A number of practical steps are included, such as execution of immediate safeguarding measures in the event of acute hazard potential becoming evident, the actual field work for repackaging within the developing country, shipment to the country of final disposal, and disposal itself. Special attention must be paid to safety aspects during execution of the field work. The last step, and an integral part of any disposal operation by the GTZ, is the establishment of prevention measures to avoid the further accumulation of obsolete pesticide stocks.

6 Strategies, Policies and Practical Steps

The problem of obsolete pesticide stockpiles and their disposal as well as the prevention of the formation of new pesticide stockpiles is embodied in a number of national and international laws, guidelines and conventions:

- Basel Convention: This is the most important convention. It is primarily concerned with the control of the international movement of hazardous waste, including obsolete pesticides, and their management. It is the legal basis for the "Return to Sender" approach.
- Directive 259/93 of the European Union: It regulates the import of hazardous waste back to the country of origin, mostly OECD countries. This was actually an instrument to prevent "waste tourism", but it is also the most important tool and the legal basis for the re-import of the obsolete pesticide stocks back to Europe
- Montreal Protocol: This counters the depletion of the ozone layer of the atmosphere and is mainly related to ozone depleting substances, such as methyl bromide, a widely used pesticide.
- Rotterdam Convention: This improves the exchange of information and the observance of import decisions in international trade with certain hazardous chemicals. The convention is internationally known as Prior Informed Consent (PIC).
- Stockholm Convention: This bans the production and use of certain Persistent Organic Pollutants (POPs) or severely restricts release. It applies to plant protection agents, industral chemicals and highly toxic by-products and incineration processes, such as dioxins and furans.

Both the PIC and the POPs conventions are mainly focused on pesticides and are important for preventing the formation of new stockpiles. Of the 12 persistent POPs currently targeted by the Stockholm Convention, 9 are pesticides.

There are also a number of guidelines which provide the basis for classification, packaging, labelling and safe transportation by sea to the point of final destruction. One is the International Dangerous Goods (IMDG) Code. Similar legislation exists for road, rail and air transportation.

The environmentally sound disposal of waste in an HTI is regulated by national laws and the European Directive on the control of dioxins and furans, the by-products of the incineration process.

6.1 The Projects and Difficulties

Administrative problems are the most time-consuming ones. From problem identification until final disposal of the obsolete stockpiles, two years and more are a realistic time frame. Very often long diplomatic negotiations are necessary for a bilateral agreement between the countries where the waste has been generated and the country of final disposal. Compared to the long planning phase, the actually fieldwork is normally carried out within weeks.

It is always difficult to find a suitable shipping company for transporting hazardous waste. They fear problems from environmental groups as well as problems during stopover in a third country. Under the Basel Convention there is an established notification process to inform third countries in advance about the transport of hazardous waste, but some governments do not want foreign waste to pass through their territory.

Climatic conditions can also delay the execution of disposal operations. It is difficult to work under tropical conditions with heavy protective gear. Political problems and unrest in the developing countries can also delay disposal operations or bring them to a complete halt. In the past the GTZ has had to cancel two disposal projects, one in Congo and another in Sudan.

6.2 Strategies and Politics

In the past different donor organisations have worked on the same problem but without co-ordination. The possible synergistic effects that would save financial resources have been ignored. Co-ordination should be taken care of by the FAO.

Developing countries do not support regional approaches, such as the idea of building a local incinerator or a disposal centre in a developing country. This idea has often been discussed. Only a few countries in Africa have enough waste to make operation economic. They have neither the technical know-how nor the financial resources to run an incinerator on their own. There is also a lack of understanding in developing countries that the environmentally sound disposal of waste costs money and must be paid for by the generator of the waste. Therefore no national donor or international organisation is willing to spend money on the realisation of a disposal facility in Africa or any other developing country elsewhere - it is more economical to send the waste back to an OECD country for disposal.

6.3 ASP Programme

During the "First African Conference on Obsolete Pesticides Stocks" in Rabat in January 2001, all African countries demanded in the Rabat Declaration directed to the OECD countries the urgent removal of the obsolete stockpiles in Africa.

The commonly used procedure – the country-to-country approach requires each country to seek funds, develop project proposals and deal with different national donor organisations, contractors or other relevant agencies individually – had limited success. A re-

gional programme, facilitated and managed by international organisations, and relevant regional partners would help to reduce or overcome many of these obstacles. This would facilitate more rapid progress in dealing with obsolete pesticides in developing countries. This was the background to the World Bank and the FAO developing the African Stockpile Programme (ASP), together with a number of other international organisations such as UNEP Chemicals, UNIDO, the Secretariat of the Basel Convention, the Organisation of African Unity (OAU) and the African Development Bank. Some non-governmental organisations such as the WWF and the Pesticide Action Network (PAN) from the UK and Africa are also involved. The target of the ASP is to remove the 48 000 tonnes of obsolete pesticide stockpiles in 53 African countries within a time frame of 13-15 years. Based on that approach, the total costs for clearing the entire continent of its obsolete pesticides would be roughly US$ 150 – 175 million. The ASP will also develop special measures to prevent new obsolete stockpile accumulation. The total budget within the ASP for the prevention measures is estimated at US$ 50 – 75 million, which brings the total ASP fund to an amount of US$ 200 – 250 million.
The funding for this programme should come from the Global Environmental Facility (GEF) of the World Bank. At this stage it is anticipated that approximately 30 % will be contributed by the GEF. The organisers of the ASP programme will also seek funds from bilateral and multilateral donor agencies, private foundations and the pesticide manufacturers. Crop Life International, the federation of the manufacturers of pesticides, is therefore also in the steering group for the ASP programme. The organisers of the programme are hoping to start with the first pilot activities in prioritised countries in spring 2003.

7 Summary

The problem of obsolete pesticide stocks in developing countries became serious at the beginning of the 90s because of the risks for human health and the environment. After a decade with a number of individual disposal operations organised and financed by national aid organisations, the international community has put this issue on their agenda, aiming to solve the problem through a number of international conventions and activities, e.g.:

- Montreal Protocol
- Basel, PIC and POPs Conventions
- African Stockpile Programme

The latest and most ambitious initiative to solve this problem is the African Stockpile Program (ASP) – a joint initiative of the World Bank and the FAO.

8 References

[1] FAO, Inventory of obsolete and unwanted and/or banned pesticide stocks in Africa and the Near East, 003/1999.

[2] IPCS (International Programme on Chemical Safety), The WHO-Recommended Classification of Pesticides by Hazard and Guidelines to Classification, 1990-1991.

[3] GTZ Publication, Obsolete Pesticides, a Dangerous Legacy. Result of a Pilot Project on the Disposal of Obsolete Pesticides, can be found under http://www.gtz.de/chs

[4] FAO Pesticide Disposal Series 2: "Prevention of accumulation of obsolete pesticide stocks", Provisional Guidelines, Rome, 1995.

[5] UNIDO/ILO Technical Report Series No. 3: "Storage of Hazardous Materials – A Technical Guide for Safe Warehousing of Hazardous Materials".

[6] UNEP/SBC/94/3: Basel Convention on the Control of Transboundary Movement of Hazardous Wastes and Their Disposal, Geneva, June 1989.

[7] EEC Directive No, 259/93: Regulation on Transboundary Waste Shipment, February 1993.

05 A Food Processor's Efforts to Minimize Pesticide Residues in the Food Chain

Hans Jöhr and Eduard Bruckner

Nestec Ltd., Avenue Nestlé 55, CH 1800 Vevey, Switzerland

1 Introduction

The key objective of the food industry is to manufacture food products that satisfy customers and consumers' expectations. Hence, product quality and safety are of primary importance and customers must be delighted with the products and the services provided by the company. While product quality may leave room for individual interpretations and priorty setting, safety of food products is not negotiable. Safety is, therefore, no sales argument; it is a precondition for bringing a product on the shelves. Customers have the right of being ensured that food products they buy are safe for their health and meet all legal safety requirements.

This chapter gives a short overview on Nestlé, highlights our business objectives and explains the importance of agricultural raw materials for our business. It describes the objectives of sourcing, the different sourcing strategies and the quality assurance programs we apply to ensure the safety of our food products.

Food companies are responsible for the products they sell; however, many safety obstacles might be caused by activities upstream in the supply chain and out of our direct control. Food quality and safety effectively begin with raw materials. It is, therefore, very important that everybody in the food chain, from farm input suppliers to the smallest retailer is aware of his responsibility and acts accordingly.

With the increasing complexity of the food chain and increasing safety requirements the importance of this collaboration between all stakeholders becomes more and more evident. Recent food scares unfortunately underline the need for more cooperation.

2 Background on Nestlé

2.1 Business Objectives

The basic business objective of Nestlé is to produce and market food products that satisfy customers and consumer expectations. It is our objective to produce and to sell good quality food and to provide our customers value for money.

In our corporate business principles we point out that in the 136-year life of Nestlé, our fundamental approach to business has been the creation of long term sustainable value for our customers, consumers, employees, shareholders and society as a whole. We state openly that we favor long term business development over short-term profit. We aim for long term relations with our customers but also with our suppliers. This includes favoring sustainable agricultural production methods because we consider them as the most reliable ones for the long-term supply of agricultural raw materials of high quality and at competitive conditions.

2.2 Nestlé History and Key Products

Henri Nestlé, a German citizen, scientist and businessman, founded Nestlé in Switzerland in 1866. His first product was a "Kindermehl" or "Farine Lactée", an infant nutrition product made out of milk, flour and sugar to help mothers who where unable to breast feed their children. Meanwhile, Nestlé has become the world's leading food company with a turnover of 47.3 billion US $ in 2001. We produce in 468 factories in 84 countries and employ almost 230'000 people. Our products are sold in practically all countries of the world.

Nestlé is the leading company in Soluble Coffee, Infant Nutrition, Confectionery, Dairy, Chocolate, Milk Beverage, Bottled Water and PetCare. We are number 2 in Ice Cream and Breakfast Cereals, and are globally present with the trademarks Nescafé, Nestlé, Nestea, Maggi, Purina and Buitoni and give local flavor to our products. Other known Brands are, to mention only a few examples: Thomy, Alete, Carnation, Perrier, Vittel, Friskies etc.

By giving this background, we underline that we work under very different conditions and with a wide range of raw materials that require individually defined sourcing strategies and quality assurance programs for raw materials.

3 Importance of Agricultural Raw Materials

The importance of agricultural raw materials for our food products is manifold and especially evident in product quality and safety as well as in production expenses.

3.1 Raw Materials and Production Expenses

The importance of raw materials in production expenses is expressed in the raw material expenditures related to our total turnover. In our Management Report 2001 we show that we spent 23.1% of our total turnover or CHF 19.7 billion for raw materials. Packaging is not included in this amount and agricultural raw materials represent the major part of this cost segment.

Although Nestlé is an important user of agricultural raw materials, we have no influence on general price developments, e.g. world market prices for coffee or cocoa or milk prices, just to mention our most important raw materials. These prices are very much market driven or influenced by political and trade regulations. However, changes in raw material prices affect our business performance because not all of those changes can be fully transferred to the final product prices.

3.2 Raw Materials and Food Quality

Food quality is a driving force in our business. It includes many aspects and can be looked at from different angles. On the one hand, quality can be judged very personally, since each consumer has his own preferences but, on the other hand, there are also common criteria. We can group them all into three quality dimensions:

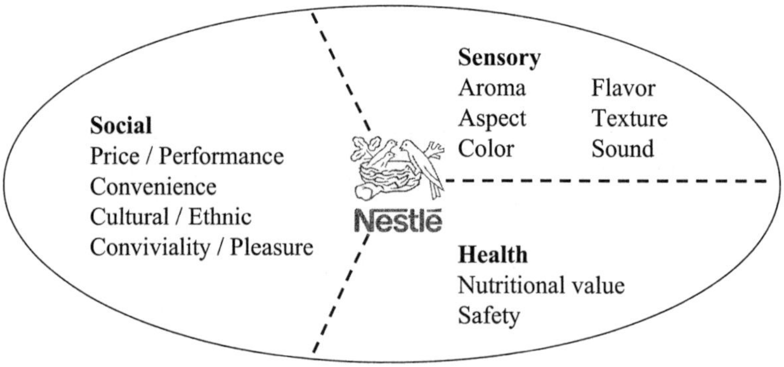

Figure 1. The three dimensions of food quality and their criteria

Agricultural raw materials play an important role in all three dimensions. Thus, in the social sector the price of food products is of key importance, but also other aspects such as ethics or pleasure are relevant. With regards to the sensorial or health dimensions, agricultural raw materials may play a role in every single aspect from aroma over color and flavor to nutritional value. The importance of raw materials on visual aspects of processed food depends on the processing steps and the presentation of the finished product which might be less effected by raw materials.

The effect of raw materials on safety aspects is again of very high importance. Poor quality or contaminated raw materials cannot be improved or "cleaned" during processing which can only make the best out of a given material. Considering the different processing methods and the range of raw materials used, we cannot expect a degradation of contaminants to non-toxic substances during the processing procedure, be these pesticides, heavy metals, mycotoxins or others.

3.3 Key Raw Materials

As already pointed out, Nestlé uses a wide range of raw materials originating from agriculture. The most important ones are milk, coffee, cocoa and cereals as far as we use them as fresh or primary processed at the farm or producer level (e.g. fermentation of cocoa beans). Beside these key raw materials we also need a range of other field crops, as well as of fruits and vegetables that we may buy as fresh or primary processed in form of juices, paste, dehydrated etc. Meat, sugar, oils, spices and other food ingredients, we buy exclusively in processed forms, as we do not have the related primary processing facilities or slaughterhouses.
Water is another "raw material" of high and even increasing importance for our water business. It is also a sensitive raw material concerning its quality, and we have to make special efforts to ensure its safety and to avoid contamination. However, as water is not an agricultural raw material, it is not further considered in this presentation.

4 Objectives of Sourcing

Sourcing of agricultural raw materials has to be in line with our overall business objectives and has to contribute to sustainable and profitable development of our business. Thus, it has to contribute to food quality and safety and to competitive pricing.

4.1 Sourcing for Competitive Overall Costs

When sourcing raw materials it is important to have the overall production costs in mind. This means raw materials have to fit our needs concerning timing, to allow the most economic utilization of our processing installations and to provide an optimized processing yield. This includes minimized processing losses and ensured quality of finished products.
The price effectively paid for a raw material is in this context important but not the only criterion. The price has to be seen in the context of the overall manufacturing expenses of a food product and the quality criteria expected from the finished product.

4.2 Sourcing for Quality Production

"Quality Raw Material" is an expression that has to be defined by the individual user of a raw material, for specific production processes and for the production objectives, i.e. for every product. The "quality", i.e. the required physical/chemical parameters of a raw material vary and are product specific; however if it comes to product safety, there is no room for interpretation. Our products must be safe and fulfill all legal requirements given in a specific country or imposed by international regulations.
We underline once more: raw materials must allow us to produce safe food products. We have no or only very limited possibilities to reduce safety hazards that have their origin in raw materials during processing.

5 Strategies of Sourcing

The sourcing strategies applied consider the objectives mentioned, as well as the proprietary characteristics and the availability of the raw materials at the required specification. Raw materials may be sourced through the trade or directly from farmers or farmers´ organizations. Nestlé has no own commercial farming activities.
Part of our overall sourcing strategy is supplier pertnering. It optimizes the value created for both the buyer and the supplier by

- concentrating on suppliers capable of maintaining competitive costs, high quality and technology leadership,
- sharing common goals of achieving lowest industry systems costs,
- seeking savings that will provide a mutual incentive to eliminate costs with the aim of attaining a sustainable competitive advantage through an integrated approach focused on supply chain opportunities.

5.1 Sourcing through Trade Channels

We source all raw materials through the trade or from primary processing companies as far as they are available or can be produced at the required specifications. We may either buy these raw materials on international commodity markets (e.g. most of our coffee, cocoa or sugar), or from “approved suppliers" selected through a cross-functional teamwork within Nestlé. The assessment procedure may begin with a commercial assessment to ensure the suppliers´ reliability and capacity to deliver the required volumes, his price and his acceptance of delivery according to agreed specifications. Quality assurance specialists assess the risks for the individual raw material and complete the approval of a company, a specific production line or a product. In the approval phase we explain suppliers amongst others our expectations regarding traceability, notification of price or of recipe changes, compliance with national or international legislation and eventual technical assistance for the production of agricultural raw materials.

The approval is product specific – to best value for the price. The assurance of product quality and safety are of key importance. The more sensitive materials are, the more in details the whole auditing procedure has to go. For instance, producers of raw materials for baby food have to ensure traceability up to the farm and to the field where a raw material is produced. They also have to prove their capacity to instruct farmers to produce raw materials allowing European Nestlé companies to comply with the newly implemented limits for pesticides in baby food, i.e. residues below 10 ppb.

5.2 Sourcing Directly from Farms

We source raw materials directly from farms if we need them in fresh conditions, if they are perishable or if the quality we need is not available on the open market. In these cases we also aim for partnering with our suppliers, and for a long-term supplier relation based on confidence and on control systems. We want to ensure the quality we need by motivating farmers and by getting their right mind set to constantly improve the quality of their produce and to optimize the quality-cost-relation.

Directly from farms we source most of our milk requirements, part of our coffee needs and most of the fruit and vegetable we need for the production of baby food. The latter are most difficult to find in the open market, because a certain percentage of these fruits and vegetables have pesticide residues above 10 ppb. These residues are in almost all cases within the legal limits for common food products and without any health risk. However, for our needs it would be too costly, too risky and logistically not feasible to test all raw materials before we could release them for the production of baby food.

With contract growing, done by ourselves or by approved suppliers, we reduce the risk of contamination with pesticides and other contaminants that play an important role for this product group, which are e.g. heavy metals, nitrates and mycotoxins.

5.2.1 Agricultural Services at Factory Level

Nestlé has a long tradition in collaborating with farmers who supply raw materials to Nestlé factories. Where required we engage agricultural staff to assist farmers achieving our quality requirements. These services are also responsible for the collection of raw materials and, in collaboration with quality assurance specialists, for the quality control of the raw materials received at a collection center or at the factory gate.

Where needed and feasible, our agricultural experts provide advice on the use of pesticides and fertilizers. They do this directly or, wherever possible, together with official advisory services and in line with local, official recommendations.

When providing this assistance to farmers, we always aim for long term collaboration, for environmentally friendly production and for correct prices based on the quality of the produce. However, with changing knowledge, regulations, market situations, technologies, customer and consumer expectations we are now in the process of re-drafting our collaboration with farmers. In future we will, as much as possible, include all three pillars of sustainable development in our collaboration with the farmers (People, Profit and Planet, or with other words, Society, Economy and Environment). This requires more

involvement of all stakeholders within the food chain, beginning from the input suppliers over the trade down to the consumers and their requirements.
We consider this new orientation of our collaboration with farmers and raw material suppliers as the "sustainable agriculture initiative of Nestlé."

5.2.2 Contract Growing for Baby Food

Contract growing for baby food is for Nestlé the most complex sourcing method with very detailed instructions for suppliers. The procedures include the selection of the growing area, of farmers who are able and willing to follow our instructions, the selection of varieties, soil tests before growing and instructions concerning the use of fertilizers and pesticides.
The selection of the pesticides that we recommend is based on many aspects and includes: the registration status, product efficiency, mode of action (systemic, contact), degradation in crops, in soils under given climates, toxicity, metabolites, application techniques, waiting periods, secondary effects (predators), risk of resistance (within control programs), costs, alternatives, future of product (pressure on use) and analytical possibilities (limits).
The list of criteria is very long and requires in many cases more knowledge than available on the labels or in published documents. A critical area in this respect is the degradation of chemicals down to a level below 10 ppb, which is usually not studied or published for normal registration of pesticides. In order to obtain this information we sometimes need close contacts with agrochemical companies and an exchange of possibly confidential information on active substances and their metabolites.
Sourcing for baby food is for us the most important and most striking example for strong efforts at the very beginning of the supply chain to minimize residues of pesticides in the food chain. We have to do this, because otherwise we would not be able to ensure the quality we require, and the production at reasonable costs.

6 Quality Assurance

Whenever feasible, the Nestlé quality assurance scheme begins at the factory gate, at the level of the suppliers and possibly beyond the direct supplier.

6.1 Quality Control at Farmers Level

Quality control at farms is only carried out in exceptional cases, i.e. if special interests or cases of doubts prevail. However, it is frequently done if farmers produce raw materials for baby food. In this case, controls of raw materials may be made even before harvesting a crop. For milk collection in developing countries, Nestlé has many collection centers at village levels, which function similar as the receiving stations at factories. There we test basic physical/chemical quality parameters of the milk, which are important for paying and motivating the farmers to produce the quality needed. Residues of pesticides are normally not controlled at this level. Their control is the subject of monitoring programs carried out at factory levels. In cases of contamination, our traceability program allows us to identify the sources of contamination and to take the necessary corrective actions.

At farmer's level it is our objective to avoid quality obstacles, instead of controlling the produce afterwards. Collaborating directly with farmers gives us the most efficient control mechanism concerning production methods of our raw materials, the use of pesticides, veterinary drugs or other chemicals that might help in the production process but could pose problems in case of misuse.

6.2 Quality Control in the Upstream Supply Chain

Quality control in the upstream supply chain is part of our supplier approval system and regular supplier auditing. The applied control systems are based, as within our own factory premises on the HACCP systems (Hazard Analysis on Critical Control Points). These quality control systems have been elaborated for each individual raw material. It is our objective to source our raw materials as far as possible from approved suppliers; these are suppliers who are able to guarantee the qualities we require.

6.3 Quality Control at the Factory Gate

Quality control at the factory gate precedes the final release of materials into the production process. Depending on the raw materials and the tests to be carried out, all factories are equipped with the necessary control systems and facilities.

As far as chemicals, especially pesticides, are concerned, they are normally controlled in our regional laboratories. The factory laboratories have only limited control possibilities, but may be able to check for the most critical products or indicators for quality obstacles. For controlling pesticide residues in raw materials for baby food in Europe, we engage a limited number of officially certified Nestlé laboratories. They network with third party certified laboratories to cross check the analytical methods applied, and to support each other at peak periods or if critical results need to be verified.

7 Conclusions

Since the possibilities of food processors to minimise pesticide residues in the food chain are very limited, the title of this chapter may be misleading. The use of crop protection chemicals in agriculture is governed by complex regulations, market forces and farmers' practices and experiences.

The business objective of Nestlé is to produce and market quality food products that satisfy consumer expectations in all respects. Our label on the products is a quality seal. Therefore, we are responsible for the safety of our products, but many safety obstacles, especially those linked to agricultural production methods, are high up in the supply chain of our raw materials and, in most cases, out of our control. Our possibilities to control pesticide residues in our products are, to a high degree, limited to a control of incoming raw materials. We rarely can control their use at the first stage but there are critical points and possible sources of obstacles.

Food safety concerns us all and potential residues of pesticides in food products are an important aspect of it. However, we have to see the whole issue in a wider context: the sustainable development of agriculture in general. Farmers need the necessary economic and a stable social environment to effectively focus their business objectives on the quality of their produce. They also need reliable sources of their inputs such as animal feed or chemicals, and they need full information on how these inputs and their production methods can affect their own produce and their natural environment on a long term.

Together with the Group Danone and with Unilever, we now have created SAI, the Sustainable Agriculture Initiative of the food industry. This initiative shall become a platform for the food industry to support sustainable agriculture in the pre-competitive field of agriculture in general. For our key raw materials we work directly with our suppliers. Concerning agriculture in general, individual food companies do not have the means to make major improvements on their own. We will integrate SAI into CIAA, the EU Confederation of the Food and Drink Industry, but at the same time search for collaboration with all stakeholders of the food chain, including the chemical industry.

Ernest Hemingway said: "No man is an island entire of himself". We can use this phrase for our subject: nobody in the food chain can work exclusively on his own. Whatever one does, it will affect the next link of the chain. The consumer, being at the end of the chain, he/she finally decides whom he/she will trust and which product he/she will buy.

Food safety scares are not only a problem for individual companies. They can affect entire industrial segments, such as the food industry as a whole together with its suppliers, farmers and even the providers of farm inputs.

Public trust in normal, everyday food products has somewhat eroded in recent years although our food has never been as safe as today. The imbalance between this reality and a mythical perception concerning the quality of our food needs to be neutralised in order to regain full confidence in food products. This needs homework in our direct fields of activities, but also efforts in ensuring an adequate political and regulatory environment for our business which must be based on sound scientific knowledge and not on short term, opportunistic political orientation.

06 The Risk of Eating: The Toxicity of Natural versus Man-Made Chemicals.

Aalt Bast

Department of Pharmacology and Toxicology, University Maastricht, Faculty of Medicine, P.O. Box 616, 6200 MD Maastricht, The Netherlands. A.Bast@farmaco.unimaas.nl

1 Introduction

Food poisoning was a major concern in ancient times with emperors and pharaohs hiring food tasters to protect themselves against criminal assaults. The bible mentions food tasters in Genesis 40, 21-22 in the well-known story of Joseph and the chief baker and chief cupbearer. Joseph meets these two employees of the pharaoh in prison where he was requested to explain their dreams. Both the cupbearer and the baker were probably food tasters of the pharaoh but the story does not tell why they were in prison. Had they been accused of making a mistake? The job of a food taster was an important one. Only a few highly esteemed priests were allowed to hand over food and drinks to the pharaoh. Such an important job could not, of course, be carried out by an ordinary man. To the toxicologist, it is appealing to think that the respected food taster in ancient times is comparable to the modern-day toxicologist. The story ends and the cupbearer's and baker's dreams come true. Genesis reads: “He (the pharaoh) restored the chief cupbearer to his position, so that he once again put the cup into Pharaoh’s hand, but he hanged the chief baker”. A food taster is apparently a risky occupation but fortunately the resemblance of the food taster to a modern toxicologist ends here.

2 Illusive Food Safety

The food taster provides their employer with a false sense of security. In fact only one grape, untested by the food taster, could contain the toxic agent. The effect of the toxin might also have a slow onset of action in which case, a food taster would not provide any protection. As a result of some recent food scandals, politicians have appealed for modern food tasters. Consumers have a right to receive absolutely safe food. Toxicologists

should control food products more thoroughly. Not only should product safety be guaranteed but even the integral food chain ('from farmer to plate') should be controlled.
The scientific food taster uses analytical chemical tools to detect unwanted chemical contamination. Thankfully the toxicologist's own life is no longer at stake. But what do we want to check, where do we look and finally what is absolute safety?
Man depends on food and wants to believe that the food he consumes is safe. That is a deep-rooted social and individual feeling. When safety is not guaranteed, this elicits an intense emotional response. For example, complaints arising from the alleged pollution of Coca-Cola cans in Belgium could not be solely attributed to the contamination of the drinks. In fact, the health complaints were thought to be largely due to a form of mass hysteria. The drinking of Cola is thought by young people to represent their freedom and independence and so is a way of making a statement about their lives. The putative contamination of the cans shook the unassailable last cultural certainty of youth: now even Coca-Cola could no longer be trusted. To further compound the problem, the report on the contaminated Coke cans was also included in a series of other reports on pollution such as BSE in cows and dioxin in poultry. Media coverage of the alleged irregularities surrounding the investigations into the contamination of the cans again fueled public concerns. This all led to the hypothesis that the health complaints reported were the result of a mass sociogenic illness [1]. Food safety is of utmost importance to the consumer. In order to attain this feeling of safety, the consumer believes all kind of toxicological myths.

3 Toxicological Myths

3.1 Homemade and Therefore Safe

The consumer generally believes that you only know what you are eating if you prepare the food yourself. However, this is by no means the case. Many new reaction products arise during food preparation. Fatty acids oxidize, proteins fall apart and the fragments react with fat oxidation products, vitamins oxidize to form new compounds, etcetera. The effects on health of most of these new reaction products are largely unknown. The duration of exposure to these new compounds is long, even an entire lifetime.
A skilled organic chemist will be able to make a new chemical entity each week. If the new compound is aimed to have a certain biological activity a lot of its properties will be known at the time of its application and testing. How different are the working conditions for the cook. Within an hour, a cook will be able to synthesize hundreds of new chemicals by mixing, heating, frying and boiling food constituents in his laboratory, which is more commonly referred to as the kitchen. It is somewhat surprising to notice that the consumer distrusts the well-characterized pure products of the organic chemist (e.g. drugs or pesticides), whereas the suspicious multitude of chemicals produced by the cook are administered orally everyday at mealtimes without any hesitation at all.

A recent study in Sweden reported relatively high quantities of acrylamide, a carcinogenic compound, to exist in food, particularly when carbohydrate-rich foods such as potatoes, rice or cereals were heated [2]. As a result, the World Health Organization convened an emergency meeting in June 2002 to evaluate this and decide what action to take.

3.2 Food has a Long Use History and Should Therefore be Safe

This is a fundamental mistake. Many crop plants have only been cultivated over the last centuries. New, previously unknown edible plants from remote areas now reach us and so we consume compounds to which we have never previously been exposed. The majority of the constituents of these new plants are unknown as are the harmful effects of these new chemical mixtures on our health.

3.3 Food is Natural and Therefore Safe

In this toxicological myth, the word 'natural' is thought to be synonymous with safe. The same school of thought associates the word chemical with danger. The conviction of this myth is widespread. It forms the basis for organic farming in which no pesticides, herbicides or artificial fertilizers are employed. The provision of extensive governmental subsidies to organic farmers does not help dispel the myth. The fact that in organic agriculture, crops with a high disease resistance are selected is also not widely reported. One of the ways to obtain disease resistance crops is to use crops with a relatively high level of natural toxins. These compounds with an insecticide action are abundantly present in plants and so in this way plants possess a natural defense against fretting/gnawing. This of course offered an evolutionary advantage as plants cannot run away if they are attacked. However, many of these natural toxins are also toxic for man [3]. Furthermore, biological or organic farming only yields around 70% of the crop produced by regular farming and so more agricultural land is needed to obtain similar yields. The chance of detrimental weeds such as nightshade in peas or moulds like ergot in rye becomes correspondingly greater.

Nevertheless, the obstinate consumer clings to the natural-is-safe myth and politicians follow this opinion. An appropriate and balanced view in which human toxicological and eco-toxicological advantages and disadvantages are weighed against each other is lacking. Unfortunately, the voice of the scientific food taster is too quiet.

4 Risk Perception

Food risks that are regarded as important by the consumer are in most cases related to man's interference. In other words, additions to food, which are perceived as unnatural, are thought by the consumer to be the greatest risk. It seems that the consumer believes chemistry to be responsible for any question marks over food safety [4].

The risks most obvious to the public and politicians are the most thoroughly investigated ones. Crop protection chemicals, for example, undergo a thorough investigation before they are allowed to the market. Why the fear? Risk is not just a strict numerical formula by which politicians and the public can make an ultimate weighting. Mere knowledge is not sufficient for public or political risk appraisal as the concern over risk also plays a key role. Therefore risk is also dependent on psychological and social-cultural factors and risk perception plays a much greater role than may have been previously thought. Also of major importance is whether the exposure to the risk is voluntary or forced. Cigarette smokers take early death as a result of smoking related diseases for granted because they determine the risk themselves. However the presence of certain crop protection chemicals in toxicologically insignificant amounts is regarded as serious because of the forced exposure. In other words, the piece of chicken that we would like to barbecue should be free from dioxin. However, the consumer ignores the carcinogenic polycyclic aromatic hydrocarbons that emerge during the barbecue to coat the chicken.

Public policy disputes over food safety is concerned with the acceptability of a given risk. However, the final judgment is more likely to depend on public outrage than on the toxicological size of the risk [5].

The consumer of today has a different risk perception then the consumer in ancient Egypt. In the western world food is in abundance. The modern consumer can permit himself to focus on the health effects of food rather than on a possible food shortage. The market for functional food and enriched food is rapidly growing. Proof for the positive health effects is rarely requested. Safety studies in relation to the benefit of these food products are not mentioned. The consumer asks and industry provides. Companies truly interested in improving the public's health will be characterized by supporting their health claims with sound scientific evidence.

The modern consumer is confronted with relatively small accidents involving only kilograms of food. When compared to the huge volume of food that is produced daily, this is an extremely small amount but trifles are considered to be newsworthy. Newspaper headlines will shout the news on the same day and in this way the putative risks enlarge and become unreal. Consequently, the distrust of consumers regarding food safety increases. The western consumer can afford this luxury. The real risks, however, are food shortage and genuine deficiencies in the third world.

Sometimes even scientists fall into this trap and start investigating ghost stories. An illustrative example is the suggestion that the decrease in the male to female ratio among newborn infants during the last decades was due to environmental factors. Environmentalists (followed by the media) claimed that this was a direct result of the extensive use of pesticides. One keen science journalist [6], however, noted that this decline in the male to female ratio at birth followed a peak in male births occurring during World War

II. A similar peak in male births had been observed during World War I (in Germany but not in the neutral Netherlands) and was also followed by a post war decline in the male proportion of newborns. He was therefore able to conclude that the decrease in the male to female birth ratio was in reality unrelated to environmental factors. However, it will undoubtedly be very difficult to stop the now nonsensical discussion on the relationship between pesticide use and decreased male to female birth ratio.

It is curious that socially accepted and favorite fears are frequently irrational and exaggerated whereas the risks that really matter are not weighed sufficiently [7].

5 Precautionary Principle

Food risks that are associated with human interference appear to receive priority in public policy. Additions to food are regarded as unnatural and therefore hazardous. From a scientific viewpoint, these most eye catching issues tend to be the most thoroughly investigated. Agents for crop protection are rigorously tested before they are admitted to the market. Nevertheless, public policy instinctively calls for a ban on or at least a very restrictive use of these agents for crop protection. Better safe than sorry seems to be the motto. This is also called the precautionary principle. Who can be against this principle in order to minimize risks? It seems simple. However, the threats associated with a precautionary ban on the use of crop protective agents are manifold.

Farming without the use of crop protection agents, so-called organic farming, retains several drawbacks: low crop yields, high levels of natural toxins, detrimental weeds or moulds. Low harvest yields will lead to increased prices. Price increases will result in replacement strategies, which may lead to less consumption of fruit and vegetables. It is, however, known that a high intake of fruit and vegetables ('five servings a day') has a positive health effect with the incidence of certain tumors shown to be reduced by 10-50%. Crop protection indirectly contributes to public health by maintaining low prices and stimulating the consumption of fruit and vegetables.

A restrictive admission policy will lead to a climate in which development of new crop protection agents, which have a more specific action and which are less damaging to the environment, is hampered.

6 Functional Food and Nutraceuticals

The health benefits of food and food constituents is currently gaining much attention. New biotechnological techniques allow the enrichment of food with micro- and macronutrients such as rice with high levels of vitamin A, tomatoes with extra flavonoids and also milk with additional calcium. These functional foods are often connected with health benefits. Specific food constituents in a pharmaceutical formulation are called nutraceuticals. These are frequently associated with medical claims.

Pharaohs regarded food to be the bearer of both divinity and vital strength. The current interest in the positive health effects of functional foods, nutraceuticals and herbal medicines seems to be a return to the philosophy of ancient Egypt. The public seems to have an impetuous optimistic view about these 'natural' products and so a realistic risk-benefit analysis for these products is not requested by the consumer. As a result these products are consumed without any toxicological hesitation, and concerns on the safety of these products are scarce.

We seem to have completely forgotten about Paracelsus (1493-1541). Paracelsus was born in Switzerland as Theophrastus Phillipus Aureolus Bombastus von Hohenheim. His adage that every compound can become toxic, as long as the dose is high enough, is still the opening sentence in many handbooks on toxicology. In fact, this adage still forms the basis for judging the safety of man-made chemicals and natural compounds today. Low levels of pesticide or herbicide residues can be completely safe whereas high doses of natural compounds should be regarded with toxicological scepticism.

7 References

[1] B. B. Nemery, M. Fisher, D. Boogaerts, The Lancet 1999, 354, 77.

[2] Detailed information under http://www.slv.se/HeadMenu/livsmedelsverket.asp, 2002.

[3] J. P. F. D'Mello, C. M. Duffus, J. H. Duffus, Toxic Substances in Crop Plants, The Royal Society of Chemistry, Cambridge, 1991.

[4] E. M. Whelan, Toxic Terror. The Truth Behind the Cancer Scares, Prometheus Books, New York, 1993.

[5] B. Craven, C. Johnson in Fearing Food: Risk, Health and Environment, (Eds.: J. Morris, R. Bate), Butterworth-Heinemann, Oxford, 1999, 141-170.

[6] J. M. van de Broek, The Lancet 1997, 349, 805.

[7] R. Bate, What Risk? Science, Politics and Public Health, Butterworth-Heinemann, Oxford, 1999.

Disease Control

Chemistry, Biochemistry, Molecular Biology

07 Molecular Diagnostics for Fungicide Resistance in Plant Pathogens

Helge Sierotzki [(a)] and Ulrich Gisi [(b)]

Syngenta Crop Protection, (a) Research Biology, WST 540, CH-4332 Stein, Switzerland, E-mail: helge.sierotzki@syngenta.com, (b) Research, Product Biology, WRO-1060, CH-4002 Basel, Switzerland, E-mail: ulrich.gisi@syngenta.com

1 Introduction

Fungicide resistance can be defined as the decreased sensitivity of an isolate of a particular pathogen species against a particular inhibitor. Field resistance is observed when the frequency of resistant individuals in a pathogen population has reached a level which results in poor disease control with the fungicide under practical use conditions. Resistant mutants created and selected in a laboratory approach do not necessarily reflect the type of resistance that may evolve under field conditions. Simultaneous resistance to different fungicides of the same mode of action (resistance) group is described as cross-resistance. If more than one biochemical or genetic mechanism confers resistance to different fungicide classes in the same individual, the pathogen is endowed with multiple resistance. Initially, the frequency of resistant individuals in a field population is extremely low and the distribution is wholly random. Therefore, early events in the evolution of resistance are difficult to detect with conventional bioassays and more sensitive and rapid tests methods are needed including molecular approaches [1].
The risk of resistance developing in plant pathogens to fungicides is related to the mode of action of compounds as well as the biology of the pathogen species [1]. In general, multi-site inhibitors confer a lower risk of evolving resistance than inhibitors with a single site of action. In many cases, systemic and curative compounds exert a higher selection pressure than contact fungicides. Moreover, the biology of the pathogen species can influence the development and spread of resistance. Important biological factors for the evolution of resistance are the mutation rate (phenotypically expressed), changes in fitness, migration of resistant individuals and the frequency of sexual and asexual reproduction. In addition, the evolution of resistance is controlled by the selection process imposed by the fungicide and is strongly influenced by factors such as number, timing and type of applications (e.g. solo use, mixtures, alternations).

Despite the precautions taken by the producers to delay the evolution of resistance such as sensitivity monitoring using bioassays and recommendations for product use strategies, resistance developed in many fungicide classes to varying degrees, which has lead in some cases to disappointing disease control under field conditions. In order to detect the first foci of resistance as early as possible and to follow the evolution of resistance as closely as possible, it is necessary to develop specific and sensitive diagnosis methods for all resistance mechanisms. Therefore, knowledge about the mode of action and resistance should be obtained as early as possible during the development of a new compound. The actual change leading to resistance can either be in the target enzyme (change in affinity to the inhibitor), in metabolic or catabolic processes, the specificity of an enzyme allowing a by-pass reaction or an altered influx/efflux balance.
The elucidation of the primary site and biochemical mode of action of inhibitors is often difficult and not necessarily associated with the mechanism of resistance in field isolates. Nevertheless, information on the mechanism of resistance can provide evidence to determine the site of action. Therefore, the classification of fungicides is based on cross-resistance reactions rather than chemical similarities of structures or proposed modes of action (Table 1). Based on available information in the literature, three categories of inhibitor classes can be made: Classes with known mode of action and known mechanism of resistance, classes with proposed mode of action and unknown mode of resistance but wide-spread field resistance, and classes in which resistance is claimed to occur in the field but both mode of action and resistance are not known.
To develop appropriate molecular methods for the detection of resistance, it is important to know the gene(s) coding for the altered target proteins and to characterize the alterations. Molecular methods of detection are only of value if there is a very high correlation with the resistant phenotype (tested in bioassays) [2,3]

2 Resistance for which the Mechanism is Known and Molecular Detection Methods are Available

2.1 Qo Site Inhibitors (QoIs)

The QoI fungicides bind to the Qo site of cytochrome b and inhibit the electron flow at the bc1 enzyme complex of the respiratory chain in the mitochondria. They are widely used because of their broad-spectrum activity [4]. However, a few years after the commercial use of QoIs resistant isolates of *Blumeria graminis* f. sp. *tritici* were detected in some regions of Northern Germany [5]. Since then, resistant isolates have been found in a range of pathogens [4].

Table 1. Fungicide classes and mechanisms of resistance

Fungicide classes[a)]	Examples of active ingredients	Target enzyme	Resistant species[b)] in field populations	Mechanism of Resistance	Molecular detection method[d)]
QoIs	azoxystrobin, kersoxim-methyl	Cytochrome bc1	8, 10, 11, 12, 17, 20, 24	Mutations in *cyt b* (G143A, F129L)	ARMS-Scorpion, AS-PCR, PCR-RFLP, DHPLC, Q-PCR
MBCs	benomyl	ß-tubulin	4, 9, 11, 19, 23, 24	Mutations in ß-tubulin gene (E198K/G/A and F200Y)	PCR-ASO, PCR-RFLP, Q-PCR
Benzamides	zoxamide	ß-tubulin		-	
N-phenyl-carbamates	diethofencarb	ß-tubulin	19, 23	Mutation in ß-tubulin gene (E198L)	
EBIs	DMIs: tridimenol, propiconazole, epoxiconazole	C14-demethylase	8, 10, 24	Muation in cyp 51 (Y136F), change of promotor of cyp51, ABC transporter	Q-PCR
	"Amines": fenpropidine	Δ14-reductase and Δ7/Δ8-isomerase		-	
	fenhexamid	3-keto reductase		-	
Dicarboximides	vinclozolin	NADPH-cyctochrome c reductase?	2, 3, 4, 5, 7, 14, 18	Mutations in kinase *os*1 (I365S and others)	PCR-RFLP
Carboxamides	carboxin	Succinate dehydrogenase (SDH)	21, 22	Mutation in *sdh* (H257L)	
Phenylpyrroles	fludioxonil	Protein kinase (in glycerol synthesis)	4	Protein kinase (*os* 2) and ABC transporter gene (*BcatrB*)	

Table 1. Fungicide classes and mechanisms of resistance (continued)

Fungicide classes[a)]	Examples of active ingredients	Target enzyme	Resistant species[b)] in field popultions	Mechanism of Resistance	Molecular detection method[d)]
Anilino-pyrimidines	cyprodinil	CBL[c)], CGS[c)] (L-methionine synthesis)	4, 24	By-pass reactions; mutation in cgs (S24P and I64V)?, ABC transporters	
Phenylamides	metalaxyl	RNA polymerase I (r-RNA)?	6, 13, 15, 17	Mutation in RNA polymerase?	
Cinnamic acids	dimethomorph	(Cell wall synthesis)	14	-	
Kasugamycin	kasugamycin	(Protein synthesis)	16	Mutation in ribosome preventing binding	
Polyoxins	polyoxin D	(Chitin biosynthesis)	1	Reduced dipeptide premease activity	
Phosphoro-thiolates	pyrazophos	(Phospholipid biosynthesis)	16	Mutational loss of ability to metabolize pyrazophos	

a) EBIs: ergosterol biosynthesis inhibitors, QoIs: Qo site inhibitors of complex III, MBCs: methyl benzimidazole carbamates

b) 1: *Alternaria kikuchina*, 2: *Alternaria alternata*, 3: *Alternaria liniola*, 4: *Botrytis cinerea*, 5: *Botrytis fabae*, 6: *Bremia lactucae*, 7: *Didymella lycopersici*, 8: *Erysiphe* (*Blumeris*) *graminis*, 9: *Helminthosporium solani*, 10: *Mycosphaerella fijiensis*, 11: *Mycosphaerella graminicola*, 12: *Magnaporthe grisea*, 13: *Pseudoperonospora cubensis*, 14: *Phoma exigua*, 15: *Phytophthora infestans*, 16: *Pyricularia oryzae*, 17: *Peronospora parasitica*, 18: *Plasmopora viticola*, 19: *Rhynchosporium secalis*, 20: *Sphaerotheca fuliginea*, 21: *Ustilago maydis*, 22: *Ustilago nuda*, 23: *Venturia nashicola*, 24: *Venturia inaequalis*

c) CBL: cystathionine ß-lyase, CGS: cystathionine γ-synthase

d) Q-PCR: quantitative polymerase chain reaction, ARMS: amplification refractory mutation sysem, AS-PCR: allele-specific PCR, PCR-RFLP: PCR-restriction fragment length polymorphism, DHPLC: denaturating high performance liquid chromatography, PCR-ASO: PCR-allele-specific oligonucleotides

The molecular mode of action of QoIs is well documented and the biochemical interactions with the enzyme have been elucidated [6]. The sequences of the cytochrome b gene *(cyt b)*, located in the mitochondrial genome, is available for many species including plant pathogens. At least 15 different mutations have been described in *cyt b* leading to resistance [7]. In field isolates of different pathogen species, several mechanisms of resistance were described: i) in most cases, a substitution of glycine with alanine at position 143, G143A, in *cyt b* was detected as a single nucleotide polymorphism (SNP) [8]; ii) a substitution of phenylalanine with leucine at position 129, F129L, in *cyt b* was detected in some cases due to 6 possible SNPs [9,10]; iii) two additional mechanisms not based on a change in *cyt b* have been described in *Venturia inaequalis* [11,12]. The actual mechanism leading to resistance is not fully elucidated, however, crystallographic and modelling studies revealed a conformational change of the binding pocket due to the change in amino acids [6]. Different methods have been developed to detect the G143A mutation in a broad range of pathogens. Based on sequence information available, species- and allele-specific PCR techniques were developed [13]:
(I) PCR RFLP (Figure 1): Species-specific fragments are amplified and then cut with the allele-specific endonuclease enzyme *Ita* 1 (*Fnu*H 4, GCN.GC). This method provides qualitative results and is suitable for diagnosis of the presence or absence of the mutation in single spore isolates [8]. It has been adapted for *B. graminis*, *Mycosphaerella fijiensis* and *V. inaequalis* [13].
(II) Allele specific amplification (Figure 2): Either a forward or reverse primer is designed with a single base pair difference at the 3' end between the primer combination detecting wild-type and mutated isolates. The amplicons can be visualized either on an agarose gel stained with ethidium bromide, or in real time PCR equipment using SybGreen as dye or by an allele specific hybridisation probe coupled to a fluorescence dye (i. e. TaqMan probe, ARMS-Scorpion). The later techniques revealed a high sensitivity for allele detection and quantification [13]. The sensitive and resistant alleles are detected in two separate amplification reactions. Specific conditions have been developed for a wide range of pathogens and both changes, G143A [13,14,15,16] and F129L [10].
(III) Allele specific hybridisation (Figure 2): Fragments amplified in an allele unspecific PCR are measured with allele specific probes (Molecular Beacons, TaqMan probes or Scorpion), which can either be used in a single PCR or in multiplex reactions for detection of both alleles in one reaction [13]. This method has minor background problems, however the dynamic range may be limited due to competition of the probes. The Molecular Beacon approach has been developed for *B. graminis* and *M. fijiensis* [13].
(IV) DHPLC (Figure 3): This method is based on species-specific amplification of a fragment containing the mutation and a separation of the heteroduplexes with HPLC. This technique is semiquantitative and suitable for high throughput and potentially multiplexing for different genes. The DHPLC method was proposed for *B. graminis*, *V. inaequalis and Plasmopora viticola* [17].
The specificity and high sensitivity of the described methods allow the detection of rare events in bulk samples without the need to produce pure cultures of the pathogens. The correlation between resistant phenotype and molecular detection of G143A was perfect

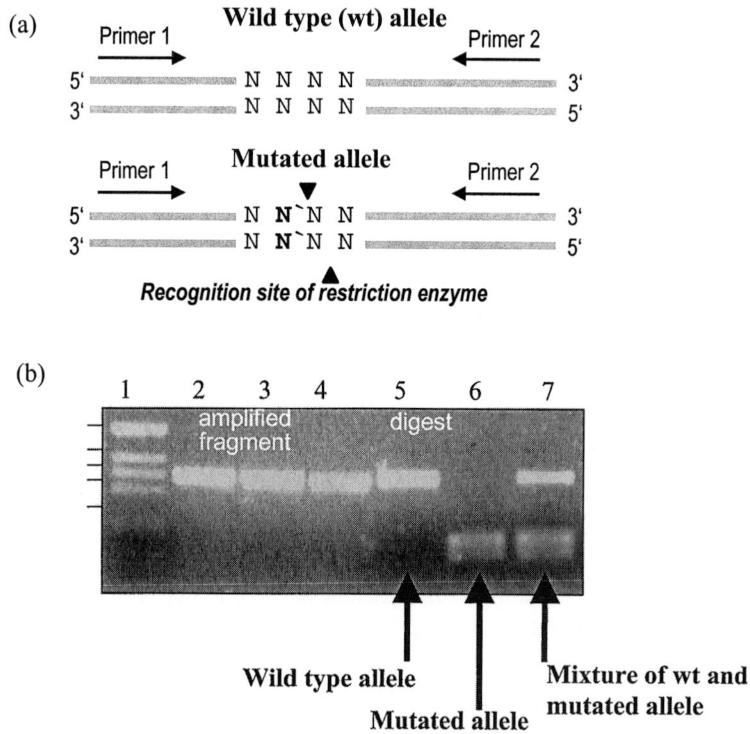

Figure 1. Principle of operation for a PCR-RFLP assay. (a) Location of the restriction site in the mutated allele of the gene of resistant isolates (R), which is absent in the *wt* allele of sensitive isolates (S). (b) Agarose gel showing the PCR products of allele unspecific amplifications (lanes 2-4) and the digestion products (lanes 5-7). Lane 1, DNA size marker.

in *B. graminis*, however in other pathogen species exceptions were observed [18]. In *Pyricularia oryzae* and *P. viticola,* the F129L change was detected recently [9,10], it is governed by several SNPs. Therefore, a method has to be developed to detect either all mutations at once or each one in independent single reactions. In some *V. inaequalis* isolates, resistance was not conferred by a mutational change in *cyt b* [11] and molecular detection methods have not yet been developed. In pathogen species with no or low resistance events such as *Mycosphaerella graminicola* and *B. graminis* f. sp. *hordei,* respectively, molecular diagnostic tools may reveal very low frequencies of resistant isolates in areas where bioassays would still indicate a sensitive response. All compounds of the QoI class of fungicides commercialised so far (azoxystrobin, picoxystrobin, pyraclostrobin, kresoxim-methyl, trifloxystrobin, metominostrobin, famoxadone, fenamidone) were reported to be cross resistant [18].

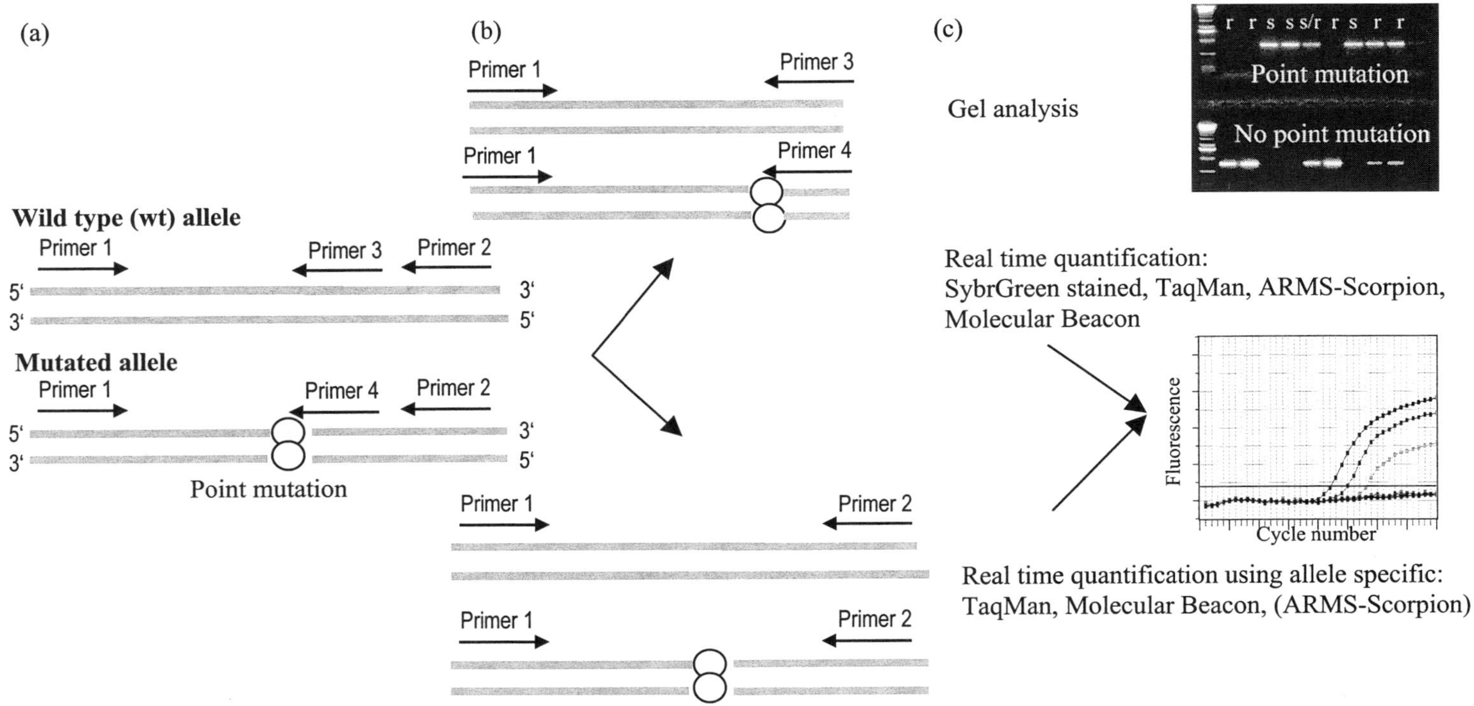

Figure 2. Allele specific determination of single nucleotide polymorphisms (SNPs). (a) The determined SNP responsible for resistance is flanked by either allele specific primers (primer pair 1 and 3 or 1 and 4, respectively) or by species specific primers (primer pairs 1 and 2). (b) The allele specific (top) or species specific (below) amplified fragments are examined by different methods: (c) yes/no answer on an agarose gel stained with ethidium-bromide, or with fragments were measured in real time PCR equipment using different chemistries: universal dye Sybr Green, fragment specific probes (top) or allele specific probes (below).

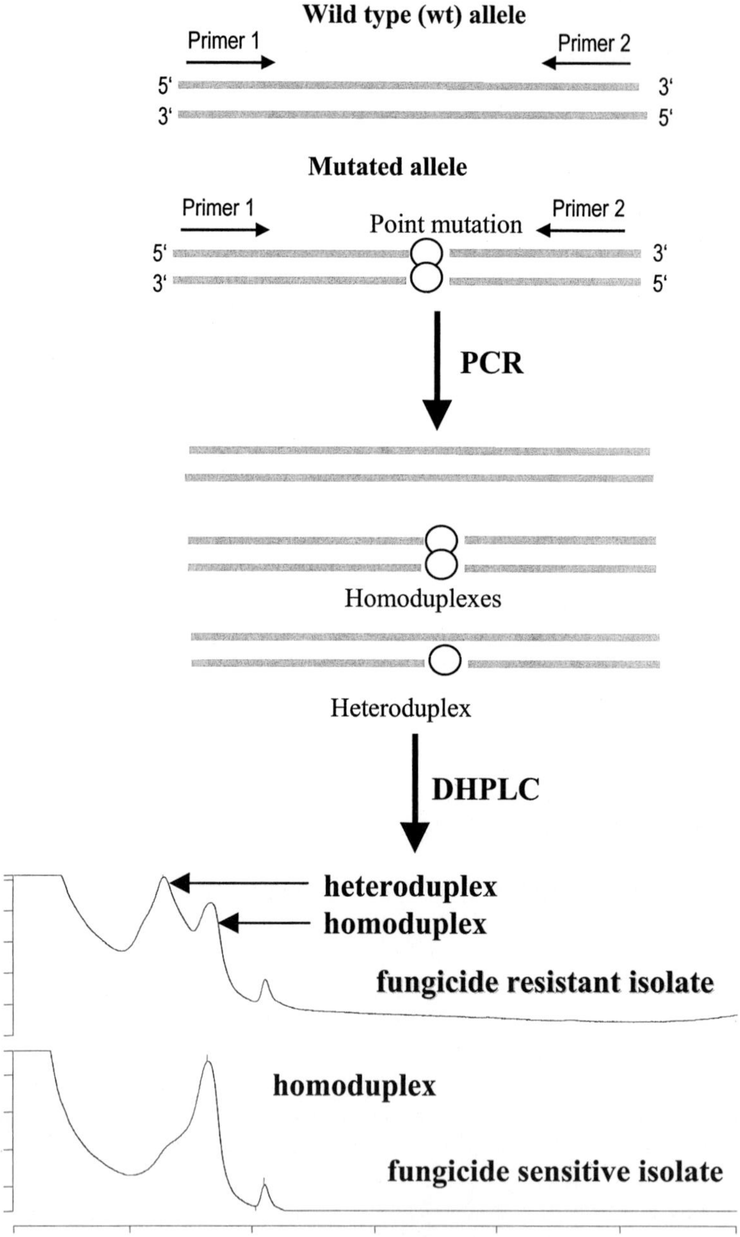

Figure 3. Denaturated high performance liquid chromatography of DNA fragments including a SNP. The fragment of a sample is always mixed with a reference isolate without SNP.

2.2 Benzimidazoles (MBCs)

The benzimidazole fungicides prevent the assembly of tubuli during cell division (mitosis) by binding to the ß-tubulin protein. Resistance is wide-spread in many pathogen species and is persistent in populations throughout the world even in the absence of fungicide applications [19]. The mechanism of resistance is well documented in the literature being point mutations in the ß-tubulin gene [20,21]. Many mutations were observed in the ß-tubulin gene leading to different degrees of resistance in a range of plant pathogens and several *Penicillium* species [19]. Many of these mutations did not influence the competitive ability of isolates. The most detrimental mutations were found at amino acid position 198 with a substitution of glutamate to glycine, E198G (*Rhynchosporium secalis, V. inaequalis*), to lysine, E198K (*Botrytis cinerea, Monilia fructicola, V. inaequalis*), or to alanine, E198A (*B. cinerea, V. inaequalis*) resulting in medium, high or very high resistance levels, respectively. At position 200, a change from phenylalanine to tyrosine, F200Y (*B. cinerea, V. inaequalis*) resulted in medium resistance levels [20].

Three different method have been applied to detect the above described changes:

(I) PCR-ASO: A 1200 bp fragment with species specificity (e.g. from *V. inaequalis*) was amplified, spotted in dots on a membrane and then hybridised with allele specific oligonucletide probes of 17 bp [22]. This method is not suitable for mixed samples, but can display all different changes on one membrane.

(II) PCR-RFLP: With *Bse* 1 a species-specific fragment (e.g. from *Helminthosporium solani*) containing the mutation was cut [23].

(III) SSCP (single stranded conformation polymorphisms): Fragments amplified from sensitive and resistant isolates of *V. nashicola* were analysed by capillary electrophoresis [24].

All benzimidazole fungicides (benomyl, carbendazim, fuberidazole, thiabendazole) and thiophanate fungicides are cross-resistant. However, the recently discovered fungicide, zoxamide, which also inhibits the ß-tubulin assembly, is not cross-resistant in oomycetes [25], although in other pathogens, the relation is not yet clear. On the other hand, negative cross-resistance between benzimidazoles and N-phenylcarbamates such as diethofencarb (and related compounds) has been described [26,27,28]. Isolates of *B. cinerea* and *Pseudocercosporella herpotrichoides* with a substitution at position 198 from glutamate to glycine or alanine were sensitive to N-phenylcarbamates (and resistant to MBCs), but a change from glutamate to lysine conferred positive cross (multiple)-resistance [24]. Within the classes of benzimidazoles and N-phenylcarbamates, different compounds may react with different intensities to the mutations and thus, positive and negative cross-resistance patterns vary within pathogen species and isolates tested [28]. The specificity and selectivity of MBCs (with insensitivity of metazoan and plant tubulins) may be related to different amino acids at position 165 as compared to alanine for sensitive fungi [29]. Recently, a second isoform of ß-tubulin has been found in knockout mutants of *R. secalis* leading to resistance [21].

2.3 Demethylation Inhibitors (DMIs)

More than 20 active ingredients (such as triadimenol, propiconazole, difenoconazole, epoxiconazole) belong to the class of DMI fungicides, which all inhibit the cytochrome P450 dependent 14α-demethylase in the biosynthesis of fungal sterols such as ergosterol [30]. Resistant isolates have been observed in several important pathogen species in the medical and agricultural area [31]. Because DMI resistance develops over many generations in a step wise manner, it is assumed to be polygenic [30]. Currently, three major mechanisms of resistance are postulated in plant pathogenic fungi: i) a point mutation in the target gene *cyp*51 at position 136 with a change from tyrosine to phenylalanine, Y136F [32]; in *Candida*, several other point mutations in *cyp*51 are known [33]; ii) overexpression of *cyp*51 due to a duplication of an enhancer in the promoter region of the gene [34,35,36]; and iii) increased efflux of inhibitor molecules due to an overexpression of specific ABC transporters [37,38]. The point mutation was found in *Uncinula necator* and *B. graminis* leading to resistance factors (RF) of up to 30. The overexpression of the *cyp*51 gene was observed in *V. inaequalis* [34] and *Penicillium digitatum* [36] conferring RF values of up to 200. The efflux mechanism through ABC transporters seems to be important especially in *Candida* spp. [35], whereas in plant pathogens such as *M. graminicola* and *B. cinerea* it may contribute only partially to resistance leading to low RF values.

The following molecular detection methods have been applied:

(I) AS-PCR: Amplification of allele-specific fragments of *cyp*51 containing the SNP, then visualization on agarose gels stained with ethidium bromide [32] or observed using real time PCR equipment using SybrGreen as dye [16].

(II) PCR: Species specific PCR amplifying the promoter region of *cyp*51, then differences in length were recorded on agarose gels [36].

DMI resistance may be governed by different resistance mechanisms co-existing in the same isolate. Molecular detection tools can help to find low frequencies of mutated individuals in populations. However, since resistance to DMIs may be a combined effect of several mechanisms, the molecular results have to be treated with care and bioassays may be always needed for verification.

The DMIs belong to the large family of ergostrol biosynthesis inhibitors (EBIs), which includes also "amines" such as morpholines (e.g. fenpropimorph), piperidines (fenpropidin) and spiroketalamines (spiroxamine) inhibiting Δ14 reductase and $\Delta 7/\Delta 8$ isomerase, as well as the hydroxyanilides (fenhexamid) inhibiting 3-keto reductase. Among the three classes of EBIs, there is no cross-resistance [39] and little information on the mechanism of resistance is available for non-DMIs.

3 Resistance for which the Mechanism is Known but Molecular Detection Methods are not Available

3.1 Dicarboximides

For the dicarboximide fungicides such as iprodione, vinclozolin and procymidone, the mode of action is not well characterized. A possible target is the inhibition of NADH cytochrome c reductase [40] leading to the formation of toxic lipids and peroxides. Laboratory isolates but not field isolates of *B. cinerea* resistant to dicarboximides can also be resistant to phenylpyrroles (e.g. fludioxonil) and are hypersensitive to osmotic stress. In *Neurospora crassa*, mutants sensitive to osmotic stress have been described, some (*os*-1) being resistant to dicarboximides and fludioxonil [41]. The *os*-1 gene codes for a two-component histidine kinase, which was described to be involved in the signal transduction for adaptation to high osmolarity of the growth medium. The homologue of *os*-1 in dicarboximide-sensitive and -resistant *B. cinerea* field isolates has been cloned and point mutations were assigned to confer resistance [42]. Three combinations of mutations were differentiated: type 1 isolates with a substitution of isoleucine to serine at position 365, I365S (previously reported as I86S, [42]); type II isolates with H225P, V368F, Q369H and T447S; type III isolates with Q369P and N373S [43]. In Japanese populations of *B. cinerea*, about 150 isolates were tested for the presence of the most abundant mutation, I365S, by PCR-RFLP, using *Taq* 1 as restriction enzyme. For the few isolates where resistance and presence of I365S did not correlate, additional mutations were found. Currently, the correlation of mutated *os*-1 gene of *B. cinerea* and dicarboximide resistance is not fully elucidated and whether all resistant isolates have a changed *os*-1 gene. Many dicarboximide resistant isolates of *B. cinerea* collected from treated fields may disappear from populations, however the correlation between SNPs and loss of fitness has not been elucidated. Molecular analyses are valid only if they include all the described mutations and if bioassays are performed in parallel.

3.2 Carboxamides

The target of carboxamide fungicides such as carboxin, benodanil, furametpyr, flutolanil and mepronil is succinate ubiquinone reductase (or succinate dehydrogenase, SDH) of complex II in the respiration chain. The mode of action is known to involve an interruption of the electron flow at the Ip (or SDH2) subunit of the enzyme [44]. The *sdh*-2 gene(s) of complex II coding for the SDH complex are nuclear, in contrast to the *cyt b* gene of complex III which are mitochondrial. Resistance to carboxamides has initially been detected in isolates of *Puccinia horiana* on *Chrysanthemum* [45]. Later, resistant isolates have been found also in *Ustilago maydis* and *U. nuda* [46]. For *U.*

maydis, the mechanism of resistance has been described as amino acid change in the target gene, *sdh*-2, leading to a conformational change in the enzyme. The substitution of histidine to leucine at position 257, H257L, is due to a single SNP [44]. The *sdh*-2 genes of *P. horiana* and *U. nuda* have not been sequenced so far. Resistance in *U. nuda* was described as monogenic, with an occasionally unusual inheritance pattern [46]. To date, molecular detection methods have not been available.

3.3 Phenylpyrroles

Phenylpyrrole fungicides such as fludioxonil and fenpiclonil inhibit a mitogen-activated protein (MAP) kinase in signal transduction of osmoregulation (glycerol synthesis). Laboratory isolates of *B. cinerea* resistant to phenylpyrroles can be cross resistant to dicarboximides, however the few available phenylpyrrole resistant field isolates did not show this behaviour [47]. Moreover, it has been demonstrated for *B. cinerea* in crossing experiments that two different genes were responsible for resistance to phenylpyrroles and dicarboximides (independent segregation pattern) [47]. In *N. crassa*, null mutants of the gene *os*-2 involved in osmoregulation encoding a *HOG1* mitogen-activated protein kinase homologue were phenylpyrrole resistant. Mutations leading to resistance were based on a frame shift or two changes of internal tryptophane to a stop codon [48]. However, while the gene putatively responsible for dicarboximide resistance (*os-1*) has been sequenced in *B. cinerea*, the gene(s) involved in phenylpyrrole resistance has not yet been mapped. Another putative resistance mechanism for phenylpyrroles was claimed to be an overexpression of ABC transpoters. Replacement mutations in the corresponding gene, *BcatrB*, yielded isolates of *B. cinerea* which were more sensitive to fenpiclonil than wild type isolates [49]. However, the corresponding mutation(s) leading to altered expression have not yet been found.

3.4 Anilinopyrimidines

Anilinopyrimidine fungicides such as cyprodinil, pyrimethanil and mepanipyrim inhibit methionine biosynthesis, but the target enzyme has not yet been identified. Complementation studies suggested that either cystathionine ß-lyase (CBL) or cystathionine γ-synthase (CGS) are the target enzymes [50]. However, isolated CBL was not sensitive to cyprodinil and CGS-deficient mutants of *N. crassa* were insensitive [51]. When the sequence of the *cbl* gene in sensitive and resistant field isolates of *B. cinerea* was compared, no mutations were detected conferring resistance. However, two different mutations (S24P and I64V) were found in the *cgs* gene of several *B. cinerea* field isolates correlating to the resistant phenotype. In two independent progeny of a cross between a sensitive and resistant single ascospore isolate of *B. cinerea*, the S24P and I64V change, respectively, co-segregated with the resistant phenotype. The mutations were located in the regulatory part of the *cgs* gene suggesting a bypass reaction in methionine biosynthesis as mechanism of resistance. However, in field isolates of *B.*

cinerea, there was no strict correlation between either mutations and the resistant phenotype [51]. Therefore, additional mechanisms of resistance beside the mutations in *cgs* might exist. Resistant field populations were described for *B. cinerea, V. inaequalis* and *P. herpotrichoides* [52,53]. To date, molecular detection tools have not been developed.

3.5 Phenylamides

The phenylamide fungicides such as metalaxyl, mefenoxam, benalaxyl, ofurace and oxadixyl inhibit the ribosomal (r)RNA synthesis [54]. When measuring the sensitivity of the three RNA species in biochemical assays, only rRNA was affected by fungicide treatment. Moreover, partially purified polymerase I enzyme was not sensitive to metalaxyl; however, the activity of endogenous RNA polymerase I in isolated nuclei was affected by phenylamides indicating that the interaction was effective only with the intact enzyme complex [54]. Phenylamide resistant isolates of *Phytophthora infestans*, *P. vitcola* and many other oomycetes are wide-spread in field populations, and fluctuate in frequency over years and within seasons but are in a dynamic equilibrium with sensitive isolates in many areas [55]. Inheritance of resistance was demonstrated to be monogenic with the participation of minor genes [56]. Biochemical studies (incorporation of radiolabeled uridine) suggested that a change of RNA polymerase I is responsible for resistance [54]. However, only recently sequence data for polymerase subunits became available in *P. infestans* [57] but molecular detection tools have not yet been developed.

4 Resistance Known in Field Populations but Resistance Mechanisms are Unknown

4.1 Cinnamic Acids and Amino Acid Amide Carbamates

The cinnamic acid compound dimethomorph is claimed to interfer with cell wall synthesis in sensitive pathogen species of the oomycetes, with the exception of *Pythium*, that is not sensitive as all Asco- and Basidiomycetes [58]. The same spectrum of activity and similar effects on cell wall synthesis were also described for the chemically unrelated amino acid amide carbamates such as iprovalicarb and benthiovalicarb, a new fungicide class [59]. Whether cross-resistance between iprovalicarb and dimethomorph exists is still speculative [60]. In artificial mutants of *Phytophthora parasitica*, a single dominant gene has been described conferring resistance to dimethomorph [61]. Field isolates of *Plasmopara viticola* resistant to dimethomorph have been reported in a few cases [62].

4.2 Kasugamycin

Kasugamycin, a hexopyranosyl antibiotic, controls *Pyricularia oryzae* in rice by inhibiting protein synthesis in 80S and 70S ribosomes, while interacting with the 30S subunit [63]. In resistant mutants, protein synthesis is not inhibited indicating a mutational change in ribosomes [63]. In a backcross progeny of *Magnaporthe grisea* a locus controlling kasugamycin resistance was identified. Bulked segregant analyses revealed three markers co-segregating with resistance; however, they were located on different chromosomes [64]. This suggests that several genes might be involved in resistance.

4.3 Polyoxins

Polyoxins such as polyoxin A, B and D which block the biosynthesis of chitin are competitive inhibitors of uridine diphosphate-N-acetylglucosamine [65]. In resistant isolates of *Alternaria kikuchiana*, no uptake of dipeptides was observed suggesting that altered dipeptide permease may be responsible for polyoxin resistance [65].

4.4 Pyrazophos

The phosphoro-thiolate compounds such as pyrazophos interfere with phospholipid biosynthesis [66]. Pyrazophos is active against several powdery mildew species by being converted to 2-hydroxy-5-methyl-6-ethoxycarbonylpyrazolo (1-5-a)-pyrimidine (PP). However, pathogen species such as *U. nuda* were incapable of metabolising pyrazophos and thus insensitive. In addition, resistant mutants of *P. oryzae* were unable to metabolise pyrazophos [67] and were cross-resistant to other molecules of the same class of fungicides such as iprobenfos (IBP) and edifenphos. No information on genes in this metabolic pathway is available.

4.5 Miscellaneous fungicides

For a range of chemically unrelated fungicides such as ethirimol, cymoxanil, quinoxyfen and aromatic hydrocarbons (e.g.chloroneb, quintozene), resistant isolates have been detected occasionaly in limited areas. For these fungicides, the mode of action and mechanism of resistance is speculative and no information is available on the genes involved in evolution of resistance.

5 Outlook

Molecular methods can only be utilised for the detection of fungicide resistance if several conditions are fulfilled:
(I) The type of change(s) in the gene(s) conferring resistance is known.
(II) Resistant isolates are collected from the field rather than from assays with artificial mutagenesis or adaptation made under laboratory conditions. Well characterized resistant isolates are needed, preferentially single spore isolates, which should represent the majority of the resistant field population.
(III) The correlation between the observed molecular changes in the genome (SNPs) and the phenotypic response (resistance) is as high as possible. A sound survey should be conducted to determine whether different mechanisms of resistance exist and how frequent they are.
(IV) The molecular technique applied detects all (or the most frequent) changes and SNPs in the presence of wild type alleles and host tissue (DNA).
(V) The molecular method is quantitative with a high sensitivity in order to detect very rare events (1 in 10000-100000).
The examples given above illustrate that different mechanisms can lead to resistance against a particular fungicide among and within a pathogen species. In general, fungal pathogens are phenotypically and genetically plastic and are able to adapt to changing environmental conditions including the exposure to fungicides. The strong selection pressure imposed by fungicide applications and the competition among individuals under field conditions will probably limit survival of resistant individuals which suffer from fitness penalties. Therefore, any resistant isolates detected in the field and the mechanisms responsible for resistance in these isolates provide the basis for the development of molecular detection tools.
For developing molecular diagnosis kits, a whole range of methods is available. The different techniques require laboratory equipment, but field versions of such methods do not currently exist. The most common method is based on PCR-RFLP using suitable restriction enzymes. Because the restriction process and the on-gel-visualisation is qualitative, the method can be used to detect the presence or absence of mutated alleles in single, fully selected or homoplasmatic isolates. Allele specific PCR-methods are more or less quantitative, especially if combined with the dye SybrGreen or a probe specific assay using a real time cycler. The wild type and mutated alleles are measured in two different PCRs. If allele specific probes are used, both alleles (potentially also alleles from a different gene) can be detected in a multiplex PCR. However, allele specific probing may encounter competition problems, which lower the sensitivity of the method. The SSCP and DHPLC techniques can potentially detect SNPs without knowing

the exact sequence of the fragment and may differentiate between different alleles of several genes.
Molecular methods for detecting resistance may be used either during the development of a new active ingredient or during the commercial use of the product. However, at an early development stage the mode of action and the mechanism of resistance are often not known and therefore in such cases, it is not possible to develop molecular methods. If inhibitors are discovered with a target based approach, the genes coding for the targets may be known and used for searching potential mutations for resistance. However, when mechanisms other than target site changes are responsible for resistance, forced selection in controlled field experiments or *in vitro* studies with the target or a model organism might be necessary to give indications of the possible mechanisms of resistance which might arise. Molecular methods can provide especially powerful tools to detect the early evolution of resistant isolates or to follow populations where resistance already exists. As a consequence, appropriate recommendations for product use can be developed. Sensitivity monitoring can give information on the migration of existing and emergence of resistant isolates. In addition, fluctuations in resistant sub-populations over time and space can be followed in cases when changes in pathogen development occur (sexual and asexual reproduction) or when product use recommendations are altered (restriction of applications or change to product free periods). In such cases, molecular methods for detecting resistant individuals will support rather than substitute sensitivity tests with bioassays. To circumvent the narrow range of detectable changes in the genome by using molecular techniques, whole genome chip microarrays, BADGE (beads array for the detection of gene expression) or proteomic methods (2-D gels MALDI-TOF, matrix-assisted laser desorption/ionisation) might be recommended, which offer high multiplex degrees and are more relevant for the actual resistant phenotypes. However, use of these techniques requires a thorough knowledge of the mechanism of action and resistance.

6 References

[1] K. J. Brent, Fungicide resistance in crop pathogens. How can it be managed? FRAC Monograph No. 1, GIFAP, Brussels, 1995, 48pp.

[2] K. J. Brent, in Pesticide Resistance: Strategies and Tactics for Management, National Academy Press, Washington, 1986, pp. 298-312.

[3] Anonymus, in Pesticide Resistance: Strategies and Tactics for Management, National Academy Press, Washington, 1986, pp. 271-278.

[4] D. W. Bartlett, J. M. Clough, J. R. Godwin, A. A. Hall, M. Hamer, B. Parr-Dobrzanski, Pest Manag. Sci, 2002, 58, 649-662.

[5] M. Reschke, DLG-Mitteilungen, 1999, 1, 51.

[6] T. Link, M. Iwata, J. Björkman, D. van der Spoel, A. Stocker, S. Iwata, in Chemistry of Crop Protection (Eds.: G. Voss and G. Ramos), Wiley-VCH Weinheim, Germany, 2002, this volume.

[7] G. Brasseur, A. S. Saribas, F. Daldal, Biochem. Biophys. Acta, 1996, 1275, 61-69.

[8] H. Sierotzki, S. Parisi, U. Steinfeld, I. Tenzer, S. Poirey, U. Gisi, Pest Manag. Sci., 2000, 56, 833-841.

[9] M. L. Farman, Phytopathology, 2001, 91, 110, abstract.
[10] C. Stanger, IUPAC Conference, Basel, Switzerland, 2002, pers. communication.
[11] U. Steinfeld, H, Sierotzki, S. Parisi, U. Gisi, in Modern Fungicides and Antifungal Compounds III (Eds.: H. W. Dehne, U. Gisi, K. H. Kuck, P. E. Russel, H. Lyr), AgroConcept, Bonn, Germany, 2002, pp. 167-176.
[12] T. Jabs, K. Cronshaw, A. Freund, in Modern Fungicides and Antifungal Compounds III (Eds.: H. W. Dehne, U. Gisi, K. H. Kuck, P. E. Russel, H. Lyr), AgroConcept, Bonn, Germany, 2002, in press.
[13] P. Wille, H. Sierotzki, C. Stanger, S. Cleer, J. Burbidge, A. A. Hall, J. Windass, U. Gisi, in Modern Fungicides and Antifungal Compounds III (Eds.: H.W. Dehne, U. Gisi, K.H. Kuck, P.E. Russel, H. Lyr), AgroConcept Bonn, Germany, 2002, 131-139.
[14] B. A. Fraaije, J. A. Butters, J. M. Coelho, D. R. Jones, D. W. Hollomon, Plant Pathology, 2002, 51, 45-54.
[15] E. A. Rohel, P. Laurent, B. A. Fraaije, N. Cavelier, D. W. Hollomon, Pest Manag. Sci., 2002, 58, 248-254.
[16] J. A. Lucas, Detection and diagnosis of plant disease, 2001, http://www.iacr.bbsrc.ac.uk/corporate/annualreport/2000-2001/3%20Detection.pdf.
[17] S. Bäumler, H. Sierotzki, A. A. Hall, U. Gisi, V. Mohler, G. F. Felsenstein, G. Schwarz, Pest Manag. Sci., 2002, in press.
[18] S. P. Heaney, A. A. Hall, S. A. Davies, G. Olaya, Proc. Brighton Crop Protection Conf. – Pests and Diseases, BCPC, Farnham, Surrey, UK, 2000, 755-762.
[19] L. C. Davidse, H. Ishii, in Modern Selective Fungicides (Ed.: H. Lyr), Gustav Fischer, Jena, Germany, 1995, pp. 305-322.
[20] H. Koenraadt, S. C. Somerville, A. L. Jones, Phytopathology, 1992, 82, 1348-1354.
[21] J. Butters, D. W. Hollomon, 7th International Congress of Plant Pathology, Edingburgh, abstract 3.3.49, 1998, http://www.bspp.org.uk/icpp98/3.3/49.html.
[22]] H. Koenraadt, A. L. Jones, Phytopathology, 1992, 82, 1354-1358.
[23] L. R. Cooke, G. J. McKay, PCR-based detection of TBZ resistance in *Helminthosporium solani*, 1999, http://www.afsni.ac.uk/News/PPS/PCR/default.htm.
[24] Anonymus, DNA-based approaches for diagnosis of fungicide resistance in phytopathogenic fungi, 1996, http://www.niaes.affrc.go.jp/annual/r1996/fr_html/24-25.htm.
[25] D. H. Young, S. L. Spiewak, R. A. Slawecki, Pest Manag. Sci., 2001, 57, 1081-1087.
[26] A. L. Jones, E. Shabi, G. R. Ehret, Can. J. Plant Pathol., 1987, 9, 195-199.
[27] P. Leroux, M. Gredt, Neth. J. Pl. Path., 1989, 95, Supplement 1, 121-127.
[28] P. Leroux, F. Chapeland, A. Arnold, M. Gredt, J. Gen. Plant Pathol., 2000, 66, 75-81.
[29] M. K. Jung , B. R. Oakley, Cell Motil. Cytoskeleton, 1990, 17, 87-94.
[30] K. H. Kuck, H. Scheinpflug, R. Pontzen, in Modern Selective Fungicides (Ed.: H. Lyr), Gustav Fischer, Jena, Germany, 1995, pp. 205-258.
[31] H. Vanden Bossche, Rev. Ibroam. Micol., 1997, 14, 44-49.
[32] C. Délye, F. Laigret, M.-F. Corio-Costet, Appl. Environ. Microbiol., 1997, 63, 2966-2970.
[33] D. Sanglard, F. Ischner, L. Koymans, J. Bille, Antim. Agents Chemother., 1998, 42, 241-253.
[34] G. Schnabel, A. L. Jones, Phytopathology, 2001, 91, 102-110.
[35] T. C. White, Antimicrob. Agents Chemother., 1997, 41, 1482-1487.
[36] H. Hamamoto, K. Hasegawa, R. Nakaune, Y. Jin Lee, K. Akutsu, T. Hibi, Pest Manag. Sci., 2001, 57, 839-843.
[37] H. Hamamoto, O. Naeata, K. Hasegawa, R. Nakaune, Y. Jin Lee, Y. Makizumi, K. Akutsu, T. Hibi, Pestic. Biochem. Physiol., 2001, 70, 19-26.
[38] L. H. Zwiers, M. A. de Waard, Fungal Genet. Biol., 2000, 30, 115-125.

[39] K.-H. Kuck, FRAC working group, SBI fungicides, 2001, http://www.frac.info/sbi_wg.html.
[40] W. Edlich, H. Lyr, in Target Sites of Fungicide Action (Ed.: W. Köller), CRC Press, London, 1992, pp. 53-68.
[41] N. Ochiai, M. Fujimura, T. Motoyama, A. Ichiishi, R. Usami, K. Horikoshi, I Yamaguchi, Pest Manag. Sci., 2001, 57, 437-442.
[42] M. Oshima, M. Fujimura, S. Banno, C. Hashimoto, T. Motoyama, A. Ichiishi, I. Yamaguchi, Phytopathology, 2002, 92,75-80.
[43] M. Fujimura, IUPAC Conference, Basel, Switzerland, 2002, pers. communication.
[44] P. L. E. Broomfield, J. A. Hargreaves, Curr. Genet., 1992, 22, 117-121.
[45] F. B. Dirkse, M. Dil, R. Linders, I. Rietstra, Med. Fac. Landbouww. Rijksuniv. Gent, 1982, 47/3, 793-800.
[46] G. Newcombe, P. L. Thomas, Phytopathology, 2000, 90, 179-182.
[47] A. Vignutelli, M. Hilber-Bodmer, W. Hilber, Mycol. Res., 2002, 106, 329-335.
[48] Y. Zhang, R. Lamm, C. Pillonel, S. Lam, J.-R. Xu, Appl. Env. Microbiol., 2002, 68, 532-538.
[49] H. Schoonbeek, G. del Sorbo, M. A. de Waard, MPMI, 2001, 14, 562-571.
[50] P. Masner, P. Muster, J. Schmid, Pestic. Sci., 1994, 42, 163-166.
[51] H. Sierotzki, J. Wullschleger, M. Alt, T. Bruyère, C. Pillonel, S. Parisi, U. Gisi, in Modern Fungicides and Antifungal Compounds III (Eds.: H. W. Dehne, U. Gisi, K. H. Kuck, P. E. Russel, H. Lyr), AgroConcept, Bonn, Germany, 2002, 141-148.
[52] P. Leroux, A. Arnold, M. Gredt, in Modern Fungicides and Antifungal Compounds III (Eds.: H. W. Dehne, U. Gisi, K. H. Kuck, P. E. Russel, H. Lyr), AgroConcept, Bonn, Germany, 2002, in press.
[53] R. Küng, K. M. Chin, U. Gisi, in Modern Fungicides and Antifungal Compounds II (Eds.: H. Lyr, P. E. Russel, H.-W. Dehne, H. D. Sisler), Intercept, Andover, UK, 1999, pp. 313-322.
[54] L. C. Davidse, in Modern Selective Fungicides (Ed.: H. Lyr), Gustav Fischer, Jena, Germany, 1995, pp. 347-354.
[55] U. Gisi, Y. Cohen, Annu. Rev. Phytopathol., 1996, 34, 549-572.
[56] A.-L. Fabritius, R. C. Shattock, H. S. Judelson, Phytopathology, 1997, 87, 1034-1040.
[57] Syngenta internal data.
[58] G. Albert, J. Curtze, C. Drandarevski, Proc. Brighton Crop Protection Conference – Pests and Diseases, BCPC, Farnham, Surrey, UK, 1988, 17-24.
[59] K. Stenzel, R. Pontzen, T. Seitz, R. Tiemann, A. Witzenberger, Proc. Brighton Crop Protection Conference - Pests an Diseases, BCPC, Farnham, Surrey, UK, 1998, 367-374.
[60] A. Suty, K. Stenzel, Pflanzenschutz-Nachrichten Bayer, 1999, 52, 1, 71-84.
[61] K. Chabane, P. Leroux, G. Bompeix, Phytopath. Medit., 1996, 35, 82-90.
[62] G. Albert, pers communication in [60].
[63] E. Cundliffe, in Ribosomes: Structure, Function and Genetics (Eds.: G. Chambliss, G. R. Craven, J. Davies, K, Davis, L. Kahan, M. Nomura), University Park Press, Baltimore, 1980, pp. 555-581.
[64] S. Fukiya, M. Kodama, H. Kito, T. Sone, F. Tomita, Biosc. Biotechnol. Biochem., 2001, 65, 1464-1473.
[65] M. Hori, K. Kakiki, T. Misato, J. Pesticide Sci., 1977, 2, 139-149.
[66] P. Braun, N. Schreiber, in Modern Selective Fungicides (Ed.: H. Lyr), Gustav Fischer, Jena, Germany, 1995, pp. 373-388.
[67] M. A. de Waard, Neth. J. Pl. Path., 1980, 86, 251-258.

08 New Antifungal Modes of Action

Matthew J. Henry [(a)] **and Gary D. Gustafson** [(b)]

(a) The Dow Chemical Company, 1500 E. Lake Cook Rd., Buffalo Grove IL 60089, USA, (b) Dow AgroSciences, 9330 Zionsville Rd., Indianapolis IN 46268, USA

1 Introduction

The development of new technologies with implications for pesticide discovery, including target site directed rational design, high-throughput screening, biotechnology and genomics, has challenged the agricultural chemistry industry to make significant breakthroughs in the identification of new fungicides with novel modes of actions and increased efficacy. To date, these new technologies have probably made a greater commercial impact through lead optimization than via the discovery of new leads and novel target sites. However, there is continued optimism that these ultimate goals can be achieved. While these new technologies have been evolving, traditional processes for the discovery of lead chemistries with pesticidal activity have also been changing. Blind screening of available organic compounds has been replaced by the use of specific criteria, such as molecular size, hydrophobicity, ionic charge, and chemical class, to preselect compounds that are more likely to have agricultural utility. Natural products also continue to be an increasingly useful source for the discovery of bioactive lead compounds.

The determination of the mode of action of new lead compounds has typically followed the discovery of their basic *in vivo* activity. In some cases, this information has provided valuable guidance for improving the pharmacokinetic properties of the molecule and the activity at the biochemical target site. Technological advancements continue to have a significant impact on our ability to rapidly determine the mechanism of action of new fungicides and fungicide leads. In particular, molecular genetic and biochemical tools are now making the elucidation of antifungal mechanisms faster and easier than ever before. This review will summarize some of the recently discovered fungicide target sites.

2 MET Complex III Qo Inhibitors (Strobilurins)

During the last 10 years, the focus of mode of action research has been concentrated on the synthetic strobilurins which are currently the most important class of fungicides in use today. These fungicides are improvements on the natural product strobilurin A and related compounds odemansin A and myxothioazole. The natural products have been extensively studied and the mode of action and resistance well documented as inhibitors of mitochondrial electron transport complex III [1,2]. The existing knowledge on the bioactivity of these natural products had a considerable impact on the development of the synthetic strobilurin fungicides. The first commercialized synthetic products were azoxystrobin (Fig. 1) and kresoxim-methyl. Since the strobilurin compounds have recently been reviewed, this update will not cover them in detail [3]. More recently, the azolone fungicides, famoxadone (an oxazolidinedione) and fenamidone (an imidazolinone), have also been identified as inhibitors of the mitochondrial electron transport cytochrome bc complex III (MET III) of the electron transport chain in eukaryotic species. MET III is a complex target of multiple polypeptides and three electron transfer sites. The strobilurins, azolones and myxothiazole are known to bind at the Qo site of this cytochrome. The Qo reaction center resides on the mitochondrially-encoded cytochrome b/c_1. This reaction center is the site of bifurcation of electron transport between a Rieske iron-sulfur cluster and a heme group, and is distinctly different from the binding site defined by the MET III inhibitor stigmatellin.

The key to the considerable commercial success of the strobilurin fungicides resides in their potent activity across various classes of fungi. However, the combination of broad-spectrum activity with a site-specific fungicide also leads to a significant increase in the potential for disruptive resistance. Indeed, mutants highly resistant to strobilurins were rapidly detected in populations of wheat powdery mildew. This resistance is known to be due to a single amino acid change in the cytochome b target. Target site mutations resulting in strobilurin resistance have more recently been identified in several additional fungal species including *Venturia* and *Plasmopara* [5,6]. When tested, such mutants have also been found to be cross-resistant to azolone fungicides. In *Venturia inaequalis* another mechanism of resistance may also be operating which may act either independently or in parallel with target site resistance. This mechanism may be based on a compensation mechanism that can bypass the cytochromes of normal electron transport [7].

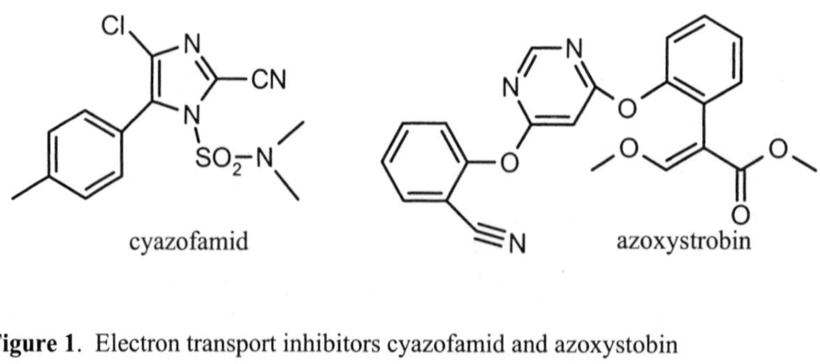

Figure 1. Electron transport inhibitors cyazofamid and azoxystobin

3 MET Complex III Qi Inhibitors

Dimefluazole and the more recently discovered compound cyazofamid (IKF 916) are, like strobilurins, specific inhibitors of the cytochrome bc complex III.[8] However, these fungicides have properties appreciably different from the strobilurins and appear to be more closely related to Antimycin A. Cyazofamid (Fig. 1) and dimefluazole do not possess the methoxyacrylate-like toxiphore that is characteristic of the strobilurins and the azolones, but are characterized instead by the presence of a dimethylsulfamoyl moiety. The selectively of these compounds is quite different from the broad-spectrum activity observed with the strobilurins as they are known to be specific inhibitors of Oomycete fungi such as *Phytophthora, Plasmopara and Pythium* [9]. There can be wide differences in sensitivity to these compounds between Oomycete species or even among isolates of a single species. For example, independant isolates of *Phytophthora infestans* have shown variances in the concentration that inhibits growth by 50% (EC_{50}) that range from 0.008mg/L to 0.02 mg/L. Other Oomycetes such as *Pythium sylvaticum* can be somewhat less sensitive with EC_{50} values in the range of 0.2 mg/L. In comparison, insensitive Ascomycete and Basidiomycete species typically have EC_{50} values for cyazofamid of more than 100mg/L.

At the molecular level, cyazofamid has been shown to have *in vitro* activity against the *Pythium* cytochrome bc complex III, while having little effect on *Botrytis*, yeast, rat liver and potato tuber cytochromes. Studies on the effect of cyazofamid on the reduction kinetics of complex III have concluded that the specific target site is the Qi reaction center. Thus, it is not entirely surprising that strobilurin resistant fungi show no cross resistance to cyazofamid. The Qi reaction center is the binding site of the well-characterized MET III inhibitor Antimycin A. Although Antimycin A has antifungal activity, it does not exhibit the narrow species specificity of cyazofamid. The Oomycete selectivity of cyazofamid and dimefluazole may be related to their proposed mechanism of inhibition. A dimethylsulfamoyl analog of dimefluazole has been used to demonstrate the formation of a covalent, irreversible adduct with the target site [10] This reactive inhibition mechanism would be unique to this class of Qi complex III inhibitors and may explain their selectivity.

4 Zoxamide

Zoxamide (Fig. 2) is a new fungicide that specifically inhibits Oomycete fungi. The encystment and germination of zoospores is not affected by zoxamide, but germ tube elongation is significantly inhibited. Oomycetes are normally coenocyctic and develop multiple nuclei per cell as the germ tube grows. However, in the presence of low levels of zoxamide, nuclear division is arrested and disruptions in the formation of microtubule filaments can be observed by immunofluorescence microscopy. These effects are identical to those observed in fungi exposed to the antitubulin compound colchicine. Biochemical studies with radiolabeled (S)-zoxamide have clearly demonstrated that zoxamide inhibits tubulin polymerization by binding to β-tubulin [11]. Thus, zoxamide has the same target site as benomyl (Fig. 2) and other benzimidazole fungicides. However, the benzimidazoles were not active against Oomycete fungi, and the mechanism of action of zoxamide may not be entirely identical to that of the benzimidazoles. The observation that zoxamide continues to remain bound to denatured β-tubulin suggests that, unlike benzimidazoles, the binding is covalent and may be a function of the α-chloroketone moiety of the compound.

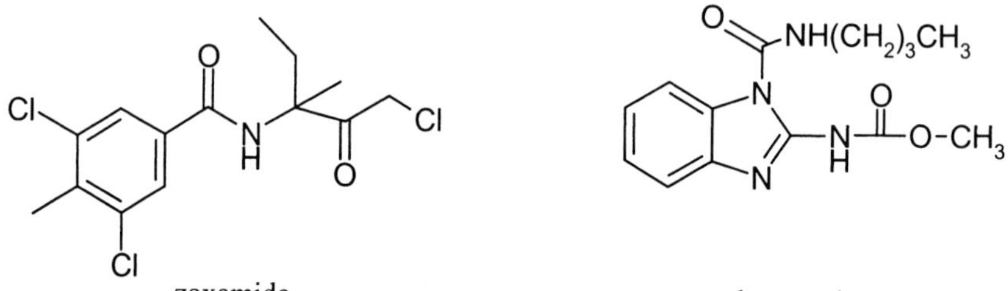

Figure 2. Tubulin polymerase inhibitors zoxamide and benomyl.

5 Phenylpyrroles

The phenylpyrrole class of fungicides includes the compounds fenpiclonil and fludioxonil. These chemistries have been developed from the natural product lead molecule pyrrolnitrin produced by *Pseudomonas pyrrocinia* (Fig. 3). The photoinstability of pyrrolnitrin was overcome in the synthetic analogs and these compounds are now used for the control of *Botrytis cinerea, Monilinia,* and *Sclerotinia* spp. [12].

The mode of action of the phenylpyrroles has been difficult to elucidate due to the lack of an effect on primary metabolic processes. Cross-resistance studies with laboratory-generated mutants suggested that dicarboximide and phenylpyrrole fungicides might share a common mechanism of action [13]. Several studies have shown that dicarboxamides and/or phenylpyrroles cause intracellular accumulation of polyols [14,15], but this accumulation of glycerol is most likely a consequence of inhibition of proteins involved in osmoregulation.[15] The direct target site of phenylpyrroles is suggested to be a protein kinase involved in the intracellular osmoregulation of glycerol [15]. A calcium- and cAMP-independent protein kinase (PK-III) isolated from *Neurospora crassa* is inhibited 50% by fenpiclonil and fludioxonil at 2 and 20 μM, respectively. Additional tests indicated that phenylpyrroles were not general inhibitors of protein kinases. While they exhibited some activity against two PK-III-related protein kinases from rat and starfish, they were also found to have no affect on many other kinases. The dicarboximide, vinclozolin, has no effect on the activity of PK-III at concentrations up to 100 μM. Thus, the phenylpyrroles appear to act at a different site than the dicarboximides within an osmoregulation signal transduction pathway that may have similarities to the osmosensing mitogen activated protein (MAP) kinase cascade known in yeast [16,17].

The ultimate cause of fungal toxicity upon treatment with dicarboximides, phenylpyrroles and perhaps aromatic hydrocarbon fungicides such as chloroneb, may be an overproduction of osmoregulating polyols. Laboratory mutants defective in the os-1 gene, which encodes an early osmosensing kinase, are resistant to all three classes of fungicides. These mutants fail to accumulate glycerol in response to osmotic stress or exposure to these three classes of fungicides. Each fungicide class may act at a different point within the osmosensing MAP kinase cascade [18].

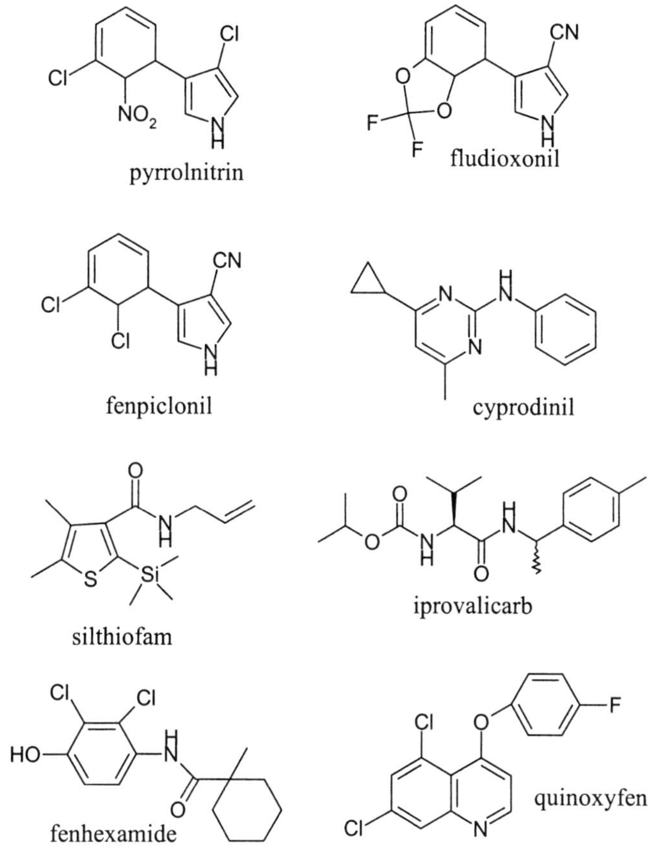

Figure 3. New Fungicides.

6 Anilinopyrimidines

Cyprodinil (Fig. 3) is an anilinopyrimidine fungicide that is effective on a variety of Ascomycete fungi [13]. The toxicity of cyprodinil can be reversed by adding methionine or homocysteine to growth media, but not with cystathionine [19]. These results suggest that cystathionine beta-lyase (CBL) is the target site of the anilinopyrimidine fungicides [20]. However, recent studies indicate that isolated CBL is not inhibited by cyprodinil. Furthermore, mutants resistant to cyprodinil were not found to have mutations in the *cbl* gene [21]. In some cyprodinil-resistant mutants, single amino acid changes have been detected in the regulatory sequences of other enzymes in the methionine biosynthetic pathway, suggesting that anilinopyrimidines may have a regulatory mechanism of action.

7 Scytalone Dehydratase Inhibitors

Melanin biosynthesis was first demonstrated to be a valid antifungal target site by the rice blast fungicide tricyclazole. Tricyclazole inhibits the naphthol reductase steps in fungal melanin biosynthesis. The inhibition of melanin production results in the inability of the appressorial infection peg to penetrate into the host [22-25]. Recently a second enzyme in the pathway, scytalone dehydratase has been identified as the target site of three new rice blast control agents, carpropamid, fenoxanil (AC 382042) and KC10017 (Fig. 4) [26,27]. These agents are effective for the control of rice blast caused by *Magnaporthe grisea* and bean anthracnose caused by *Colletotrichum lagenarium*. The target site from *M. grisea* has been extensively investigated at the molecular level to understand the mechanism of inhibitor binding [28]. Scytalone dehydratase converts the melanin intermediate scytalone to 1,3,6-trihydroxy napthalene. This dehydration is essentially identical to a subsequent step in the pathway where vermelone is dehydrated to 1,8 dihydroxynapthalene (Fig. 4). The inhibition of melanin biosynthesis results in the incomplete development of the fungal appressorial cell wall. This weakened wall is unable to maintain the turgor pressure required for direct the penetration of the infection peg into the host. The action of these disease control agents is targeted at a process involved in the pathogenicity of the fungus, and is not directly inhibitory to the *in vitro* growth of the fungus. Thus, scytalone dehydratase inhibitors and other inhibitors of melanin biosynthesis are referred to as antipathogenic compounds.

8 Silthiofam

Silthiofam (MON 65500, Fig. 3) is a new fungicide developed primarily for the control of "take-all" disease in wheat caused by the pathogen *Gaeumannomyces graminis* var. *tritici*. This class of chemistry has been referred to as the "silyl-hindered" amides. A variety of cellular processes have been investigated with respect to the action of silthiofam. It has been suggested that the mechanism of action for silthiofam is inhibition of the export of ATP from the mitochondria to the cytosol. Although the total cellular ATP levels are not appreciably reduced in intact mycelia of *Gaeumannomyces* treated with silthiofam, the levels of ATP produced from isolated mitochondria are reduced. These experiments are difficult to interpret because the inhibition of transport must be deduced by inference rather than direct measurement. However, the inhibition of ATP transport could explain why a variety metabolic processes that require ATP are rapidly affected by silthiofam [29].

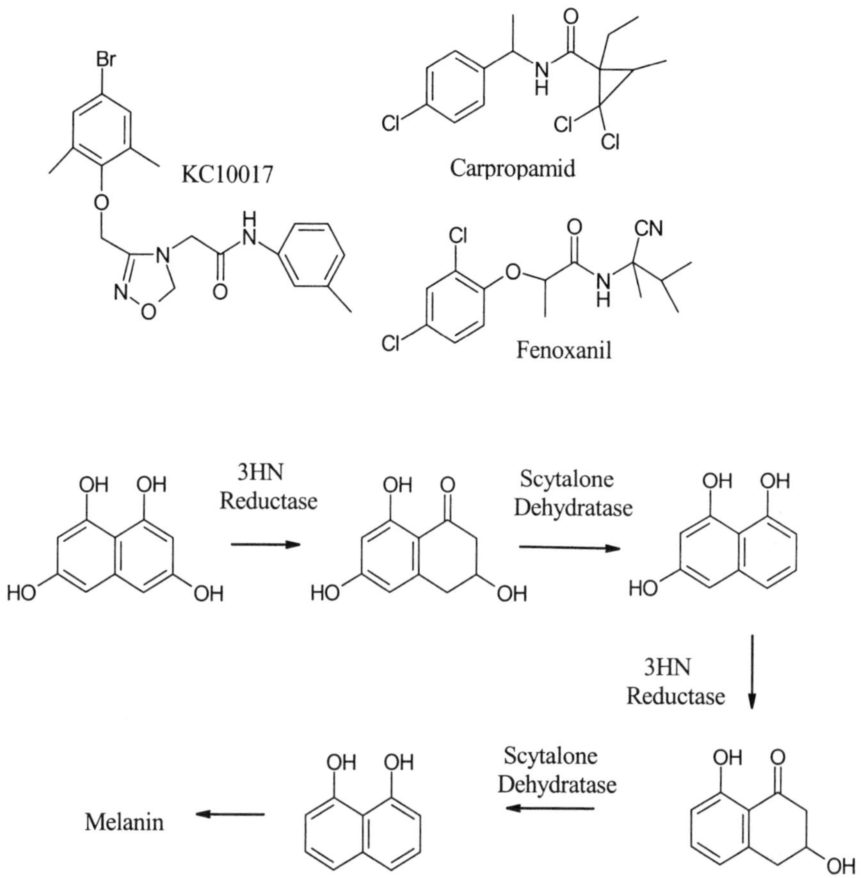

Figure 4. Melanin Biosynthesis Inhibitors and Pathway

9 Fenhexamid

Fenhexamid (Fig. 3) is a recently developed fungicide that is used for the control of gray mold, caused by *Botrytis cinerea*, on many fruits, vegetables and ornamental crops. Recent reports have provided evidence that fenhexamid is a sterol biosynthsis inhibitor [30]. At concentrations of 0.2 to 1 mg/L, fenhexamid does not inhibit germination of conidia, but elongating germ tubes are distorted, bulged and granular. This germ tube morphology is consistent with other sterol biosynthesis inhibitors, including azoles and morpholines. Mycelial growth is inhibited about 50% at a concentration of 0.1 mg/L, and there is a corresponding reduction in ergosterol and an accumulation of 3-keto-sterols. The C-4 dimethyl-sterol precursors of ergosterol are decarboxylated by a different mechanism than the 14α-methyl group of ergosterol precursors. The C-4 methyl groups are oxidized to the alcohol, aldehyde and ultimately the carboxylic acid (Fig. 5). This decarboxylation occurs via a 3-keto-steroid intermediate that is then reduced to the alcohol. The accumulation of 3-keto-sterol intermediates in mycelium treated with fenhexamid points directly to the 3-ketoreductase as the probable target site of fenhexamid.

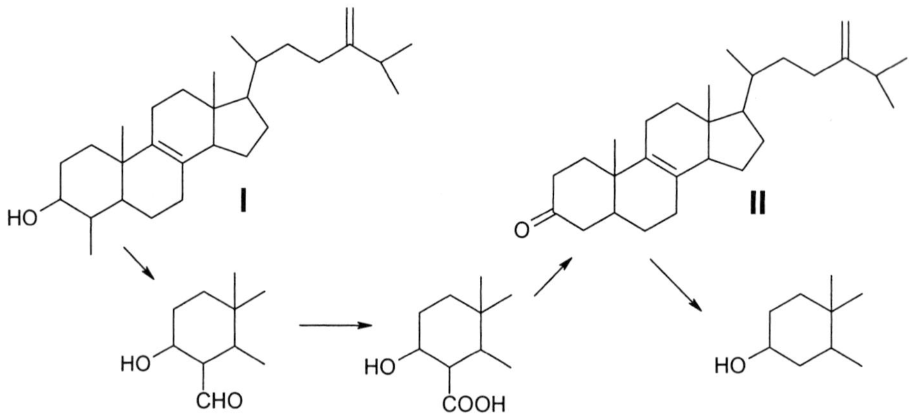

Figure 5. Sterol 3-keto reductase. I) 14α-methyl fecosterol, II) fecosterone.

10 Iprovalicarb

Iprovalicarb (Fig. 3) is a new Oomycete-specific fungicide and part of a novel class of amino acid amide carbamates. These fungicides are effective inhibitors of zoospore and sporangia germination as well as mycelial growth. Biochemical studies have demonstrated that respiration, electron transport, nucleic acid metabolism, and basic lipid metabolism are not primary targets of iprovalicarb [31]. Although incorporation of glucose into cell wall polymers is unaffected by iprovalicarb treatment, irregular tubular extensions are observed during regeneration of cell walls in protoplasts of *Phytophthora infestans* exposed to iprovalicarb. These morphological aberrations might suggest that structural components of cell wall assembly are being affected. his could involve microtubules or actin filaments associated in directing the cell wall biosynthetic enzymes in the cell membrane.

11 Quinoxyfen

Quinoxyfen (Fig. 3) is a protectant fungicide that controls powdery mildew diseases by interfering with germination and/or appressorium formation. Because quinoxyfen is only active against obligate powdery mildew pathogens, the mechanism of action of quinoxyfen has been difficult to investigate using biochemical techniques. In barley powdery mildew (*Blumeria graminis* f.sp. *hordei*), quinoxyfen interferes mainly with appressoriun formation. Mutants of barley powdery mildew resistant to quinoxyfen have been isolated in the laboratory and from quinoxyfen-treated field plots [32]. These mutants produce morphologically abnormal colonies with few conidia and are unable to undergo sexual recombination [33]. Conidia from quinoxyfen-resistant mutants germinate and form appressoria on artificial substrates more promiscuously than wild-type spores[34], suggesting that resistance bypasses normal host recognition signals. A gene transcript has been identified in wild-type conidia that is absent, or much less abundant, in conidia from quinoxyfen-resistant mutants [34]. This mRNA, which is no longer detectable 24 hours after wild-type conidia are inoculated onto barley, has been shown to code for a GTPase activating protein (GAP). In the presence of quinoxyfen, this GAP mRNA fails to dissipate during the infection stages leading up to appressorium formation. This effect is specific to quinoxyfen as it is not observed with ethirimol, a fungicide that also inhibits appressorium formation in *B. graminis*. The involvement of a GAP in resistance to quinoxyfen suggests that quinoxyfen inhibits mildew infection by disruption of early cell signaling events. However, the biochemical target of quinoxyfen has not yet been identified.

12 Summary

Antifungal mechanisms for the control of plant pathogenic fungi are diverse in scope. The targets range from structural components like tubulin to energy and biosynthetic targets as well as regulatory pathways. These examples indicate that there are still new opportunities to identify novel targets to assists and support the discovery of new disease control agents. Fundamental mode of action studies are essential to continue this discovery process.

13 References

[1] H. Brandt, H. Schagger, and G. Von Jagow. FEBS Lett. 1988, 201,499-506.

[2] T. Tron, M. Crimi, A. M. Colson, and M. D. Esposti. Eur. J. Biochem. 1991, 199, 753-60.

[3] D. W. Bartlett, J. M. Clough, J. R. Godwin, A. A. Hall, M. Hamer, and B. Parr-Dobranski. Pest Management Sci. 2002, 58, 649-662.

[4] D. Jordon, R. Livingston, J. Bisaha, K. Duncan, S. Pember, M. Picollelli, R. Schwartz, J. Sternberg and X. Tang, Pest. Sci. 1999, 55, 105-118.

[5] H. Ishii, B. A. Fraaije, T. Sugiyama, K. Noguchi, K. Nishimura, K. Takeda, T, Amano, and D. W. Holloman, Phytopathology 2001, 91, 1166-1171

[6] S. P. Heaney, A. A. Hall, S. A. Davies, and G. Olaya, Brighton Crop Prot. Conf. Pest Dis. 2000, 2, 755

[7] U. Steinfeld, H. Sierotzki, S. Parisi, S. Poirey, and U. Gisi. Pest Management Sci. 2001, 57, 787-796.

[8] S. Mitani, S. Araki, T. Yamaguchi, Y. Takii, T. Ohshima, and N. Matsuo. Pestic. Biochem. Physiol. 2001, 70, 92-99.

[9] S. Mitani, S. Araki, Y. Takii, T. Ohshima, N. Matsuo, and H. Miyoshi. Pestic. Biochem. Physiol. 2001, 71, 107-115.

[10] C. Pillonell, Pesticide Science 1995, 43, 107-113.

[11] D. H. Young and R. A. Slawecki, Pestic. Biochem. Physiol. 2001, 69, 100-111.
[12] K. Gehmann, R. Nyfeler, A. J. Leadbeater, D. Nevill, and D. Sozzi, Proceedings of the British Crop Protection Conf. Pests and Diseases, 1990, pp. 399-406.
[13] P. Leroux, C. Lanen and R. Fritz, Pestic. Sci. 1992, 36, 255-261.
[14] A. B. K. Jespers, L. C. Davidse and M. deWaard, Pestic. Biochem. Physiol. 1993, 45, 116-129.
[15] C. Pillonel and T. Meyer, Pestic. Sci. 1997, 49, 229-236.
[16] T. Maeda, S. M. Wurgler-Murphy, and H. Salto, Nature 1994, 369, 242-245.
[17] M. C. Gustin, J. Albertyn, M. Alexander and K. Davenport, Mol Biol. 1998, Rev. 64, 1264-1300.
[18] N. Ochiai, M. Fujimura, T. Motoyama, A. Ichiishi, R. Usami,K. Horikoshi, and I. Yamaguchi, Pest Manag Sci. 2001, 57, 437-442.
[19] P. Masner, P. Muster, and J. Schmid, Pestic. Sci. 1994, 42, 163-166.
[20] R. Fritz, C. Lanen, V. Colas, and P. Leroux, Pest. Sci. 1997, 49, 40-46.
[21] H. Sierotzki, J. Wullschleger, M. Alt, T. Bruyere, C. Pillonel, S. Parisi, and U. Gisi, Proceeding of 13th Int. Reinhardsbrunn Symp., Modern Fungicides and Antifungal Compounds, 2001.
[22] C. P. Woloshuk, H. D. Sisler, M. C. Tokousbalides, and S. R. Dutky. Pestic. Biochem. Physiol. 1980, 14, 256-264.
[23] C. P. Woloshuk, H. D. Sisler, and E. L. Vigil. Physiol. Plant Pathol. 1983, 22, 245-259.
[24] Y. Kurahashi, Pesticide Outlook 2001, 12,32-35.
[25] J. Thompson, G. S. Basarab, A. Anersson, Y. Linqvist, and D. B. Jordan, Biocchemistry 1997, 36,1852-1860
[26] G. Tsuji, T. Takeda, I. Furusawa, O. Horino, and Y. Kubo, Pestic. Biochem. Physiol. 1997, 57, 211-219.
[27] J. C. Kim, J. Y. Min, H. T. Kim, B. S. Kim, Y. S. Kim, B. T. Kim, S. H. Yu, I. Yamaguchi and K.Y. Cho, Pestic. Biochem. Physiol. 1998, 62, 102-112.
[28] D. B. Jordan, G. S. Basarab, J. J. Steffens, R. S. Schwartz, and J. Doughty. Biochemistry, 2000, 39, 8593-8602.
[29] T. Joseph-Horne, C. Heppner, J. Headrick, and D. W. Hollomon. Pestic. Biochem. Physiol. 2000, 67, 168-186.
[30] D. Debieu, J. Bach,M. Hugon, C. Malosse and P. Leroux, Pest Management Sci, 2001, 57, 1060-1067.
[31] A. Mehl. Proceeding of 13th Int'l Reinhardsbrunn Symp. Modern Fungicides and Antifungal Compounds, 2001
[32] D. W. Hollomon, I. E. Wheeler, K. Dixon, C. Longhurst, and G. Skylakakis, Pestic Sci. 1997, 51, 347-351.
[33] D. W. Hollomon, I. E. Wheeler, 2002. Controlling powdery mildews with Chemistry. In *The powdery mildews: a comprehnsive treatise*. (R.R.Belanger, A.J. Dik, T.L.W. Carver, and W. R Bushnell, eds.) American Phytopathological Society Press, St. Paul, USA.
[34] I. E. Wheeler, D. W. Hollomon, E. Green, C. Longhurst, 2000. Quinoxyfen signals a stop to infection by powdery mildews. *Proc. Brighton Crop Protect. Conf- Pests and Diseases*, 841-846.

09 Elucidating Pathways Controlling Induced Resistance

Jurriaan Ton and Brigitte Mauch-Mani*

University of Neuchâtel, Institute of Botany, Biochemistry Unit, Rue Emile Argand 9, PO Box 2, CH-2007 Neuchâtel, Switzerland; E-mails: jurriaan.ton@unine.ch, brigitte.mauch@unine.ch (* for correspondence)

1 Introduction

During most of their lifetime, plants are exposed to a wide range of potential pathogens, pests, and abiotic stresses. Therefore, they had to develop a number of passive and active resistance mechanisms to protect themselves. These defense mechanisms, especially when biotic stresses are involved, have been defined broadly as avoidance, tolerance and resistance [1]. Avoidance results in a strongly reduced contact between the plant and the potential parasite. This type of resistance is mainly found against animal pests and herbivores where bad smells, tastes or special structures such as thorns are the reason for a reduced contact between the partners of the interaction. Tolerance of a plant against a given pathogen does not interfere with the development of the invading pathogen itself but leads to a noticable reduction of the damage resulting from the pathogen's presence in the host.

Phenomenologically, resistance is well known to manifest itself at different levels. For reasons of clarity, resistance can be subdivided into different classes that, from a biochemical point of view, might well be identical. The diversity of responses that occurs at the cellular level in both host and non-host plants suggests many different stages during an attempted infection, at which an interaction between the parasite and the host can take place. Each of these stages might represent a "switching point", during which the response determines the further development and outcome of the interaction [2].

Probably, the most ubiquitous type of resistance is non-host resistance. Plants are constantly in contact with many different microbial organisms, of which only a very restricted number is able to successfully infect a given plant species. This type of resistance protects the plant entirely from infection by most potential pathogens, and is manifested as an inability of the pathogen to cause disease upon contact with any individual of a particular plant species. In such an interaction, the pathogen is considered to be non-pathogenic. Non-host resistance is therefore the most common and the most effective form of resistance in higher plants.

Basal resistance, also called field resistance, quantitative, or horizontal resistance is usually not race-specific, and is commonly assumed to be polygenic. Plants expressing this type of resistance can be infected by pathogens but the rate of disease progress and corresponding symptom expression is significantly reduced compared to susceptible plants. Basal resistance thus confers a partial, quantitative protection. In contrast, race-specific resistance, also called gene-for-gene resistance or vertical resistance is based on the specific interaction between the products of avirulence (*avr*) genes in the pathogen and resistance (*R*) genes in the host [3]. Certain individuals within populations of the species will be susceptible to some races of a pathogen but resistant to other races of the same pathogen whereas in compatible interactions, the pathogen is able to colonize the tissues and cause disease. In incompatible interactions, the pathogen is rapidly arrested at the site of infection although it is still capable of initiating an infection. Usually, this early recognition of the so-called avirulent pathogen gives rise to a hypersensitive response (HR). The HR involves a range of active defense mechanisms, including a form of programmed cell death at the site of infection [4].

In contrast to the above-mentioned constitutive types of resistance, plants also possess adaptive defense mechanisms to defend themselves against pathogen or insect attack. Following an appropriate stimulation, they are capable of developing an enhanced defensive capacity, commonly referred to as induced resistance [5,6,7,8].

2 Basal Resistance to Plant Pathogens: Involved Signalling Pathways

Investigations dealing with the signal transduction pathways controlling defense reactions of plants have repeatedly revealed a prominent role for the plant hormones salicylic acid (SA), jasmonic acid (JA) and ethylene (ET). In most pathosystems, infection by both avirulent and virulent pathogens is associated with enhanced production of these regulators, and exogenous application of these compounds often results in an enhanced level of resistance [9,10,11,12,13,14]. Additionally, inhibition or reduction of either of these signals can lead to an increased susceptibility of the plants to certain pathogens or insects [15,16,17,18,19,20,21,22]. The prominent role of SA in plant defense was elegantly demonstrated with the use of NahG transformants. These transgenic plants constitutively express a bacterial salicylate hydroxylase, the *NahG* gene, which converts SA into catechol, a chemical with no resistance-conferring activity. In tobacco and Arabidopsis plants, constitutive expression of the *NahG* gene leads to an increased disease susceptibility to a large range of oomycetous, fungal, bacterial and viral pathogens [15,23]. The role of SA has also been demonstrated via another approach. A screen for mutants impaired in the accumulation of SA following infection by an avirulent bacterial pathogen yielded two mutants, *sid1* and *sid2* (salicylic acid induction defficient), affected in pathogen-induced biosynthesis of SA [24]. The two *sid* mutants show an increased susceptibility to the virulent pathogens *Pseudomonas syringae* pv. *tomato* and

Peronospora parasitica, emphasizing the importance of SA in the basal resistance against both the bacterial and the oomycetous pathogen. However, the involvement of SA in basal resistance against oomycetes is not ubiquitous. Arabidopsis plants carrying the *NahG* gene as well as the *sid2* mutants both only showed a very slight shift towards suceptibility when infected with another oomycete, *Phytophthora porri* [25]. Mutants *sid1* and *sid2* are allelic with the enhanced disease susceptibility mutants *eds5* [24] and *eds16* [26], respectively. These display enhanced susceptibility to a virulent strain of the bacterial pathogen *Xanthomonas campestris* pv. *raphani* [27] and the powdery mildew fungus *Erysiphe orontii* [26]. Both mutated genes have been cloned, *EDS5* coding for a member of the MATE transporter family [28] and *EDS16* for an isochorismate synthase [26].

Most of the data available concerning the role of JA in resistance of plants against pathogens stem from investigations with Arabidopsis mutants affected in the biosynthesis of, or in their responsiveness to, JA. The *coi1* mutant [29] is affected in response to jasmonate and shows an enhanced susceptibility to the necrotrophic fungi *Alternaria brassicicola* and *Botrytis cinerea* [20], and the bacterial pathogen *Erwinia carotovora* pv. *carotovora* [30]. In the Arabidopsis *jar1* mutant [31], which is also JA-insensitive, an increased colonization by the virulent pathogen *P. syringae* pv. *tomato* was measured in the leaves [12]. Furthermore, the enhanced disease susceptibility mutant *eds8-1* of Arabidopsis that displays enhanced susceptibility to different pathovars of *P. syringae* [32], was recently found to exhibit reduced sensitivity to JA [33]. In addition, the *jar1* mutant, as well as the *fad3*, *fad7*, *fad8* triple mutant that is defective in JA biosynthesis, exhibit susceptibility to normally non-pathogenic soil-borne oomycetes of the genus *Pythium* [18,34]. The above-mentioned examples point to a considerable involvement of JA in the expression of basal resistance against various pathogens. However, the story seems more complicated. The jasmonic acid deficient mutant *opr3* exhibits an unaltered strong resistance to the dipteran *Bradysia impatiens* and the fungus *Alternaria brassicicola* [35], suggesting that the various components of the JA response pathway contribute differentially to basal resistance.

The role of ET in plant basal resistance seems even more controversial. Although in some cases ET promotes disease development, in other cases it is associated with disease resistance. ET-insensitive tomato genotypes allowed wild-type levels of growth of virulent *P. syringae* pv. *tomato* and *X. campestris* pv. *vesicatoria*, but the observed disease symptoms were atenuated [36,37]. This clearly points to a regulation of symptom expression by ET. Conversely, Knoester *et al.* [16] reported that ET-insensitive tobacco plants transformed with the mutated ET receptor gene *etr1-1* from Arabidopsis (Tetr tobacco), displayed susceptibility to the normally non-pathogenic oomycete *Pythium sylvaticum*. In addition, several ET-insensitive mutants of Arabidopsis exhibit enhanced susceptibility to *B. cinerea* [38], *P. syringae* pv. *tomato* [12] and *E. carotovora* [30]. These observations support the notion that ET-dependent defenses contribute to basal resistance against these pathogens. The relative contribution of SA, JA and ET to basal resistance seems to be specific for a given host-pathogen interaction. In the model plant Arabidopsis, containment of some pathogens has been shown to depend mainly on SA-dependent pathways, i.e. *P. parasitica* and turnip crinkle virus (TCV), whereas others, such as *A. brassicicola*, *B. cinerea*, and *E. carotovora,* were predominantly controlled through JA-and ET-dependent resistance mechanisms.

3 Induced Resistance

By ways of induced resistance, a genetically susceptible plant can be induced to defend itself more efficiently against a broad range of virulent pathogens including oomycetes, fungi, bacteria and viruses, thus displaying phenotypic resistance. This form of resistance is reminiscent of the immunization system found in animals. Induced resistance has been known for a long time [39] but it was only with the studies by Ross [40] and Kuc and coworkers [41,42] that the resistance could be linked to the activation of defence mechanisms. The induced state is achieved by treating the plant locally with specific biotic or abiotic agents. This results in an increased resistance in distal plant parts. Induced disease resistance can be subdivided into three categories depending on the initial inducing treatment.

A localized predisposing infection with a necrotizing pathogen usually results in a type of resistance referred to as systemic acquired resistance (SAR). SAR has been described for a large number of plant species [5,6,42].

A phenotypically similar type of resistance is observed following the colonization of plant roots by selected strains of non-pathogenic *Pseudomonas fluorescens* bacteria. This type of induced resistance is called rhizobacteria-mediated induced systemic resistance (ISR) [7].

Finally, plants can also be transferred into an induced state by exogenous application of selected chemical agents [43]. The best investigated chemical inducers are salicylic acid (SA) and its functional analogues 2,6-dichloroisonicotinic acid (INA) and benzothiadiazole (BTH) [43]. INA and BTH are thought to induce a resistance type similar to SAR because they activate the same set of defense-related genes and are effective against the same spectrum of plant pathogens [5]. The non-protein amino acid β-aminobutyric acid (BABA), a powerful chemical inducer of broad-spectrum resistance in different plant species [8,44], on the contrary seems to be able to operate independently from an initial SAR gene induction [45,46].

3.1 Mechanisms of Induced Resistance

During the last decade, the development of many pathosytems with Arabidopsis as the host has allowed a better insight into the molecular mechanisms underlying the different types of induced resistance. Numerous signal transduction mutants have been isolated and characterized, leading to the notion that plants are capable of differentially activating specific signalling pathways depending on the attacker they are confronted with.

3.1.1 SAR

One of the first alterations that can be measured in plants expressing SAR is an increase in the production of SA [47,48]. At the same time a set of genes coding for specific pathogenis-related proteins (PR) [49,50], now often referred to as SAR genes, is activated. Interestingly, exogenous application of SA itself induces resistance accompanied by the transcriptional activation of the same set of SAR genes [39,50]. A strong support for a crucial involvement of SA in SAR (as in basal defense capacity of plants) became evident with the availability of the SA-non accumulating NahG plants. Neither NahG tobacco nor NahG Arabidopsis are capable of developing SAR and activiating SAR genes following infection with a pathogen [51,52]. The Arabidopsis mutants *sid1* and *sid2*, both affected in SA accumulation [24], do not express SAR against *P. parasitica*, which further emphasizes the crucial role of SA in this type of induced resistance. The earlier notion that SA might also be the translocated signal for SAR seems quite unlikely today. By grafting wild type tobacco onto NahG tobacco and vice versa, Vernooij et al. [53] showed, that SA is essential for the induction of SAR, but that a signal other than SA is translocated to the upper parts of the plant and there induces resistance. More recent evidence indicates that the mutant *dir1-1*, defective in induced resistance, is affected in a gene that has a prominent role in the production or transmission of an essential mobile signal from the inoculated leaf to the rest of the plant [54].
A further key factor in the SAR signalling pathway is encoded by the *NPR1* gene (also known as *NIM1, SAI1*). The Arabidopsis *npr1* mutant accumulates wild-type levels of SA when treated with avirulent pathogens but fails to develop biologically or chemically induced SAR [55, 38,56]. This indicates that NPR1 acts downstream of the accumulation of SA in the SAR pathway. Another factor implicated in the final stages of SAR signalling is the SNI1 gene (suppressor of *npr1-1*, inducible 1) that encodes a leucine-rich nuclear protein [57]. The recessive *sni1* mutation restores INA-induced SAR expression and *PR-1* transcription in the background of the *npr1* mutant. This indicates that SNI1 functions as a transcriptional repressor of SAR that can be counteracted by NPR1 after activation of the SA-dependent SAR pathway.

3.1.2 ISR

Large populations of rhizobacteria are usually present in the rhizosphere, mainly because of the ideal environmental conditions they encounter there due to a constant supply of nutrients provided through root exudates and lysates [58]. Many of the present strains exert a disease controlling activity simply by directly antagonizing soil-borne pathogens [59,60]. In order to investigate disease suppression conditioned by ISR, it is therefore important to spatially separate the bacteria from the roots.
Certain rhizobacterial strains produce SA when exposed to iron-limiting conditions [61, 62]. This is for example the case in strain CHA0 of *Pseudomonas fluorescens*. The observed ISR in tobacco plants might well be due to an induction by SA [63], especially when considering that treatment of the plants‘ roots with the CHA0 bacteria triggered accumulation of SA-inducible PRs in the leaves [63]. An additional fact pointing to an involvement of SA in ISR is presented in the following example. When *P. fluorescens* strain P3 is transformed with the SA-biosynthetic gene cluster from CHA0 its ISR-

inducing capacity is strongly improved [64]. Despite examples of induction of ISR mediated by SA, there is a large body of evidence pointing to a strong involvement of a SA-independent pathway for the establishment of ISR [11,12, 65].

The first evidence for a SA-independent pathway controlling ISR became apparent in experiments where the rhizobacterium *P. fluorescens* strain WCS417r was used to induce resistance against the plant pathogenic bacteria *P. syringae* pv *tomato* in Arabidopsis [11]. It was shown that WCS417r-mediated ISR was fully maintained in NahG plants, and not associated with the transcriptional activation of genes encoding SA-inducible PRs. However, in the JA-insensitive *jar1* [31] or in the ET-insensitive *etr1* [66] mutants, the WCS417r bacteria failed to trigger ISR [12]. Further studies revealed that both systemically and at the site of bacterial application, JA content and the level of ET evolution remained unaltered upon ISR induction [67,68]. This indicates that JA and ET dependency of ISR is not based on an enhancement of JA and ET production, but rather on an enhanced sensitivity to these hormones. At the molecular level, it became evident that ISR in Arabidopsis is not associated with major changes in defense-related gene expression and neither is it with changes in the expression of several JA-and/or ET responsive genes [22]. Apparently, WCS417r-mediated ISR in Arabidopsis is achieved through a sensitization for JA- and ET- dependent defences.

Interestingly, there seems to be a common signalling component shared by both the IRS and the SAR pathway. The *npr1* mutant of Arabidopsis is not only impaired in the expression of SAR, but also fails to express ISR after induction with WCS417r bacteria 12]. Thus, NPR1 is required for the establishment of both SA-dependent SAR and JA- and ET-dependent ISR. This dual involvement of NPR1 suggests that the protein is able to differentiate between different defence responses, depending on the signalling pathway that is activated upstream of it.

4 Priming

Induced disease resistance is associated with a heightened capacity to induce cellular defense responses following an infection. This has been called the primed state of the plant [69]. The physiological and molecular mechanisms underlying this type of reaction are widely unknown. Induced resistance has mostly concentrated on direct effects of a given treatment on gene transcription and priming was overseen in most of the cases, since it only becomes apparent when the plant is actually attacked by a pathogen. Characteristic for primed plants is a more rapid and effective defence reaction upon contact to a wide range of biotic and abiotic stresses.

4.1 Priming in ISR and SAR

Priming has been observed in various pathosystems as an integral part of both SAR and ISR as well as in chemically induced resistance [69]. Early reports on priming were purely descriptive. The response of the plants was accelerated and they proved also to be more resistant [70,71,72,73]. In parsley cell cultures low doses of the SAR inducers SA,

INA, and BTH did not directly induce various assayed cellular defense responses but primed the cells for potentiated activation of defense responses [74,75,76,77,78,79]. SA has been shown to strongly augment defense gene activation, the HR and H_2O_2 accumulation in soybean suspension cells following a treatment with avirulent *P.syringae* pv *glycinea* [80]. In tobacco plants, pre-treatement with SA enhances the expression of defence genes that are usually not directly responsive to SA itself [81]. Methyl jasmonate also shows a conditioning activity in tobacco cell cultures. Here, priming leads to a significant increase in lipoxygenase (LOX) activity [82].

In ISR expressing Arabidopsis, it was shown that there are no local and systemic changes in the levels of JA and ET, although an intact response to both JA and ET is required [12,68]. This suggests that ISR is based on an enhanced sensitivity to these plant hormones rather than on an increase in their production. This is supported by ISR expressing plants which show a primed expression of the jasmonate responsive gene *Atvsp* [22] upon infection with *P. syringae* pv. *tomato*.

4.2 β-Aminobutyric Acid-Induced Priming

Several synthetic resistance inducers have been shown to induce priming. BABA–mediated priming in Arabidopsis has proven to be an attractive model system to investigate the molecular mechanisms behind this phenomenon. This is due to the fact that BABA concentrations sufficient to induce resistance do not induce defense gene expression in Arabidopsis [46]. This opens the possibility to separate priming from direct effects of the inducing agent on gene activity that cause enhanced resistance.

The induction of resistance by BABA in Arabidopsis is not primarily based on a previous accumulation of defense gene transcripts but on an additional mechanism of resistance induction present in BABA-treated Arabidopsis plants. This is supported by the observation that BABA induces resistance against the oomycete *P. parasitica* in wild-type Arabidopsis plants as well as in plants that are impaired in defense gene expression [46], such as the *npr1, jar1* or *etr1* mutants [55,31,66], and *NahG* transgenic Arabidopsis plants [15]. In this case, the resistance functions independently of SA, *PR* gene expression, or other defence gene activation. The explanation for the resistance observed in all of these different defence mutants is a more rapid and stronger deposition of callose-containing papillae at the site of infection by *P. parasitica* [46] making further defense responses obsolete as ingress by *P. parasitica* has already been stopped at this point. Apparently, this primed callose deposition is controlled by a yet unidentified signalling pathway.

When challenged with virulent bacteria, priming of BABA-treated Arabidopsis plants becomes apparent as a strong potentiation of *PR-1* gene expression [46]. In this case however, priming by BABA depends on an intact SA signalling pathway, but operates independently of the JA- and ET-dependent pathways, as evident from experiments with the same defense response mutants as described above [46]. Interestingly, in the Arabidopsis-*B. cinerea* interaction, it is the *PR-1* gene that shows strong potentiation [45] and not *PDF1.2* [20] that is commonly assumed to play a dominant role in defense against *B. cinerea*.

During our research on BABA-induced resistance we noticed that the priming effect of this chemical was not restricted to the plant's reaction to biotic stresses. BABA-treated plants were also sensitized to react faster and more effectively to abiotic stresses. BABA-treated Arabidopsis, for example showed a 75% survival rate following a treatment of two days at –5 °C, whereas control plants were all killed [83]. We also observed faster reactions at the molecular and phenotypical level to high salt, heat and drought treatment [8,69]. All these observations point to an important involvement of BABA in the expression of general priming mechanisms of stress tolerance.
To gain more insight in the molecular mechanisms behind BABA-mediated priming, a mutant screening was performed based on the observation that higher doses of BABA induce sterility in Arabidopsis [8]. This led to the rescue of a number of BABA insensitive mutants (*bai* mutants) that all showed formation of siliques after repeated treatments with BABA, whereas the wild type displayed sterility [84]. Three of the *bai* mutants have been preliminarily characterized [84]. The *bai*38 mutant is affected in a zeaxanthine epoxidase gene playing a role in ABA biosynthesis. Interestingly, ABA is well-known to affect drought resistance of plants [85]. The *bai*24 mutant has an insertion in a putative phosphoinositide phosphatase gene, a member of a cellular signalling pathway which has been associated among others to osmo-sensing in plants and to cold tolerance in yeast [86]. The *bai*65 mutant has an insertion in a putative kinase gene. Further phenotypic analysis of these three mutants has shown that they are impaired in distinct defense pathways, controlling the primed defence activation during BABA-induced resistance against biotic and abiotic stresses. This suggests that there is a general priming mechanism that controls the plants' answer to biological, chemical and physical stresses and since this general priming mechanism has a genetic basis, it is also accessible to mutational analysis.

5 Conclusions

The capacity of a plant to express induced resistance depends on multiple signal transduction pathways well adapted to support a plant's struggle in very defined host-pathogen stress situations. The emerging picture today is that priming accelerates and increases a plant's ability to induce the pathway best adapted to help it cope with a defined situation. Therefore, a plant's capacity for priming is an important ecological and agronomical parameter. Efficient priming mechanisms are likely to play a role in the adaptation of natural and agronomical plant populations to rapidly changing environmental conditions. Moreover, priming-inducing agents confer enhanced resistance with minimal inhibitory effects on commercially important traits such as plant growth and fruit setting. A better knowledge of the molecular mechanisms underlying priming is the basis to engineer or select for crop plants with an increased ability to better cope with biotic and environmental stress.

6 Acknowledgements

We gratefully acknowledge financial support from the Swiss National Foundation (Grant 3100-064024) and from the National Center of Compentence in Science (NCCR) „Plant survival in agricultural and natural ecosystems".

7 References

[1] J. E. Parlevliet in Plant Breeding in the 1990s, Proceedings of the symposium on plant breeding in the 1990s, (Eds.: H. T. Stalker, J.P. Murphy) CAB International, Wallingford, 1992.
[2]M. C. Heath Physiol. Plant Pathol., 1974, 4, 403-414.
[3]H. Flor, Ann. Rev. Phytopathol., 1971, 9, 275-296.
[4] R. Mittler, E. Lam , Trends Microbiol.,1996, 4, 10-15.
[5] J.A. Ryals, U.H., Neuenschwander, M.G Willits, A.. Molina, H.-Y. Steiner, M.D. Hunt, Plant Cell, 1996, 8, 1809-1819.
[6] Sticher, L. B. Mauch-Mani, J.-P. Métraux, Annu. Rev. Phytopathol. 35, 235-270.
[7] Van Loon, L.C. et al. (1998) Systemic resistance induced by rhizosphere bacteria. Annu. Rev. Phytopathol. 36, 453-483.
[8] G. Jakab, V.Cottier, V. Toquin, G. Rigoli, L. Zimmerli, J. P Métraux, B. Mauch-Mani, Eur. J. Plant Pathol, .2001, 107, 29-37.
[9]T. Boller in The plant hormone ethylene, (Eds. : A.K. Mattoo, J.D. Suttle), CRS Press. Boca Raton, USA, pp. 293-314.
[10] D.A. Dempsey, J. Shah, D.F. Klessig , Crit. Rev. Plant Sci. 1999, 18, 547-575.
[11] C.M.J. Pieterse, S.C.M. Van Wees, E. Hoffland, J.A. Van Pelt, L.C. Van Loon, Plant Cell 1996, 8, 1225-1237.
[12] C.M.J. Pieterse, S.C.M. Van Wees, J.A. Van Pelt, J.A. Knoester, R. Laan, H. Gerrits, P.J. Weisbeek, L.C. Van Loon, Plant Cell, 1998, 10, 1571-1580.
[13] C.M.J. Pieterse, J.A. Van Pelt, J. Ton, S. Parchmann, M.J. Mueller, A.J. Buchala, J.P. Metraux, L.C. Van Loon, Physiol. Mol. Plant Pathol. 2000, 57, 123-134.
[14] B.P.H.J. Thomma, K. Eggermont, W.F. Broekaert, B.P.A. Cammue, Plant Phys. Biochem. 2000, 38, 421-427.
[15] T.P. Delaney, S. Uknes, B. Vernooij, L. Friedrich, K. Weymann, D. Negretto, T. Gaffney, M. Gut-Rella, H. Kessmann, E. Ward, J. Ryals, Science, 1994, 266, 1247-1250.
[16] M. Knoester, L.C. Van Loon, J. Van den Heuvel, J. Hennig, J.F. Bol, H.J.M. Lindhorst, Proc. Natl. Acad. Sci. USA, 1998, 95, 1933-1937.
[17] M. McConn, R.A. Creelman, E. Bell, J.E. Mullet, J. Browse, Proc. Natl. Acad. Sci. USA, 1997, 94, 5473-5477.
[18] P.E. Staswick, G.Y. Yuen, C.C. Lehman, Plant J., 1998, 15, 747-754.
[19] M.J. Stout, A.L. Fidantsef, S.S. Duffy, R. M. Bostock, Physiol. Mol. Plant Pathol., 1999, 54, 115-130.
[20] B.P.H.J. Thomma, K. Eggermont, I.A.M.A. Penninckx, B. Mauch-Mani, R. Vogelsang, B.P.A. Cammue, W.F. Broekaert, Proc. Natl. Acad. Sci. USA, 1998, 95, 15107-15111.
[21] J. Ton, S. Davison, J.A. Van Pelt, L.C. Van Loon, C.M.J. Pieterse, Plant Physiol., 2001, 125, 652-661.
[22] S. C. M. Van Wees, M. Luijendijk, I. Smoorenburg, L. C. van Loon, C. M. J. Pieterse, Plant Mol. Biol., 1999, 41, 537-549.

[23] P. Kachroo, K. Yoshioka, J. Shah. H.K. Dooner, D.F. Klessig, Plant Cell, 2000, 12, 677.690.
[24] C. Nawrath, J.P. Metraux, Plant Cell, 1999, 11, 1393-1404.
[25] A. Roetschi, A. Si-Ammour, L. Belbahri, F.Mauch, B. Mauch-Mani, Plant Journal, 2001, 28, 293-305.
[26] M.C. Wildermuth, J. Dewdney, G. Wu, F.M. Ausubel, Nature, 2001, 414, 562-565.
[27] E.E. Rogers, F.M. Ausubel, Plant Cell, 1997, 9, 305-316.
[28] C. Nawrath, S. Heck, N. Parinthawong, J.P. Metraux, Plant Cell, 2002, 14, 275-286.
[29] B. J. F. Feys, C. E. Benedetti, C. N. Penfold, J. G. Turner Plant Cell , 1994, 6, 751-759.
[30] C. Norman-Setterblad, S. Vidal, T.E. Palva, Mol. Plant-Microbe Interact., 2000, 13. 430-438.
[31] P.E. Staswick, W. Su, S.H. Howell, Proc. Natl. Acad. Sci. USA, 1992, 89, 6837-6840.
[32] J. Glazebrook, E.E. Rogers, F.M. Ausubel, Genetics, 1996, 143, 973-982.
[33] J. Ton, M. De Vos, C. Robben, A. Buchala, J.P. Metraux, L.C. Van Loon, C.M. Pieterse, Plant J., 2002, 29,11-21.
[34] P. Vijayan, J. Shockey, C.A. Lévesque, C.A. Cook, J. Browse, Proc. Natl. Acad. Sci. USA, 1998, 95, 7209-7214.
[35]A. Stintzi, H. Weber, P. Reymond, J. Browse, E.E. Farmer, Proc. Natl. Acad. Sci. USA, 2001, 98, 12317-12319.
[36]A. Ciardi, D.M. Tieman, S.T. Lund, J.B. Jones, R.E. Stall, H.J. Klee, Plant Physiol., 2000, 123, 81-92.
[37] B.P.H.J. Thomma, K. Eggermont, F.M.J. Tierens, W.F. Broekaert, Plant Physiol., 1999, 121, 1093-1101.
[38] T.P. Delaney, L. Friedrich, J.A. Ryals, Proc. Nat. Acad. Sci. USA , 1995. 92, 6602-6606.
[39] K. S. Chester, Quart. Rev. Biol., 1933, 8, 275-324.
[40] A. F. Ross, in Viruses of Plants, (Eds.: A.B.R. Beemster, J. Dijkstra), North-Holland Publishing, Amsterdam, 1966, 127-150.
[41] N. R. Madamanchi, J.Kuc, in The fungal spore and disease initiation in plants and animals, eds. G. T. Cole, H.C. Hoch, Plenum Press, New York, 1991, 347-362.
[42] R.Hammerschmidt , J.Kuc, Induced resistance to disease in plants, Kluwer Dordrecht.
[43] www.scri.sari.ac.uk/TiPP/SAR/Sar0.htm.
[44] Y. Cohen, Plant Dis., 2002, 86, 448-457.
[45] L. Zimmerli, J.P. Metraux, B. Mauch-Mani, Plant Physiology, 2001, 126, 517-23.
[46] L. Zimmerli, G. Jakab, J.P. Metraux, B. Mauch-Mani, Proc. the Nat. Acad. Sci. USA, 2000, 97, 12920-12925.
[47] J. Malamy, J.P. Carr, D.F. Klessig, I. Raskin, Science, 1990, 250, 1002-1004.
[48] J.P. Metraux, H. Signer, J. Ryals, E. Ward, M. Wyss-Benz, J. Gaudin, K. Raschdorf, E. Schmid, W. Blum, B. Invardi, Science, 1990, 250, 1004-1006.
[49] L.C. Van Loon, Eur. J. Plant Pathol., 1997, 103, 753-765.
[50] E.R. Ward, S. Uknes, S.C. Williams, S.S. Dincher, D.L. Wiederhold, D.C. Alexander, P. Ahl-Goy, J.P. Metraux, J.A. Ryals, Plant Cell, 1991, 3, 1085-1094.
[51] Gaffney, T., L. Friedrich, B. Vernooij, D. Negrotto, G. Nye, S. Uknes, E.Ward. J. Ryals, Science, 1993, 261, 754–756.
[52] K. Lawton, K Weimann, L. Friedrich, B. Vernooij, S. Uknes, J. Ryals, Mol. Plant Microbe Interact., 1995, 8, 863-870.
[53] B. Vernooij, L. Friedrich, A. Morse, R. Reist, R. Kolditz-Jowhar, E. Ward, S. Uknes, H. Kessmann, J. Ryals, Plant Cell, 1994, 6, 959–965.
[54] A. M. Maldonado, R.K. Cameron, P. Doerner, R.A. Dixon, C. Lamb, XIII Iterntional conference on Arabidopsis research, Sevilla, 2002.
[55] H. Cao, S.A. Bowling, A.S. Gordon, X.Dong, Plant Cell, 1994, 6, 1583-1592.
[56] J. Shah, F. Tsui, D.F. Klessig, Mol. Plant-Microbe Interact., 1997, 10, 69-78.
[57] X. Li, Y.L. Zhang, J.D. Clarke, X. Dong, Cell, 1999 98, 329-339.

[58] J.M. Lynch, J.M. Whipps in The rhizosphere and plant growth, eds. D.L. Keister, P.B. Cregan, Kluwer Academic Press, Dordrecht, The Netherlands, 1992, 15-24.
[59] P.A.H.M. Bakker, R. Van Peer, B. Schippers, in Biotic interactions and soilborne diseases, Eds. A.B.R. Beemster, et al., Elsevier Scientific Publishers, Amsterdam, The Netherlands,1995, 23, 217-230.
[60] G. Wei, J.W. Kloepper, S- Tuzun, Phytopathology, 1996, 86, 221-224.
[61] J.M. Meyer, P. Azelvandre, C. Georges, Biofactors, 1992, 4, 23-27.
[62] P. Visca, A. Ciervo, V, Sanfilipino, N. Orsi, J. Gen. Microbol., 1993, 139, 1995-2001.
[63] M. Maurhofer, C. Hase, P. Meuvly, J.P. Metraux, G. Défago, Phytopathology, 1994, 84, 139-146.
[64] M. Maurhofer, C. Reimmann, P. Schmidli-Sacherer, S.D. Heeb, G. Défago, Phytopathology, 1998, 88, 678-684.
[65] C.M. Press, M. Wilson, S. Tuzun, J.W. Kloepper, Mol. Plant-Microbe Interact., 1997, 10, 761-768.
[66] A.B. Bleeker, M.A. Estelle, C, Sommerville, H. Kende, Science, 1988, 241, 1086-1089.
[67] M. Knoester, C.M.J. Pieterse, J.F. Bol, L.C. Van Loon, Mol. Plant-Microbe Interact., 1999, 12, 720-727.
[68] C.M.J. Pieterse, J.A. Van Pelt, J.A. Ton, J. Parchmann, M.J. Mueller, A.J. Buchala, J.P. Metraux, L.C. Van Loon, Physiol. Mol. Plant Pathol., 2000, 57, 123-134.
[69] U. Conrath,. C.M.J. Pieterse, B. Mauch-Mani, Trends Plant Sci., 2002, 5, 210-216.
[70] R.A. Skipp, B.J. Deverall, Physiol. Plant Pathol., 1973, 2, 357-374.
[71] K. Kovats, A. Binder, H.R. Hohl, Planta, 1991, 183, 491-496.
[72] K. Kovats, A. Binder, H.R. Hohl, Planta. 1991, 183, 484-490.
[73] P. Silvermann, M. Seskar, D. Kanter, P. Schweizer, J.-P. Metraux, I. Raskin, Plant Physiol., 1995, 108, 633-639.
[74] H. Kauss, E. Theisinger-Hinkel, R. Mindermann, U. Conrath, Plant J, 1992, 2, 655-660.
[75] H. Kauss, K. Krause, W. Jeblick, Biochem. Biophys. Res. Comm., 1992, 189, 304-308.
[76] H. Kauss, W. Jeblick, Plant Physiol, 1995, 108, 1171-1178.
[77] O.U. Thulke, U. Conrath, Plant J., 1998, 14, 35-42.
[78] V.A. Katz, O.U. Thulke, U. Conrath, Plant Physiol ,1998, 117, 1333-1339.
[79] V.A. Katz, A. Fuchs, U. Conrath, FEBS Lett., 2002, 520, 53-57.
[80] K. Shirasu, H. Nakajima, V. Krishnamachari Rajasekhar, R.A. Dixon, C. Lamb, Plant Cell, 1996, 9: 261-270.
[81] L.A.J. Mur, G. Naylor, S.A.J. Warner, J.M. Sugars, R.F. White and J. Draper, Plant J., 1996, 559-571.
[82] I.A. Dubery, L.G. Teodorczuk, A.E. Louw, Mol. Cell. Biol. Res. Commun., 2000, 3, 105-110.
[83] L. Zimmerli, PhD. Thesis, Fribourg, 2000.
[84] V. Toquin, G. Jakab, M.N. Maeder, B. Mauch-Mani, in Induced Resistance in Plants against Insects and Diseases (I. Baldwin, M. Dicke, E. Haukioja, B. Mauch-Mani, A. Schmitt, eds), IOBC/wprs Bulletin, 2002, in press.
[85] J.-K. Zhu, Annu. Rev. Plant Biol. 2002. 53:247–73.
[86] T. Munnik, Trends Plant Sci., 2001, 6, 227-233.

10 Molecular Modeling of Inhibitors at Q_i and Q_o Sites in Cytochrome bc_1 Complex

Thomas A. Link[(a)], **Momi Iwata**[(b)], **Joakim Björkman**[(c)], **David van der Spoel**[(c)], **Andreas Stocker**[(d)] **and So Iwata**[(b)]

(a) Institut für Biophysik, Johann-Wolfgang-Goethe-Universität Frankfurt, Theodor-Stern-Kai 7, Haus 75, D-60590 Frankfurt/Main, Germany, E-mail: Link@biophysik.uni-frankfurt.de
(b) Imperial College of Science, Technology and Medicine, Department of Biochemistry and Division of Biomedical Sciences, Exhibition Road, London, SW7 2AZ, UK, E-mail: s.iwata@ic.ac.uk
(c) Uppsala University, Department of Biochemistry, Biomedical Center, Box 576, SE 751 23, Uppsala, Sweden, E-mail: spoel@xray.bmc.uu.se
(d) Institut für Molekulare Genetik, Johann-Wolfgang-Goethe-Universität Frankfurt, Marie-Curie-Str. 9, D-60439 Frankfurt/Main, Germany, E-mail: astocker@stud.uni-frankfurt.de

1 Introduction

The ubiquitous cytochrome bc_1 complex is the central part of electron transfer chains in mitochondria and many bacteria. The related b_6f complex is found in chloroplasts, algae, and some gram-positive bacteria. Together, they form the family of *bc* complexes; they all contain a two heme *b* centers, a high-potential heme center (heme c_1 or heme *f*), and a "Rieske" iron sulfur protein comprising a high potential [2Fe-2S] cluster. The *bc* complexes oxidize hydroquinones (ubihydroquinone, plastohydroquinone, or menahydroquinone) and transfer electrons to a water-soluble redox protein (cytochrome *c* for bc_1 complexes or plastocyanin for b_6f complexes). This exergonic transfer of electrons ($\Delta E^{0'} \approx +300$ mV) is coupled to the translocation of protons across the respective membrane (inner membrane in mitochondria, plasma membrane in bacteria) to generate the electrochemical membrane potential that is required, e.g., for ATP synthesis and the active transport of metabolites.

1.1 The 'Q-cycle' mechanism

The coupling of electron transfer and proton translocation is described by the proton-motive 'Q-cycle' mechanism first proposed by Peter Mitchell [1]. In the bc_1 complex, hydroquinone is oxidized at a reaction site which is at the positive side of the membrane, i.e., the outer side of the mitochondrial inner membrane; therefore, this site has been named Q_o or Q_P site (Figure 1). The two electrons delivered by hydroquinone go to two different redox centers which are both close to the Q_o quinone binding site: the first electron goes to the "Rieske" iron sulfur cluster and from there via cytochrome c_1 to the electron acceptor, cytochrome *c*. The second electron goes to the low-potential heme b_L and from there across the membrane dielectric to the high-potential heme b_H. Heme b_H is close to a quinone reaction site at the negative side of the membrane, i.e., the inner side of the mitochondrial inner membrane; therefore, this site has been named Q_i or Q_N site. At the Q_i reaction site, oxidized quinone is reduced by electron transfer from heme b_H. One complete cycle requires the oxidation of two molecules of hydroquinone; one molecule of quinone is reduced at the Q_i site by the two electrons transferred through cytochrome *b*, while two molecules of reduced cytochrome *c* are formed at the outside of

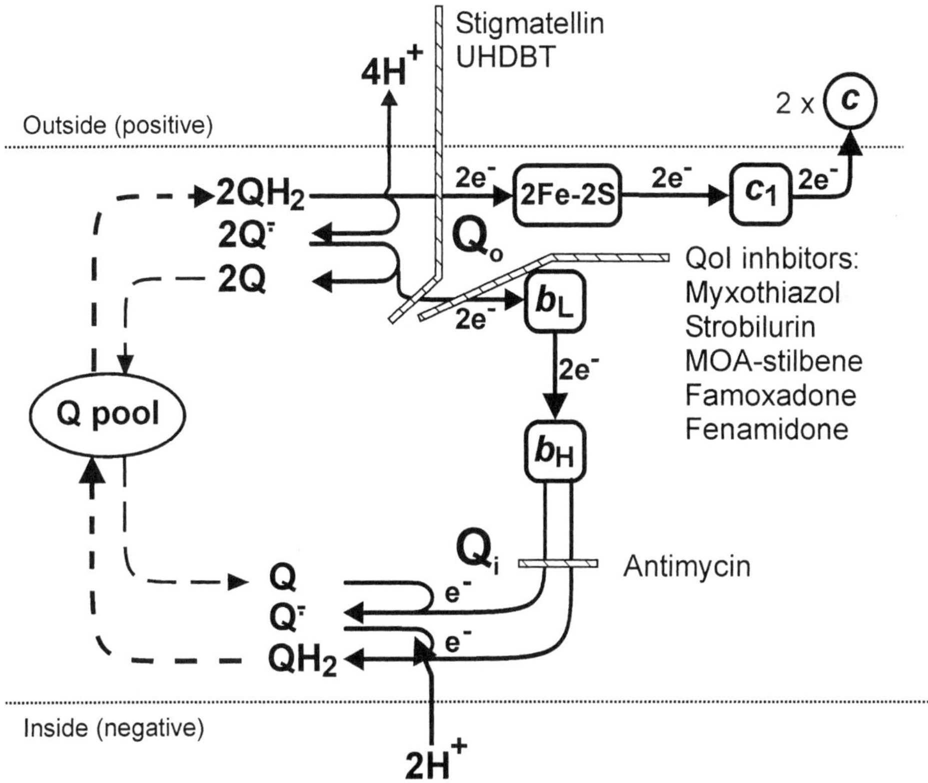

Figure 1. The 'Q-cycle' mechanism of the cytochrome bc_1 complex (cf. section 1.1). The points of inhibition of three classes of inhibitors are indicated.

the inner mitochondrial membrane. Therefore, the net reaction is:

$$QH_2 + 2\text{ cyt } c(Fe^{3+}) \longrightarrow Q + 2\text{ cyt } c(Fe^{2+}) \quad (1)$$

During this reaction, four protons are released upon oxidation of hydroquinone at the outside of the membrane while two protons are taken up during the re-reduction of quinone at the inner side of the membrane.

The 'Q-cycle' mechanism requires the presence of two separate quinone binding sites that are in contact with different sides of the membrane: the hydroquinone oxidation (Q_o or Q_P) site at the positive P side of the membrane and the quinone reduction (Q_i or Q_N) site at the negative N side of the membrane. These sites have first been characterized by their different inhibitor binding properties [2] (see below); the existence of two distinct quinone binding sites was confirmed by the X-ray structure of the bc_1 complex [3-6].

2 Inhibitors of the Cytochrome bc1 Complex

The cytochrome bc_1 complex is inhibited by a variety of organic compounds that can be grouped into three inhibitor classes [2] (Figure 1):

1. Inhibitors binding at the Q_o site exclusively to cytochrome *b*, e.g., myxothiazol as well as E-ß-methoxyacrylate (MOA) and related inhibitors. These inhibitors prevent the reduction of both the "Rieske" center and heme b_L through the Q_o site but heme b_H can still be reduced through the Q_i site with hydroquinone acting as a reductant (reverse electron flow).
2. Inhibitors binding at the Q_o site to both cytochrome *b* and the "Rieske" protein, e.g., stigmatellin and 5-undecyl-6-hydroxy-4,7-dioxobenzothiazol (UHDBT). These compounds bind tightly to the reduced "Rieske" protein and shift its midpoint potential to more positive values [7]. The binding to the "Rieske" protein can be monitored specifically by EPR spectroscopy of the reduced iron sulfur cluster. Binding of class 2 inhibitors is competitive to the binding of class 1 inhibitors although both classes of inhibitors have separate binding sites within the Q_o reaction site.
3. Inhibitors binding at the Q_i site close to heme b_H, e.g., antimycin A and hydroxyquinoline-N-oxides. Q_i site inhibitors prevent the reduction of quinone and the re-oxidation of cytochrome *b* through the Q_i site; due to the coupled reaction at the Q_o site, this will also inhibit the steady state reduction of cytochrome c_1 but not the pre-steady state reduction of oxidized bc_1 complex.
4. In addition to the three classes of organic compounds, zinc ions reversibly inhibit the bc_1 complex. Zn^{2+} binds at the Q_o site and prevents the release of protons which are formed during hydroquinone oxidation [8]. Binding of Zn^{2+} is not competitive to the binding of other Q_o site inhibitors.

2.1 MOA and Related Qo Inhibitors

Q_o site inhibitors of class 1 have been discovered during a systematic search for new antibiotic agents from basidiomycetes which was started in 1976 [9]. The fungicidal activity of strobilurin A and the localization of its site of action in the respiratory chain [10] led to a systematic exploration of strobilurin-like compounds, first by BASF and ICI and then by many other agrochemical companies. In summer 2002, six companies have registered Q_o site inhibitors as fungicides for agrochemical use, in particular for fruit, grapes, cereals, vegetables and other crops: BASF (Kreosoxim-methyl, Pyraclostrobin), Syngenta (Azoxystrobin, Picoxystrobin), DuPont (Famoxadone), Aventis (Fenamidone), Shionogi (Metominostrobin), and Bayer (Trifloxystrobin) [11]. Since the E-ß-methoxyacrylate (MOA) toxophore which is present in the natural product, strobilurin, is covered by patents to ICI (now merged into Syngenta), many of the compounds contain a 'modified methoxyacrylate' toxophore which was obtained by replacing carbon atoms with nitrogen (Figure 2). The oxazolidinone toxophore which is present in the inhibitor famoxadone (Figure 2) and the related imidazolinone toxophore of fenamidone belong to a different chemical class compared to strobilurin related inhibitors but occupy the same binding pocket and form similar binding valences with the protein.

It is important to recognize that none of the Q_o site inhibitors blocks electron transfer

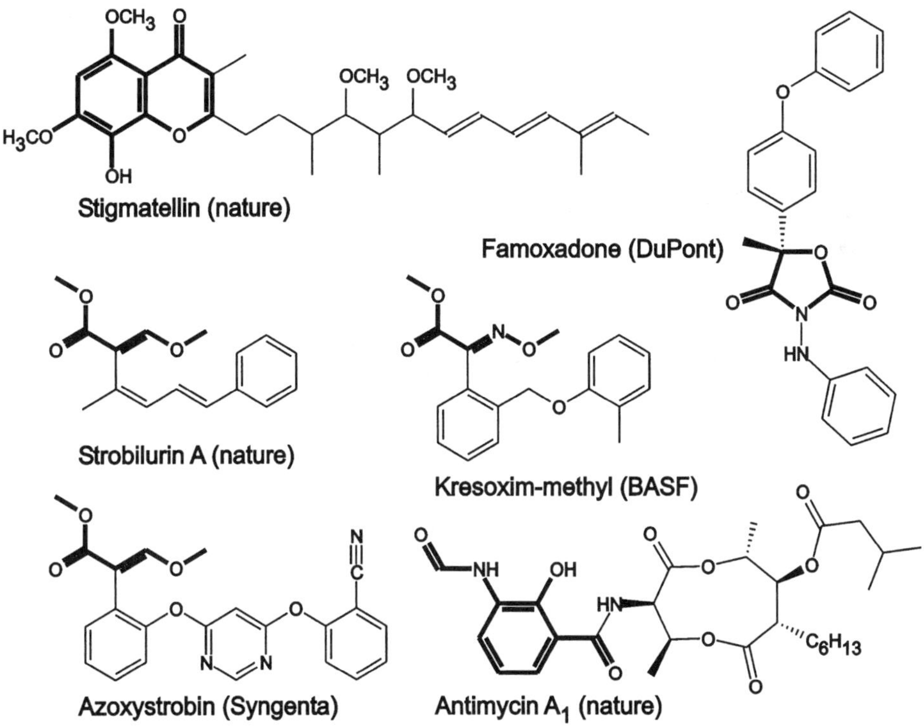

Figure 2. Representative structures of inhibitors of the cytochrome bc_1 complex (cf. section 2). The toxophore moiety has been indicated.

"*between cytochrome b and cytochrome* c_1" as is often stated since there is no electron transfer between these two redox centers; instead, they block electron transfer between hydroquinone and the bc_1 complex by preventing the binding of the quinone (except for zinc ions). There was a debate whether Q_o site inhibitors, in particular MOA inhibitors, act by displacing hydroquinone since Brandt et al. [12] have measured noncompetitive binding kinetics for these compounds. However, the same group has recently measured the binding of quinones to bovine heart mitochondrial bc_1 complex directly using high resolution magic angle spinning NMR spectroscopy and has shown the displacement of quinone by myxothiazol [13]. Furthermore, they have shown that *two* molecules of quinone are bound at the Q_o site in addition to one molecule of quinone at the Q_i site, thereby confirming previous results obtained in bacteria using EPR spectroscopy [14]. The functional consequences of this result are not yet clear as high resolution structural information about the arrangement of the two quinone molecules in the Q_o site is not yet available; the Q_o site was found to be devoid of quinone in all X-ray structures of the bc_1 complex obtained so far while electron density for quinone is clearly detectable at the Q_i site (see below).

2.2 Stigmatellin

The inhibitor stigmatellin which binds to both the "Rieske" protein and to cytochrome *b* was isolated from the myxobacterium *Stigmatella aurantiaca* [15]. It contains a 8-hydroxy-5,7-dimethoxy-chromone ring system as toxophore (Figure 2) which resembles semiquinone in its hydrogen bonding pattern, i.e., hydrogen bond donor on one side of the aromatic ring and hydrogen bond acceptor on the other side. Therefore, the binding of stigmatellin between the "Rieske" protein and cytochome *b* is thought to indicate the position where hydroquinone oxidation occurs and where the semiquinone intermediate is formed. Stigmatellin binds also to the secondary quinone (Q_B) site in bacterial photoreaction centers where it mimicks the structure of the $Q_B^{\cdot-}$ semiquinone [16].

2.3 Antimycin

The Q_i site inhibitor antimycin contains a 3-formamidosalicylic acid toxophore which is bound via an amide to a complex dilacton moiety (Figure 2). Antimycin is highly specific and binds only to the Q_i site of mitochondrial and most bacterial bc_1 complexes but not to the Q_i site of the b_6f complex while class 1 Qo inhibitors are much more promiscuous with some of them acting at the Q_i *site* of the b_6f complex or at complex I (NADH:quinone oxidoreductase) [17].

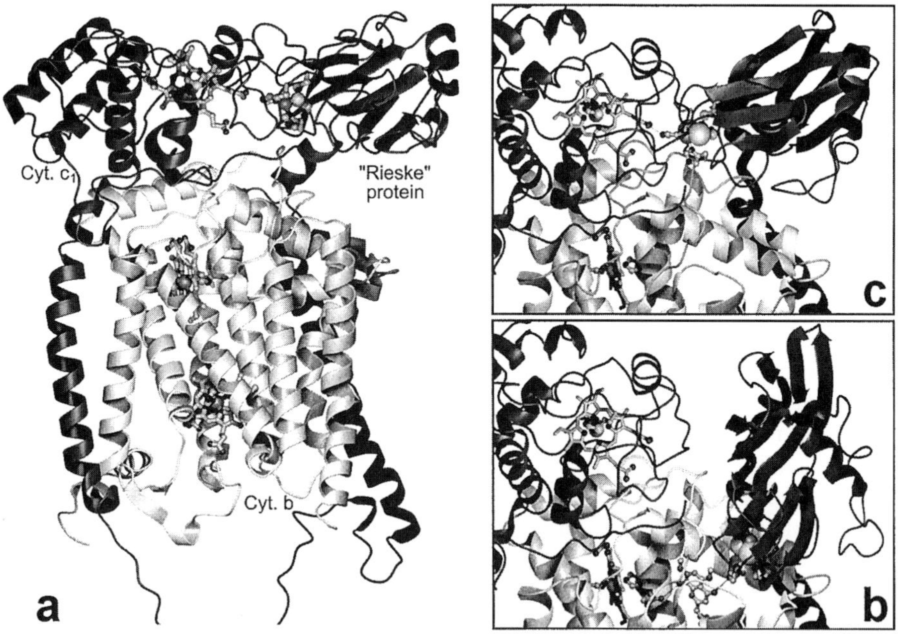

Figure 3. (a) Schematic diagram of the structure of the three catalytic subunits of the cytochrome bc_1 complex: cytochrome b (light grey), cytochrome c_1 and "Rieske" iron sulfur protein (dark grey). The redox centers are represented as space filling models. (b) Schematic diagram of the position of the "Rieske" protein in the 'b positional state' (see section 3.3.1). (c) Schematic diagram of the position of the "Rieske" protein in the 'c_1 positional state' (see section 3.3.1). Prepared with the program Molmol [41].

3 Structure of the Cytochrome bc_1 Complex

The bc complexes from mitochondria, chloroplasts, and bacteria all contain three catalytic subunits harboring the four redox centers: cytochrome b, the high-potential cytochrome c_1 or f, and the "Rieske" iron sulfur protein. These subunits are required and sufficient to support electron transport since most bacterial bc_1 complexes only consist of these three subunits. However, some bacterial bc_1 complexes contain a fourth subunit with yet unknown function. Mitochondrial bc_1 complexes contain in addition to the three catalytic subunits 7-8 subunits without redox centers: two large "core" proteins which are peripherally located and which are members of the family of matrix processing peptidases (MPP), and 5-6 small subunits. In cytochrome b_6f complexes, cytochrome b is split into cytochrome b_6 and subunit IV containing the C-terminal part of cytochrome b; in addition, 3 small hydrophobic subunits are present [18].

The bc_1 complexes form dimers in the membrane with molecular masses of approximately 480 kDa (mitochondria) and 130 kDa (bacteria), respectively. Each monomer has 10 – 13 membrane spanning helices, depending on the number of noncatalytic subunits. The membrane spanning helices of cytochrome *b* are in the center of the structure and form the dimer interface while the other membrane spanning helices are located around cytochrome *b*. Cytochrome c_1 and the "Rieske" iron sulfur protein both have water soluble domains containing the redox centers, heme c_1 and the [2Fe-2S] cluster, respectively. These domains are at the outside of the inner mitochondrial membrane, i.e., in the intermembrane space, and bound to the membrane via membrane spanning helices acting as membrane anchors.
The X-ray structures of the bc_1 complexes from bovine heart, chicken heart, and the yeast *Saccharomyces cerevisiae* have been determined at 3.0 – 2.3 Å resolution [3-6]; therefore, it is possible to explore the interaction between the complex and quinones and inhibitors at molecular level. First, we will give an overview of essential structural features of the bc_1 complex. As the non-catalytic subunits are not involved in the coordination of the redox centers or the formation of quinone binding sites, only the structure of the three catalytic subunits will be discussed in the following section (Figure 3); for a more detailed description see [19]. In order to avoid confusion, the residue numbering of the yeast bc_1 complex will be used even if other organisms have been studied.

3.1 Cytochrome *b*

Cytochrome *b* is a hydrophobic membrane protein containing both heme *b* centers in a transmembrane arrangement. In bc_1 complexes of mitochondria and bacteria like *Paracoccus* or *Rhodobacter*, cytochrome *b* has approximately 400 amino acid residues and contains eight membrane spanning helices (A – H). Cytochrome b_6 of the b_6f complex contains only the first four transmembrane helices (A – D) coordinating both heme centers. These helices form an all-antiparallel 4-α-helical bundle; the helices are tilted approximately 20° from the membrane normal.
Within the 4-α-helical bundle, the heme centers are arranged on top of each other with an iron-iron distance of 20 Å; the closest edge-to-edge distance between the hemes is 8 Å. This geometry enables rapid electron transfer between the hemes across the membrane dielectric. The two heme *b* centers are coordinated by two pairs of histidine residues which are 13 residues apart in helices B and D: His83 and His182 coordinate heme b_L close to the outer surface of the membrane while His97 and His196 coordinate heme b_H at the matrix side of the membrane. Helices A and C both contain a pair of glycine residues spaced, like the histidine residues, 13 residues apart; these residues point inside the 4-α-helical bundle and are critical for heme packing.
The membrane spanning helices of cytochrome *b* are connected by four long (ab, cd, de, and ef) and three short (bc, fg, gh) loops (loops are named according to the membrane spanning helices that they connect). The loops ab, cd, and ef at the outer side of the membrane contain peripheral helices which are oriented parallel to the membrane (αab, αcd1, αcd2, αef); the loops ab and de lie directly above or below the 4-α-helical bundle while the loops cd and ef form a lid on top of the Q_o quinone binding site which is outside the 4-α-helical bundle (Figure 4).

3.2 Cytochrome c_1

Cytochrome c_1 shows no significant sequence homology to other cytochromes or even to to cytochrome *f*, its counterpart in the b_6f complex; however, its structure is related to mitochondrial cytochrome *c* and to cytochrome c_2, the prototypes of Ambler's class I cytochromes [20]. In Ambler's class I cytochromes, the heme is surrounded by five helices ($\alpha 1$ - $\alpha 5$); three of the five helices ($\alpha 1$, $\alpha 3$, $\alpha 5$) are conserved in cytochrome c_1. Helix $\alpha 2$ of class I cytochromes is absent; this results in the exposure of the heme propionates at the protein surface. Thus, in contrast to cytochrome *c*, cytochrome c_1 has two electron transfer pathways: one through the heme propionates at the 'back face' (where cytochrome c_1 receives electrons from the "Rieske" protein), and the other at the front face through the exposed pyrrole C corner of the heme (where electrons are passed on to cytochrome *c*).

Immediately after helix $\alpha 2$ follows the long membrane spanning helix $\alpha 6'$ which forms the membrane anchor. The connection between the water soluble domain and the membrane anchor is not flexible but stabilized by a loop from the N-terminal part of the protein.

3.3 "Rieske" iron sulfur protein

The "Rieske" protein has an N-terminal membrane anchor followed by a C-terminal water soluble domain of approximately 14 kDa; between these two elements, there is a 12-residue flexible linker which allows movement of the water soluble domain (see section 3.3.1). The folding of the water soluble domain is the ''Rieske fold' consisting of three antiparallel ß-sheets [21]. Sheet 3 is formed by the ß strands ß5 – ß8; these strands and the loops connecting them form the 'cluster binding subdomain' comprising 44 residues.

The "Rieske" [2Fe-2S] cluster is coordinated by 2 histidine and 2 cysteine residues, in contrast to the 4-cysteine coordination observed in [2Fe-2S] ferredoxins. The inner iron (Fel) is coordinated by Cys159 and Cys178, while the (redox active) outer iron (Fe2) is coordinated by His161 and His181. The histidine coordination is essential for the unique properties of the "Rieske" [2Fe-2S] cluster, in particular for its high redox potential and for the redox dependent protonation/deprotonation [22].

The "Rieske" cluster is located at the tip of the cluster binding subdomain with the histidine ligands fully exposed at the surface; both histidine residues are coordinated through the ND1 nitrogen atoms while the NE2 nitrogen atoms and, if protonated, the associated hydrogens are freely accessible.

3.3.1 Movement of the catalytic domain of the "Rieske" protein

In the X-ray structures of the cytochrome bc_1 complex, the water soluble catalytic domain of the "Rieske" protein was found in different locations which have been grouped into three 'positional states' [4-6,23] (Figure 3b):

1. a 'c_1 positional state' where the exposed NεH group of His181 can form a hydrogen bond with a propionate group of heme c_1 (Figure 3);
2. an 'intermediate state' where the ligands of the "Rieske" cluster do not interact with any other subunit of the bc_1 complex;
3. a '*b* positional state' where the exposed histidine ligands of the "Rieske" cluster point towards the ubiquinone binding site in cytochrome *b*.(Figure 3c).

In the '*b* positional state', the distance is > 30 Å between the "Rieske" cluster and cytochrome c_1 whereas in the 'c_1 positional state', the Rieske [2Fe-2S] cluster is far away from the ubiquinone binding site and points in the opposite direction (Figure 3). Therefore, the "Rieske" protein has to move between these positional states in order to transfer an electron from ubihydroquinone to cytochrome c_1. The movement involves a rotation of the entire water soluble domain of 57°; this rotation requires stretching of the flexible linker.

The movement of the catalytic domain of the "Rieske" protein is related to the binding of Q_o site inhibitors. Stigmatellin or 5-undecyl-6-hydroxy-4,7-dioxobenzothiazol (UHDBT) bind directly to the exposed His181 of the "Rieske" protein [6] and fix the "Rieske" protein in the '*b* position', while binding of MOA-stilbene or myxothiazol favors the 'c_1 position' [23].

3.4 Phospholipids

Phospholipids are important structural components of the bc_1 complex. In bovine heart bc_1 complex, eight to nine molecules of cardiolipin are bound to each bc_1 monomer; their removal by extended detergent washes results in irreversible inactivation. In addition, approximately 100 molecules of neutral phospholipids per dimer are required for full electron transfer activity [24].

In the X-ray structure of yeast bc_1 complex, five phospholipid molecules and one detergent (undecylmaltoside) molecule have been identified in each monomer [25]. The five visible lipid molecules are cardiolipin (CL), phosphatidylcholin (PC), phosphatidylethanolamin (2)(PE), and phosphatidylinositol (PI); all but PI are located on the inner (matrix) side of the membrane. The PI molecule is bound between membrane spanning helices and stabilizes the packing of the membrane anchor of the "Rieske" protein to the transmembrane core of the complex. One of the PE lipids (L2) sits at the dimer interface in a large hydrophobic cleft through which the Q_o and Q_i quinone reaction sites are accessible from the membrane; this lipid might represent a larger population of lipids in the cleft which support diffusion of quinones and are therefore essential for activity.

4 Quinone and Inhibitor Binding Sites

The regions of cytochrome *b* involved in the interaction with inhibitors have been identified before X-ray structures were available by the use of random or site directed mutagenesis in combination with inhibitor binding studies. It has been concluded that Qo inhibitors bind in a pocket formed by the C-terminus of helix C, the beginning of the cd

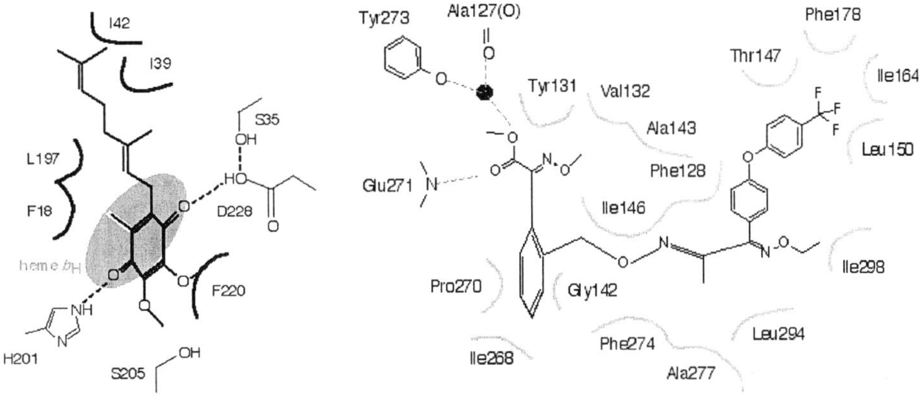

Figure 4. (a) Schematic view of the binding of ubiquinone (UQ2) in the Q_i binding site (cf. section 4.1). (b) Schematic view of the binding of a Qo inhibitor (cf. section 4.2).

loop, the "PEWY region" of the ef loop containing the highly conserved sequence Pro-Glu272-Trp-Tyr, and the N-terminus of helix F [26]. In the first X-ray structure of the *bc*$_1$ complex, electron density for the Qo inhibitor myxothiazol was found underneath the peripheral helix αcd1 and surrounded by helices C, F, and αef [3]. Unlike the Q_i binding site, the Qo binding site is not in direct contact with the nearby heme (heme b_L).

The Q_i binding site is located between the C-terminus of helix D and the N-terminus of helix E and surrounded by the N-terminal peripheral helix αa' and the N-terminus of helix A [3].

The Q_o and the Q_i binding site are located on opposite sides of the central 4-α-helical bundle of cytochrome *b* (Figure 3a); this indicates that the two sites of one cytochrome *b* molecule are strictly independent. Access of quinones and inhibitors from the membrane bilayer may occur through two large cavities located at the dimer interface. The Q_o site of one monomer and the Q_i site of the other monomer of the dimer are accessible through the same cavity; this suggests a preferred diffusion path from the Q_o site of one monomer to the Q_i site of the other monomer.

4.1 The Q_i binding site

Density for ubiquinone (UQ10 in chicken and beef heart, UQ6 in yeast) has been observed in several crystal forms. The quinone ring is oriented in a right angle to heme b_H; the closest distance to the pyrrole A ring is approximately 4 Å (Figure 4a). In the bovine complex, residue Phe220 lies on the opposite side of the quinone ring at a distance of 3.5 Å.; the corresponding residue in yeast, Met221, is approximately 7 Å away. The carbonyl oxygen atoms of ubiquinone are coordinated to His202 (201 in beef) on one side and to Asp229 (228 in beef) on the other side; in the yeast complex but not in the bovine or chicken complexes, the interaction between His202 and oxygen O5 is mediated by a water molecule. Ser34 (35 in beef) is linked to Asp229 (228) since aspartate has to be protonated to act as hydrogen bond donor to ubiquinone. Ser206 (205 in beef) from the de loop is close to the oxygen of the 3-methoxy group; in yeast, there is a hydrogen bond between Ser206 and the O3.

The Q_i site inhibitors mimick the binding of ubiquinone. Antimycin is bound primarily via the hydroxy group of the salicylic acid moiety to the OD2 oxygen atom of Asp229 (228 in beef), but also through an extended hydrogen bond (N-O distance 3.8 Å) from His202 (201 in beef) to the amide carbonyl group. Thus, the aromatic ring is not in the position of the quinone ring but overlapping the methoxy groups of the quinone ring. The formamido group of antimycin forms additional hydrogen bonds with the OD1 oxygen atom of Asp229 (228 in beef) and with Asn31 (32 in beef), while the dilacton moiety is in a similar position as the isoprenoid side chain of ubiquinone and extensively involved in van-der-Waals interactions. Other Q_i site inhibitors also bind in the same position as ubiquinone by hydrogen bonding to His202 and Asp229.

The hydrogen bond pattern of the Q_i site residues (hydrogen bond donor + hydrogen bond acceptor) matches the hydrogen bonding pattern of semiquinone; similar motifs have been found in other semiquinone stabilizing quinone reaction sites, e.g., in the bacterial photoreaction center [16] or in quinol-fumarate reductase of the bacterial respiratory chain [27]. The high specificity of antimycin for the Q_i site of the bc_1 complex is therefore likely to be due to the dilacton moiety which will prevent binding to other sites. The fact that the b_6f complex does not bind antimycin is related to the split nature of cytochrome b_6: the split between cytochrome b_6 and subunit IV is in the de loop of cytochrome b which contributes to the formation of the Q_i site. Therefore, the Q_i site of the b_6f complex is significantly different from the Q_i site of the bc_1 complex so that antimycin cannot bind.

4.2 The Q_o binding site

Unlike in the Q_i site, the relationship between the quinone binding pocket and the Qo inhibitor binding pocket is not clear, particularly due to the lack of structural information since ubiquinone is not present in the Q_o reaction site in any X-ray structure of the bc_1 complex. Binding of two molecules of ubiquinone at at the Qo site has been demonstrated [13,14]; both quinone molecules are displaced on binding of either stigmatellin or the Qo inhibitor myxothiazol. One quinone binding site is most likely at the stigmatellin binding site between residue His181 of the “Rieske” cluster and Glu272 of cytochrome *b* [6]. This site lies outside the membrane above the peripheral helix αef and has all the characteristic features of a semiquinone binding site.

The binding site for Qo inhibitors is distinct from the stigmatellin binding site within the membrane below the peripheral helix αcd1. The binding pocket can be divided into three regions: toxophore, linker region, and hydrophobic tail (Figure 4b). The toxophore is in a predominantly aromatic pocket formed by residues Phe129 (128 in beef), Tyr132 (131 in beef), 133 (Val132 in beef, Cys133 in yeast), Glu272 (271 in beef), Tyr274 (273 in beef), and 275 (Phe274 in beef, Leu275 in yeast). The key feature is a hydrogen bond between the carbonyl group of the toxophore and the backbone nitrogen of Glu272 (271 in beef). The N-O distance of this H bond varies between 2.66 Å and 3.15 Å in 8 Qo inhibitor complexes investigated, only two compounds are bound through a weak H bond with a length > 3.5 Å. The binding of the toxophore is further stabilized by a water molecule which is hydrogen bonded to the ester oxygen or amide nitrogen and which is held in place by the OH group of the conserved Tyr274 (273 in beef) and by the carbonyl oxygen of the conserved Ala128 (127 in beef) at the C-terminus of helix C. The sidechain of Glu272 (271 in beef) which points towards the outside of the membrane in the stigmatellin complex and coordinates to the inhibitor is pointing into the Qo pocket and hydrogen bonded to the OH oxygen of Tyr274 (273 in beef). This requires a conformational change of the "PEWY region" of the ef loop and is likely to be the mechanism by which binding of Qo inhibitors induces the positional change of the "Rieske" protein (see section 3.3.1).

The hydrophobic tail of the Qo inhibitors extends into a cylindrical hydrophobic pocket which is approximately 6 Å in diameter and more than 15 Å long. This hydrophobic pocket is lined mainly by residues from the ef loop right after the "PEWY" loop (residues 276-279), by residues at the N-terminus of helix F (290-298), and by residues from the C-terminus of helix C (residues 125-132).

Between the toxophore and the hydrophobic tail is the linker region of the inhibitors; all synthetic 'modified methoxyacrylates' (including those containing nitrogen) contain an aromatic linker which replaces the double bond of the natural component, strobilurin (Figure 2). The aromatic linker increases the stability as well as the binding affinity of the inhibitor; this is explained by the fact that the linker region is in van-der-Waals contact with residues Gly143 (142 in beef), Ile147 (146 in beef), Ile269 (268 in beef), and Pro271 (270 in beef). This region is of particular importance for the development of inhibitor resistance (see below).

5 Molecular Basis of Inhibitor Resistance

The study of the molecular basis of inhibitor binding and resistance has at least three aspects:

1. The analysis of inhibitor resistant mutants of model organisms (yeast, *Rhodobacter*) was used immediately after the sequence of cytochrome *b* had been established to identify those parts of the protein that are involved in the interaction with inhibitors and, possibly, quinone. This approach led to models of the structure of cytochrome *b* that were in generally good agreement with the X-ray structure, e.g., [26,28,29].
2. The analysis of bc_1 complexes from a great number of organisms has revealed species which show natural resistance to various inhibitor classes. Comparison of the cytochrome *b* sequence with a large number of sequences from sensitive organisms gave clues about the amino acid residues responsible for resistance [30]. A well studied example of natural resistance is found in strobilurin producing basidiomycetes, *Strobilurus tenacellus* and *Mycena galopoda*. Although both organisms are phylogenetically closely related, resistance is due to different amino acid exchanges (compared to sensitive organisms) in both fungi [31].
3. With the success of Qo inhibitors as fungicides in agriculture, inhibitor resistance has become a possible threat for both farmers and chemists. After a short time of

fungicide use, resistant isolates of various important plant pathogens have been found [32]. In many cases, analysis of resistant plant pathogenic fungi revealed a mutation in the cytochrome *b* gene that can be attributed to the resistance; however, there are also resistant isolates where no point mutation in the cytchrome *b* gene can be detected [33].

The prevalent mutation in phytopathogenic fungi is the exchange of Gly143 to alanine [32]. Recently, also the mutation Phe129Leu has been detected in *Pyricularia grisea* [34]. Both amino acid exchanges have previously been reported in studies of model organisms. The amino acid exchange Gly143Ala in mouse leads to a very strong resistence against myxothiazol ($K_{I,\ resistant}/K_{I,sensitive} \approx 10\ 000$) [35]; the exchange Phe129Leu was reported both in *Rhodobacter capsulatus* [36] and yeast [37] and results in a ratio $K_{I,resistant}/K_{I,sensitive}$ for myxothiazol of approximately 500 (15 for strobilurin). Thus, these mutations have strongest effect on myxothiazol binding of all mutants studied so far.

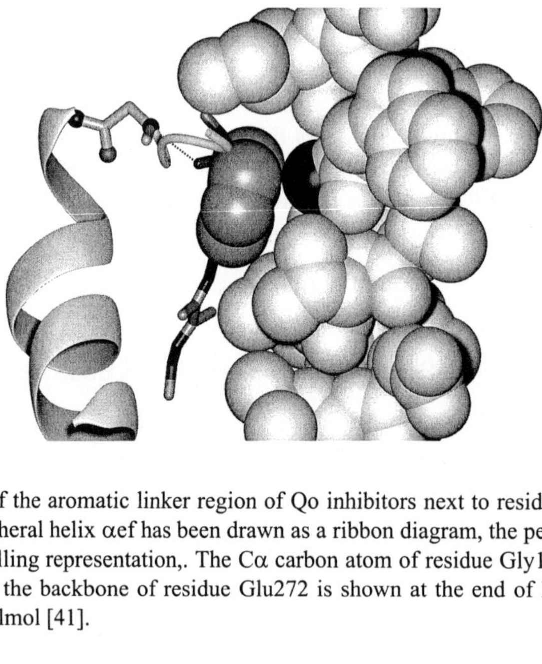

Figure 5. Binding of the aromatic linker region of Qo inhibitors next to residue Gly143 (cf. section 5.1.1). The peripheral helix αef has been drawn as a ribbon diagram, the peripheral helix αcd1 is shown in a spacefilling representation,. The Cα carbon atom of residue Gly143 in helix αcd1 is highlighted in black, the backbone of residue Glu272 is shown at the end of helix αef. Prepared with the program Molmol [41].

5.1 Structural Basis of QoI Resistance

5.1.1 Mutation Gly143Ala

Residue Gly143 (142 in beef) is from the peripheral helix αcd1 which forms the lid of the Qo binding site; its Cα atom is in van-der-Waals contact with the aromatic linker region of commercial Qo inhibitors (see section 4.2, Figure 5). When Qo inhibitors are added to the site, the helix αcd1 is forced to move approximately 2 Å; this causes the binding to be highly unfavorable. Qo inhibitors containing the aromatic linker are likely to be more affected than the natural homologue, strobilurin; this was measured in the basidiomycete *Strobilurus viridimarginata* which has alanine in position 143 [31]. Thus, an important factor for the binding of Qo inhibitors proves to be a significant structural problem as far as resistance is concerned.

The mutation Gly143Ala is unlikely to cause a structural perturbation in the absence of inhibitors, consequently, the activity of the bc_1 complex is not greatly reduced. We have created a series of mutations in position 143 to test the effect of the size of the amino acid in position 143 (Table 1). The table shows that the mutation of glycine to alanine gives high resistance to methoxyacrylate-stilbene with high activity (approximately 60% of the wild type level) while in the other mutants (Ser, Cys, Pro) the activity is 10% of the wild type level or less; with valine, almost no activity was observed. Despite, the differences in the residual activity, the I_{50} for MOA-stilbene is the same in all mutants but proline.

5.1.2 Mutation Phe129Leu

Residue Phe129 (128 in beef) is from the C-terminus of the membrane spanning helix C; it is involved in the binding of the toxophore by interacting with the acrylate part (see section 4.2, Figure 5). The distance between the phenyl ring and the oxygen of the methoxy group is 3.6 Å. The methoxy group is present in the acrylate part of all commercial fungicides of the 'modified methoxyacrylate' type except in pyraclostrobin; therefore, it will be interesting to see whether in phytopathogens carrying the mutation Phe129Leu pyraclostrobin might be affected similarly as other Qo inhibitors. As in the case of the mutation Gly143Ala, no large structural perturbation is likely to occur upon mutation of residue Phe129.

Table 1. Properties of mutants in position 158 of cytochrome *b* in *Paracoccus denitrificans* (corresponding to Gly143 in yeast).

Amino acid	Activity [% of wt]	I_{50} [nM]	$I_{50}/I_{50}(Gly)$
Gly	100	4.4	1
Ala	62	14 000	3200
Ser	11	10 600	2400
Cys	10	15 100	3400
Pro	5	3 800	900
Val	0.4	14 700	3300

Activity of the isolated bc_1 complex was measured using a synthetic analogue of ubihydroquinone and horse heart cytochrome *c* as substrates. I_{50} values were measured with MOA-stilbene.

6 Molecular Modeling of the Q_o Binding Site

The determination of X-ray structures of bc_1 complexes has been an important step towards understanding the details of quinone and inhibitor binding. However, X-ray structures are available only from mammals, chicken, and the yeast *Saccharomyces cerevisiae*; no X-ray structures are yet available for bacteria (although many crystallization attempts have been undertaken) or for phytopathogenic fungi. Therefore, we have decided to use molecular modeling to perform binding studies with fungal bc_1 complexes *in silicio* using the program package GROMACS [38].

6.1 Setting up of the system

Water

In order to get a realistic view of the protein dynamics, it was necessary to include an environment of membrane and waters. The program package DOWSER [39] was used to incorporate the first layer of waters having favorable interactions with the protein (≈ -10 kcal/mol). These, as well as good crystallographic water molecules were further refined using a program in the same package. The distribution of waters from this layer clearly indicates the position of the membrane; almost no waters are found in the putative membrane region.

Membrane

For the incorporation of the membrane, the protein was oriented such that the C2 axis of the dimer was oriented along the z-axis; the lipid bilayer was in the xy plane. A periodic and equilibrated lipid (POPC) solvent system with dimensions 193 x 183 x 60 Å was used. The size of the membrane is sufficient to cover the protein (which has dimensions 135 x 150 x 155 Å) in the xy directions. The position of the lipid in the z-direction was correlated to the distribution of tryptophane residues, the first 1 solvent layer, and surface charge density along the z-axis. As a final check a "penalty curve" was constructed that shows the summation of surface charge over a region corresponding to the length of the hydrophobic membrane part. This function should have a minimum at the optimal membrane position. All these four methods were in agreement with each other. Once the proper membrane position was determined, a program in the GROMACS package was used to exclude lipids that overlapped with the protein. Minor overlaps (lipids that were in conflict with atoms being 1-2 atoms away from the surface of the protein) were checked graphically and then either discarded or carefully adjusted.

Bulk solvent

The periodic box was extended to cover the protein in the z-dimension (box dimension 195 x 182 x 174 Å) and equilibrated bulk solvent on both sides of the membrane was added yielding a total of ~500 000 atoms (including protein and lipids). Then, the number of atoms was reduced by cutting away waters more than 12 Å away from protein or lipids, 25 Å from the principal z-axis and 25 Å away from inhibitor sites giving a remaining total of 230 000 atoms. At the water-vacuum interface water molecules can and will evaporate. In principle this could be prevented by imposing a restraining potential, but we have not done this because such a potential can introduce new artifacts. Since we have a relatively large distance between the active site and the nearest vacuum boundary

(> 25 Å) it is reasonable to assume that neither a restraint potential nor a real vacuum boundary will have large impact on the free energy calculations.

Molecular dynamics (MD) calculation: Equilibration of the model
The system was equilibrated for more than 100 ps; the equilibration was checked by monitoring energy trajectories and root mean square (rms) deviations from the X-ray structure. Using 8 parallel processors, it took 1 hour to calculate a 2 ps simulation. The restraints on the system during the molecular dynamics calculation were released in a stepwise mode, first on side chains and then on other atoms. The rms deviation of all Cα atoms between the cytochrome *b* subunit in the X-ray structure and the unrestrained model after 120 ps was 1.38 Å.

6.2 Modeling of the Qo site of *Mycosphaerella fijiensis*

As a model system, we chose cytochrome *b* from *Mycosphaerella fijiensis*. The sequence of cytochrome *b* from *M. fijiensis* was determined by the group of U. Gisi; this organism was of special interest as it acquires resistance by the mutation Gly143Ala [40]. Starting from the equilibrated models of inhibitor complexes of the bc_1 complex, each residue within 6 Å of the Qo site was replaced by the corresponding residue of the cytochrome *b* subunit of the bc_1 complex from *M. fijiensis*. Some residues further away were also replaced if they caused steric problems with the closer substitutions. In total, 14 residues were replaced including Gly143 which was changed to Ala143. The remaining residues of the entire complex were kept unchanged. In order to reduce computation time, only residues in a region of 14-18 Å from the inhibitor site were allowed to be free to move; residues beyond 18 Å were restrained. A MD calculation was performed on this model to calculate the *in silicio* structure of inhibitor complexes of *M. fijiensis* cytochrome *b*.

6.3 Calculation of Binding Constants

The direct calculation of binding constants of hydrophobic compounds to the bc_1 complex from free energy calculations is not possible since it is impossible to include the unbound inhibitor in the energy calculation. Therefore, we have performed a calculation of the relative binding energies of Qo inhibitors in wild type *M. fijiensis* cytochrome *b* (Gly143) and in the mutant Ala143. This was done by calculating the free energy changes upon changing Ala143 to Gly in free cytochrome *b* ($\Delta G\{A \rightarrow G\}$) and in cytochrome *b* with Qo inhibitor bound ($\Delta G\{AI \rightarrow GI\}$). The difference of the inhibitor binding energies of the glycine variant ($\Delta G\{G \rightarrow GI\}$) and of the alanine variant ($\Delta G\{A \rightarrow AI\}$) can then be obtained by the equation:

$$\Delta G\{A \rightarrow AI\} - \Delta G\{G \rightarrow GI\} = \Delta G\{A \rightarrow G\} - \Delta G\{AI \rightarrow GI\} \quad (2)$$

The calculations showed that the result depends very much on the restrictions applied on the systems. When all atoms in the region of 14-18 Å from the inhibitor site were allowed to move, the difference of the ligand interaction energy between the Ala and the Gly variant became negligible. If only Gly143, Pro271, and the inhibitor were allowed to move while the rest of the protein was fixed, the ligand interaction energy was 27 kJ/mol lower for the Gly compared to the Ala variant corresponding to a 50 000 fold difference in the binding constant. Although this setup is unrealistic, it provides an estimate of the maximal energy differences that could be obtained. When other restraints were used, the ligand interaction energies were between the extreme values.

One problem with the simulations that was apparent on close inspection of the structures was the presence of hydrophobic cavities which are not properly filled with lipids. Although approximately 90% of the membrane surface were covered with lipids, the remaining 10% might effect the MD simulations. It appears that the surrounding lipids have a great effect on the bc_1 complex not only *in vivo* and *in vitro* (see section 3.4), but also *in silicio*.

7 Acknowledgements

We thank Prof. Dr. Gisi, Dr. Girardet, Dr. Ziegler, Dr. Cederbaum and the late Dr. Moesinger (all from Syngenta, formerly Novartis, Basel) for their cooperation. The work was supported by a fund from Novartis Crop Protection AG, Basel. A.S was supported by a fellowship from the research training group "Protein structures, dynamics and function" of the Deutsche Forschungsgemeinschaft (DFG).

8 References

[1] P. Mitchell, J. Theor. Biol. 1976, 62, 327-367.
[2] G. von Jagow, T.A. Link, Methods Enzymol. 1986, 126, 253-271.
[3] D. Xia, C.-A. Yu, H. Kim, J.-Z. Xia, A.M. Kachurin, L. Zhang, L. Yu, J. Deisenhofer, Science 1997, 277, 60-66.
[4] Z. Zhang, L. Huang, V.M. Shulmeister, Y.-I. Chi, K.K. Kim, L.-W. Huang, A.R. Crofts, E.A. Berry, S.-H. Kim, Nature 1998, 392, 677-684.
[5] S. Iwata, J.W. Lee, K. Okada, J.K. Lee, M. Iwata, B. Rasmussen, T.A. Link, S. Ramaswamy, B.K. Jap, Science 1998, 281, 64-71.
[6] C. Hunte, J. Koepke, C. Lange, T. Roßmanith, H. Michel, Structure 2000, 8, 669-684.
[7] G. von Jagow, T. Ohnishi, FEBS Lett. 1985, 185, 311-315.
[8] T.A. Link, J. Biol. Chem. 1995, 270, 25001-25006.
[9] T. Anke, F. Oberwinkler, W. Steglich, G. Schramm, J. Antibiot. (Tokyo) 1977, 30, 806-810.
[10] W.F. Becker, G. von Jagow, T. Anke, W. Steglich, FEBS Lett. 1981, 132, 329 – 333.
[11] D.W. Bartlett, J.M. Clough, J.R. Godwin, A.A. Hall, M. Hamer, B. Parr-Dobrzanski, Pest Manag. Sci. 2002, 58, 649-662.
[12] U. Brandt, H. Schägger, G. von Jagow, Eur. J. Biochem. 1988, 173, 499-506.
[13] S. Bartoschek, M. Johansson, B.H. Geierstanger, J.G. Okun, C.R.D. Lancaster, E. Humpfer, L. Yu, C.-A. Yu, C. Griesinger, U. Brandt, J. Biol. Chem. 2001, 276, 35231-35234.
[14] H. Ding, D.E. Robertson, F. Daldal, P.L. Dutton, Biochemistry 1992, 31, 3144-3158.
[15] B. Kunze, T. Kemmer, G. Höfle, H. Reichenbach, J Antibiot (Tokyo) 1984, 37, 454-61.
[16] C.R.D. Lancaster, H. Michel, Structure 1997, 5, 1339-1359.

[17] M. Degli Esposti, Biochim. Biophys. Acta 1998, 1364, 222-235.
[18] G.M. Soriano, M.V. Ponamarev, C.J. Carrell, D. Xia, J.L. Smith, W.A. Cramer, J. Bioenerg. Biomembr. 1999, 31, 201-213.
[19] T.A. Link in Handbook of Metalloproteins (Eds.: A. Messerschmidt, R. Huber, T. Poulos, K. Wieghardt), Wiley, Chichester, 2001, pp. 402-423.
[20] R.P. Ambler, Biochim. Biophys. Acta 1991, 1058, 42-47.
[21] S. Iwata, M. Saynovits, T.A. Link, H. Michel, Structure 1996, 4, 567-579.
[22] T.A. Link, FEBS Lett. 1997, 412, 257-264.
[23] H. Kim, D. Xia, C.-A. Yu, J.-Z. Xia, A.M. Kachurin, L. Zhang, L. Yu, J. Deisenhofer, Proc. Natl. Acad. Sci. USA 1998, 95, 8026-8033.
[24] H. Schägger, T. Hagen, B. Roth, U. Brandt, T.A. Link, G. von Jagow, Eur. J. Biochem. 1989, 190, 123-130.
[25] C. Lange, J.H. Nett, B.L. Trumpower, C. Hunte, EMBO J. 2001, 20, 6591-6600.
[26] T.A. Link, U. Haase, U. Brandt, G. von Jagow, J. Bioenerg. Biomembr. 1993, 25, 221-232.
[27] T. Iverson, C. Luna-Chavez, G. Cecchini, D.C. Rees, Science 1999, 284, 1961-1966.
[28] A.R. Crofts, B. Hacker, B. Barquera, C.-H. Yun, R.B. Gennis in Research in Photosynthesis, Vol. II (Ed.: N. Murata), Kluwer Academic Publishers, Dordrecht, 1992, pp. 463-470.
[29] T.A. Link, H. Wallmeier, G. von Jagow, Biochem. Soc. Trans. 1994, 22, 197-203.
[30] M. Degli Esposti, S. de Vries, M. Crimi, A. Ghelli, T. Paternello, A. Meyer, Biochim. Biophys. Acta 1993, 1143, 243-271.
[31] P. Kraiczy, U. Haase, S. Gencic, S. Flindt, T. Anke, U. Brandt, G. von Jagow, Eur. J. Biochem. 1996, 235, 54-63.
[32] U. Gisi, H. Sierotzki, A. Hall, A. McCaffery, Pest Manag. Sci. 2002, 58, in press.
[33] U. Steinfeld, H. Sierotzki, S. Parisi, S. Poirey, U. Gisi, Pest Manag. Sci. 2001, 57, 787-796.
[34] M.L. Farman, Phytopathology 2001, 91, S110.
[35] N. Howell, Biochemistry 1990, 29, 8970-8977.
[36] D.E. Robertson, F. Daldal, L. Dutton, Biochemistry 1990, 29, 11249-11260.
[37] T. Tron, M. Crimi, A.-M. Colson, M. Degli Esposti, Eur. J. Biochem 1991, 199, 753-760.
[38] H.J.C. Berendsen,, D. van der Spoel, R. van Drunen, Comp. Phys. Comm. 1995, 91, 43-56.
[39] L. Zhang, J. Hermans, Proteins 1996, 24, 433-438.
[40] H. Sierotzki, S. Parisi, U. Steinfeld, I. Tenzer, S. Poirey, U. Gisi, Pest Manag. Sci. 2000, 56, 833-841.
[41] R. Koradi, M. Billeter, K. Wüthrich, J. Mol. Graphics 1996, 14, 51-55.

Weed Control

Chemistry, Biochemistry, Molecular Biology

11 News from Old Compounds: The Mode of Action of Auxin Herbicides

Klaus Grossmann

BASF Aktiengesellschaft, Agricultural Center, PO Box 120, D-67114 Limburgerhof, Germany, E-mail: klaus.grossmann@basf-ag.de

1 Introduction

Synthetic compounds which interfere with the phytohormonal auxin system in plants include the first selective organic herbicides to be developed for agricultural and horticultural use (for review see [1,2,3,4,5,6,7]. These so-called growth regulator or auxin herbicides started a new era of weed control in modern crop production due to their ability to be translocated systemically in the plant and to exert selective action, primarily against dicot weeds in cereal crops. Some of these compounds have also gained practical applications as plant bioregulators for yield improvement [7]. Throughout the six decades since their initial discovery, auxin herbicides have been among the most successful herbicides used in agriculture and remain so to this day [5]. By ranking the worldwide herbicide sales in 2000 for different sites/modes of action, auxin herbicides comprised *ca.* 9% (1.3 billion US $) of the total sales, in fourth position after inhibitors of amino acid biosynthesis, photosynthesis and lipid biosynthesis [8].

Various chemical classes of auxin herbicides with different weed spectra and types of selectivity have been synthesized and commercially introduced over the years. They include phenoxycarboxylic acids, benzoic acids, pyridine carboxylic acids, aromatic carboxymethyl derivatives and a relatively new category, the quinolinecarboxylic acids (Figure 1). These compounds basically act as synthetic mimics of indole-3-acetic acid (IAA), which is the principal natural auxin in higher plants [2,3,5,7].

Phytotropins are a further group of practically relevant, selective herbicides which interfere with the plant's auxin system [9,10]. Phytotropins prevent polar auxin transport in sensitive plants, which causes stunting and loss of tropic responses. They are members of a subset of auxin transport inhibitors which include benzoic acids ortho-linked to an aromatic ring system (e.g. naptalam, NPA) and a new class of chemistry, the semicarbazones, exemplified by diflufenzopyr (Figure 1). When applied in combination, diflufenzopyr interacts with auxin herbicides and synergistically enhances their activity and the range of weed species affected [11].

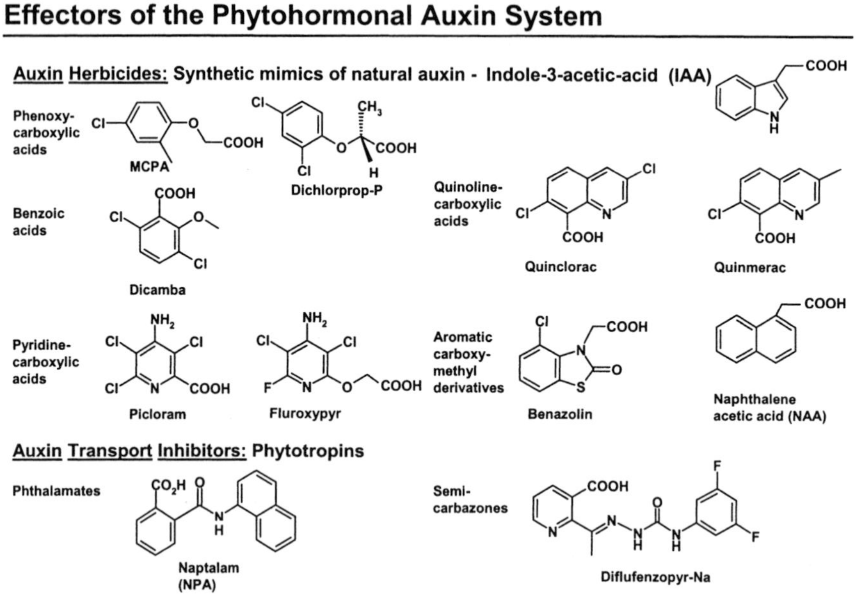

Figure 1. Structures of natural auxin, indole-3-acetic acid, and synthetic auxin herbicides and auxin transport inhibitors (phytotropins) belonging to the different chemical families.

Although, to date, the exact molecular target sites of auxin herbicides and auxin transport inhibitors have yet to be identified, considerable progress has recently been made in the elucidation of their herbicidal mode of action. In the course of this article, progress in the understanding of the succeeding series of biochemical and physiological events leading to phytotoxicity will be reviewed. During the past decade, these studies have been particularly focused on the effects of the latest discovered chemical family of auxin herbicides, the quinolinecarboxylic acids, exemplified by quinclorac and quinmerac [7,12]. Both compounds have been commercially introduced worldwide from 1988 onward. They possess a novel type of selectivity. While quinmerac effectively controls important dicot weeds, such as cleavers (*Galium aparine*), in sugarbeet, oilseed rape, and small-grain cereals, quinclorac is used particularly in rice against important dicot and grass weeds [7].

2 Auxin Herbicide Action in Dicots: Mimic of Auxin Overdose

2.1 The Phenomenon of Deregulation of Plant Growth

Auxins belong to the classical five types of plant hormones, together with ethylene, gibberellins, abscisic acid (ABA), and cytokinins [10]. In addition to their extreme chemical diversity, the effects of natural and synthetic auxins are biphasic. Depending on the concentration, opposite effects can be observed [10,13]. When present at low concentrations at the cellular sites of action, growth and developmental processes, such as cell division and elongation, differentiation of vascular tissue, root formation and apical dominance, are usually promoted. With increasing concentration, growth deformation and phytotoxic inhibition of growth are induced. Such biphasic dose-response curves are typical of phytohormones. In the extreme case of auxins, this Janus-faced characteristic of physiological action allows their use as both bioregulators and herbicides in agriculture [2,5,7,14]. Indole-3-acetic acid (IAA), even when applied exogenously, is subjected to rapid inactivation through conjugation and degradation by multiple pathways in the plant [10]. In contrast, applied synthetic auxins are long-lasting and more effective than IAA. They are stable and not inactivated by the plant as rapidly as the endogenous phytohormone. Besides the biokinetic properties, the spectrum of biological activities of the different structural types of auxin analogs depends on tissue sensitivity (type of tissue, physiological stage, plant species), most probably mediated by differentially elicited signal transduction pathways [2,4,5,7,15].

When applied as herbicides, synthetic auxins cause the same deformative and growth-inhibiting effects as those elicited by IAA applied at high concentrations [2, 3, 4] and those observed in transgenic, auxin-overproducing plants [16]. This phenomenon has been described as an auxin-overdose or as an effect of supraoptimal endogenous auxin concentrations [2,4]. There is broad consensus in the literature that this great imbalance in auxin homeostasis and interactions with other hormones in the tissue provide the basis of auxin herbicide action in susceptible dicots. Since Gilbert´s statement from 1946 [17] that auxin herbicides cause susceptible plants „to grow themselves to death“, the hypothesis prevailing until recently mainly seized on the observed growth abnormalities, while the ultimate cause of death remained open [2,3,4,5,18]. A continuous overreaction and stimulation of plant metabolism, including nucleic acid and protein synthesis, was thought to elicit a deregulation of growth through distorted cell division and expansion, leading to the collapse of the correlating plant growth structure. Normal apical growth of shoot and root is inhibited. Abnormal growth is induced in other tissues at an inappropriate time and position, giving tissue swelling and epinastic deformations, accompanied by transiently initiated cell division and differentiation in meristematic and cambial cell regions. These newly formed tissues were assumed to act as „metabolic sinks“ leading to mobilization and attraction of carbohydrate and protein reserves at the expense of the functional tissues, which senesce to power it [2,4,18,19]. However, the high level of species selectivity of most auxin herbicides, their rapid and in some cases enantioselective activity (e.g. herbicidally active (+)-D-enantiomer as dichlorprop-P, Figure 1) and the relatively low application rates suggest a more specific mechanism and mode of action causing growth inhibition and death.

Indeed, with increasing concentrations in the tissue and distribution (gradient development) in the plant, the time course of events induced by auxin herbicides or IAA at high concentrations can be divided into three phases [2,3]. In Figure 2, this is exemplified for the dicot weed *Galium aparine* after root-treatment in hydroponics against the background of other reported data in the literature [2,3,5,7,20]. (1) A stimulation phase, within the first hours after application. This includes stimulation of ethylene biosynthesis through induction of 1-aminocyclopropane-1-carboxylic acid (ACC) synthase in the shoot tissue (1-2 h), followed by symptoms of abnormal growth (3-4 h), including initiation of stem curling, tissue swelling and the downward curvature of leaves (leaf epinasty). Subsequently, abscisic acid (ABA) accumulates, initially detectable in the shoot tissue after 5-8 h. (2) An inhibition phase, within 24 h, which includes growth inhibition of root and, to a greater extent, of shoot, with decreased internode elongation and leaf area, and intensified green leaf pigmentation. Concomitantly, reductions in stomatal aperture, transpiration, carbon assimilation and starch formation occur. (3) Senescence / decay phase, which is accompanied by accelerated foliar senescence with chloroplast damage and progressive chlorosis and by the destruction of membrane and vascular system integrity, leading to desiccation, localized tissue death (necrosis) and finally to plant death .

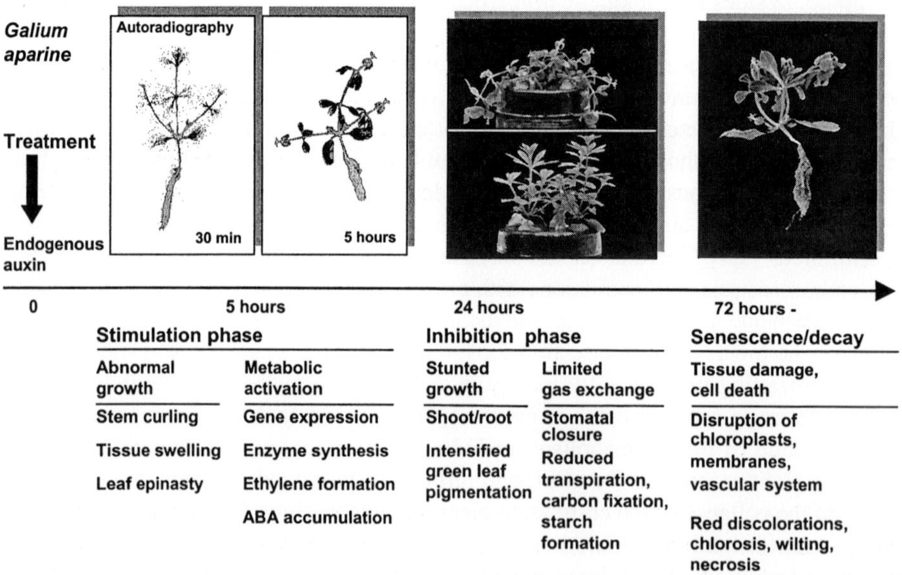

Figure 2. Three-phase-process of auxin herbicide action exemplified for the dicot weed *Galium aparine* after root-uptake and translocation in the plant (shown in autoradiographs using labelled compound).

2.2. Current Molecular Perspective

From the molecular perspective, there are two major concepts as to the site of action of auxin herbicides. One of these centers on cell elongation and the cell wall as the site of action, with impairment of its stability, while the other concept focuses on the overexpression of specific auxin-induced genes and protein synthesis, leading to aberrations in RNA synthesis with consequent metabolic changes [2,3,5 7]. In both cases, extensive auxin perception and signaling are assumed to be the primary mechanism of action.
According to a current model, the cell elongation response to auxin includes the activation of the plasma membrane H^{+}-ATPase. The enzyme triggers proton efflux into the apoplast, leading to acidification and loosening of the cell wall, which rapidly increases its extensibility („acid growth phase“, [21]). Activation of genes encoding wall lytic and modifying enzymes and proteins and subsequently, cell wall synthesis contribute to the long-term growth response („synthesis growth phase“, [3,22]. Guanosine triphosphate (GTP) binding proteins, release of phosphatidylinositol (PI)-derived molecules, rapid fluctuations in cytosolic calcium and protons involving ion channel regulation, activation of calmodulin and protein kinases, and phospholipase A-generated lipid second messengers have all been discussed in the literature as important components in the network of auxin signaling (for review see [5,7,15,23,24]). The latter signals are assumed to activate the plasma membrane H^{+}-ATPase by a phosphorylation-dependent mechanism.
At present, five major classes of early auxin-response genes are known which are induced within a few minutes of auxin treatment, mostly independently of *de novo* protein synthesis. Control of gene expression by auxin seems to be initiated by pre-existing transcription factors (auxin response factors, ARFs) that bind to auxin response domains within the promoter and become activated through degradation of repressor proteins by auxin-induced ubiquitin-proteasome system (for review see [25,26]). It is suggested that auxin promotes the ubiquitination of repressor proteins, such as Aux/IAA, by targeting them to a SCF-type ubiquitin-protein ligase (SCF^{TIR1}) [26]. The early auxin-responsive genes include genes of, as yet, unknown function (*SAUR*, *GH3* gene families), those encoding glutathione S-transferase (*GH2/4*), short-lived transcription factors (*Aux/IAA*) functioning as repressors of auxin-responsive genes, and 1-aminocyclopropane-1-carboxylic acid (ACC) synthase (*ACS*) in ethylene biosynthesis [27,28]. The last gene family encodes the key regulatory enzyme in ethylene biosynthesis that catalyzes the conversion of S-adenosylmethionine into ACC, the direct ethylene precursor [14]. Isoforms of ACC synthase genes are differentially expressed or post-transcriptionally regulated in different species and tissue types during senescence, in response to environmental stress conditions, and by auxins within a few minutes of treatment [27, 29].
It is postulated that before a hormone signal is transmitted and a downstream series of molecular, biochemical and physiological events in plant growth and development is elicited, the signal has to be first recognized by a receptor molecule [10]. Accordingly, auxin perception is thought to involve auxin binding to a plasma membrane-associated and/or soluble receptor at an internal site that targets the nucleus and regulates gene expression [2,15,23,24]. However, no auxin-binding protein has yet been unambiguously identified and characterized as a specific receptor for IAA or synthetic auxin perception [30]. As a possible candidate for an auxin receptor, auxin binding protein 1 (ABP1) is

currently discussed, which is assumed to function on the outer face of the plasma membrane, associated with a membrane-spanning docking protein (for review see [23, 30, 31]). The three-dimensional structure of the auxin-binding site has been predicted and a role in embryogenesis and auxin-induced cell expansion has been proposed [32]. Characterization of the *ABP* gene family from auxin herbicide resistant and susceptible biotypes of wild mustard showed differences in the deduced amino acids [15].

2.3 Ethylene Burst, Abscisic Acid (ABA) Biosynthesis and Hydrogen Peroxide as the Phytotoxic Principle

The most prominent example of hormonal cross-talk in growth regulation is the stimulation of ethylene production when auxins are applied to sensitive species [5, 6, 7, 14] or overproduced in transgenic plants [16]. This early and ubiquitous auxin response was described as early as 1935 by Zimmermann and Wilcoxon [33]. Ethylene is a gaseous hormone involved in plant responses to stress and the regulation of senescence and plant growth [14,34]. It promotes lateral cell expansion leading to swelling of stems and roots by reorienting microtubules from a transverse to a more longitudinal orientation [10]. Ethylene has also been discussed as a second messenger for auxin to regulate processes which are not controlled by auxin directly, such as leaf abscission and epinasty, and the regulation of auxin levels locally through ethylene-inhibited auxin transport [14]. Therefore, phenomena such as epinasty of leaf lamina or petiole and stem curvature could be attributed to auxin-stimulated ethylene or auxin activity alone [14,16,35]. Although ethylene itself is not the primary agent by which auxin herbicides cause phytotoxicity [13], former investigations have made it plausible that enhanced ethylene production is involved in this process (for review see [2,5,6,7]. In a multitude of dicot species and biotypes of differing sensitivity, stimulation of ethylene formation by auxin herbicides has been found to correlate with the induced growth effects, which could be alleviated by adding inhibitors of ethylene biosynthesis [13,14,36,37,38,39]. Accordingly, inactive auxin analogues did not promote ethylene production [5]. However, conclusive evidence for a causal role of ethylene synthesis in auxin herbicide action has been presented only recently. This progress has been achieved with the identification of auxin-induced *de novo* synthesis of ACC synthase and plant mutants and transformants modified in ethylene biosynthesis, perception and signaling [5,6,7,16,27,34]. In experiments using transgenic tomato plants containing the ACC synthase antisense gene *LE-ACS2* [38], auxin-resistant or ethylene-insensitive lines of *Arabidopsis,* wild mustard [29] and tomato [39], herbicide-induced phenotypical alterations could be alleviated or fully reversed.

In this context, we recently discovered close correlations between auxin-induced ethylene synthesis, massive accumulations of ABA in root and even more in shoot tissue and inhibition of their growth. This phenomenon, which has been reported for the first time, was demonstrated for auxin herbicides from different chemical classes (Figure 1), such as quinmerac, quinclorac, MCPA, naphthalene acetic acid, dicamba and picloram, and IAA at high concentration, in a variety of dicot species, including members of the Solanaceae, Umbelliferae, Fabaceae, Scrophulariaceae and Brassicaceae [35, 39, 40]. ABA accumulation and concomitant epinastic effects and reduced shoot growth were also observed when plants were exposed to ethylene-releasing ethephon [7,35]. Studies with isolated shoots and roots revealed that auxin-induced increases in ethylene and ABA occur primarily in the shoot tissue. From there, ABA appears to be translocated to the roots [6,20]. As shown in *Galium aparine*, auxin treatment via the root stimulates ethylene biosynthesis through induced ACC synthase activity in the shoot tissue within 2 h (Figure 3). This induction was inhibited by cordycepin or cycloheximide, indicating *de novo* enzyme synthesis [6]. Subsequently, increases in ABA and its biosynthetic precursor xanthoxal were detectable after 5-8 h [39]. ABA accumulation reached 70 times the level in controls after 48 h and closely correlated with reductions in stomatal aperture, with consequent inhibition of transpiration, carbon assimilation, plant growth and progressive foliar tissue damage [35,39,40,41] (Figure 3). The latter effect was accompanied by an overproduction of reactive oxygen species (ROS) such as hydrogen peroxide [41]. This accumulation of hydrogen peroxide appears to be triggered by the decline of photosynthetic activity due to ABA-mediated stomatal closure which leads to higher leakage of electrons from the photosystems to O_2 in the chloroplasts [41]. In addition, the amounts of Cu/Zn-superoxide dismutase increased in the leaf tissue (S. Tresch, K. Grossmann; unpublished result). The enzyme converts superoxide radicals to hydrogen peroxide and thus promotes hydrogen peroxide accumulation. Exogenously applied ABA mimicked the action of auxin on these processes [40,41]. An accumulation of hydrogen peroxide is generally considered to contribute to oxidative tissue damage through membrane lipid peroxidation and probably process signaling in senescence [42]. In addition, a role of reactive oxygen species (ROS) in auxin-induced cell-wall loosening has been suggested [43]. The causality between these processes was established more firmly by molecular dissection of the target pathways of ethylene and ABA using inhibitors of biosynthesis and tomato mutants defective in ethylene perception or ABA synthesis [39]. It is also in accord with the recognized hormonal role of ABA in stomatal closure, growth inhibition and promotion of leaf senescence [34]. In addition, quantification of xanthophylls, xanthoxal, ABA and its catabolites revealed that auxin-induced ethylene stimulates ABA biosynthesis, mainly by increasing xanthophyll cleavage to xanthoxal, catalyzed by the key regulatory enzyme 9-*cis*-epoxycarotenoid dioxygenase (NCED) [20,39,45] (Figure 3). Therefore, we concluded that ABA functions, together with ethylene, as a hormonal second messenger in the mode of action of auxin herbicides [20, 35,40]. The overproduction of ABA and consequently of hydrogen peroxide are the long-searched for missing links between induced ethylene synthesis and herbicide effects leading to growth inhibition, senescence and tissue decay (Figure 3). Dicot crop species with natural tolerance to quinmerac, such as oilseed rape and sugarbeet, showed only slight changes in ACC synthase activity and the subsequent events [44]. This suggests selectivity based on target site, the induction of ACC synthase.

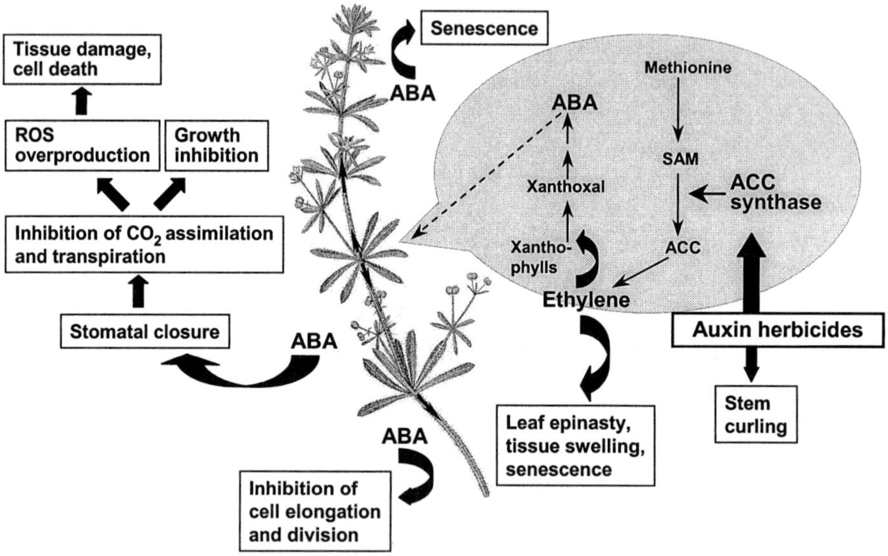

Figure 3. Model of the mode of action of auxin herbicides in the induction of growth inhibition, tissue damage and senescence in dicot plant species, as illustrated for cleavers (*Galium aparine*). ABA, abscisic acid; ACC, 1-aminocyclopropane-1-carboxylic acid; ROS, reactive oxygen species; SAM, S-adenosylmethionine. Modified from [20, 41].

2.4 Synergistic Cross-Talk between Auxin Transport Inhibitor and Auxin Herbicide

According to a recent model, the predominant naturally occurring auxin indole-3-acetic acid (IAA, Figure 1) is synthesized primarily in apical growing regions of plants, the apical shoot meristems and the young leaves [10]. From there, IAA is transported polarly through a parenchymatous cell-to-cell pathway towards basally located shoot tissues and the root system. The polar transport is mediated by auxin efflux carriers which presumably belong to the PIN protein family [46,47]. They are asymmetrically localized in the plasma membrane on the basal side of cells. Polar auxin transport is ubiquitous among higher plant species and specific for IAA and synthetic auxins. It contributes to the formation of an auxin gradient from the shoot to the root, which choreographs plant growth and development [10,46,47].

Auxin transport inhibitors, such as the phytotropins, disrupt the efflux of IAA and synthetic auxins from the cell and consequently their polar transport away from plant meristems. This causes an abnormal accumulation of IAA and exogenously applied auxin herbicides to be trapped in the growing tissues and meristematic shoot and root regions. Stunting, severe inhibition of root growth, loss of tropic responses and synergistic herbicide effects are the consequences. Diflufenzopyr, in combination with the benzoic acid dicamba, delivered more effective dicot weed control and selectivity in corn [11]. Concomitantly, diflufenzopyr synergistically enhanced ethylene-induced ABA accumulation and hydrogen peroxide overproduction by dicamba in cleavers and waterhemp (*Amaranthus* sp.) [48]. Diflufenzopyr also increased the activity of other synthetic auxin herbicides, including phenoxycarboxylic acids, pyridine carboxylic acids and quinclorac [11,48]. Phytotropins are not transported polarly [10]. Although the mechanism by which they control auxin efflux is unknown, they appear to interact with a plasma membrane and actin cytoskeleton associated protein (NPA-binding protein) which regulates the activity of the auxin efflux carrier [47]. However, the NPA-binding protein has not yet been identified [47]. Diflufenzopyr has a high affinity for the NPA-binding site, as shown by an auxin efflux inhibition value (I_{50}) of 19 nM and displacement of bound NPA [9,11]. Inhibition of the auxin efflux carrier blocks polar transport of both IAA and auxin herbicides.

3 Auxin Herbicide Action in Grasses: The Role of Cyanide

Unlike other auxin herbicides, quinclorac additionally controls important grass weeds, such as *Echinochloa*, *Digitaria*, *Setaria* and *Brachiaria* spp. [6,7]. Phytotoxic symptoms were characterized by rapidly inhibited shoot growth with progressive leaf chlorosis, followed by wilting and necrosis of the entire shoot. Only in sensitive grass species, quinclorac induced 1-aminocyclopropane-1-carboxylic acid (ACC) synthase activity in the root tissue, leading to subsequent increases in ACC as early as 1 h after treatment [49,50]. The accumulated ACC appears to be transported to the shoot, where it induces more synthesis of ACC which is converted to ethylene and cyanide. Treatment of isolated shoots with ACC *via* the vascular system stimulated ACC synthase activity and ethylene and cyanide production [7,37,49,51]. Cyanide was formed as a coproduct of ethylene in the oxidation of ACC, catalyzed by ACC oxidase [14,52]. While quinclorac-induced increases of ABA and hydrogen peroxide are not enough to elicit herbicidal effects [7], cyanide levels in the grass shoot tissue accumulated according to the herbicide concentration and application time and closely correlated with phytotoxicity [37,51]. In *Echinochloa crus-galli*, endogenous cyanide accumulated up to 30 µM. Applied KCN mimicked quinclorac effects at comparable endogenous cyanide concentrations [51]. Auxin herbicides from other chemical classes including dicamba and naphthalene acetic acid caused lower ACC and cyanide levels and correlated lower phytotoxicity [50]. Cyanide is an inhibitor of many important plant enzymes including cytochrome-c oxidase, ribulose-bisphosphate carboxylase and nitrate/nitrite reductase [52]. Therefore, we concluded that the accumulation of cyanide, derived ultimately from quinclorac-induced ACC synthase activity in ethylene biosynthesis, plays a key role in the herbicidal action of quinclorac in sensitive grasses [7,37,51] (Figure 4). In addition, Koo et al. [53] suggested that quinclorac inhibits cellulose (cell wall) biosynthesis in susceptible grasses. This was concluded from experiments using glucose incorporation into root wall fractions [53]. However, this effect of quinclorac could not be confirmed in studies using cytochemical techniques for microscopic examination (S. Tresch, K. Grossmann; unpublished result).

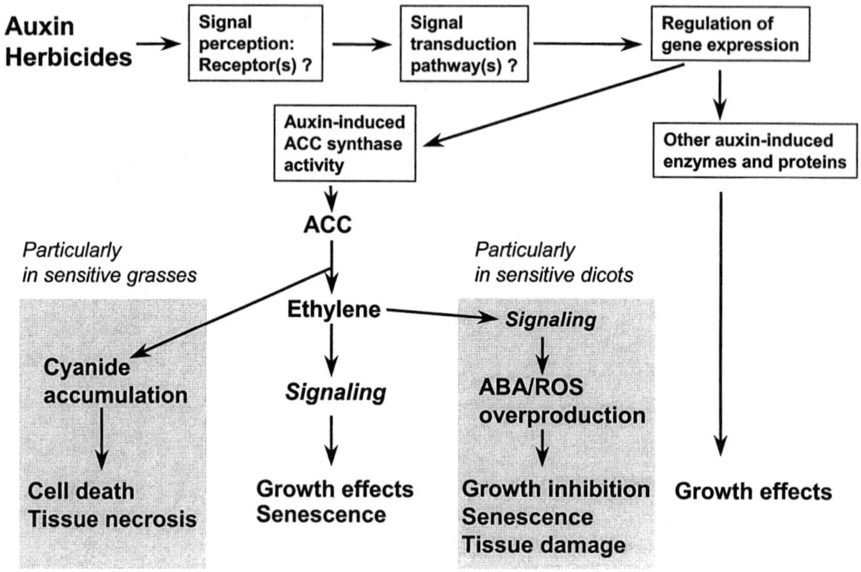

Figure 4. Proposed model of molecular interactions in the herbicidal mechanism of action of auxin herbicides in sensitive dicot and grass weeds. ABA, abscisic acid; ACC, 1-aminocyclopropane-1-carboxylic acid; ROS, reactive oxygen species. Modified from [7].

4 Conclusion

Extensive affinity for auxin receptor(s) or subsequent signal transduction processes are assumed to be the primary mechanism of herbicide action as a consequence of an auxin-overdose. Deduced from the plethora of developmental and morphological effects, it is suggested that one or several classes of auxin receptors and auxin signal transduction pathways exist (Figure 4). These receptors and pathways are thought to be heterogeneously distributed among different cell types and tissues and differentially affected by the different structural types of auxin herbicides [6,7,15]. However, a specific receptor for auxin perception has not yet been characterized unambiguously and only sparse information of components in auxin signal pathways are available. Recent studies showed that a newly discovered sequential hormone interaction plays a decisive role in auxin herbicide action. The induction of 1-aminocyclopropane-1-carboxylic acid (ACC) synthase in ethylene biosynthesis is a primary response directly following auxin herbicide signaling (Figure 4). Organ-specific gene expression or post-transcriptional *de novo* enzyme synthesis could be involved. In sensitive dicots, overproduction of ethylene causes epinastic growth, tissue swelling and leaf abscission and stimulates abscisic acid

(ABA) biosynthesis (Figure 4). ABA directly affects cell division and expansion and induces stomatal closure. This limits photosynthetic activity and biomass production, accompanied by an overproduction of reactive oxygen species. Growth inhibition, senescence and tissue decay are the consequences [20,41] (Figure 4). Current evidence indicates that ethylene-triggered ABA is not restricted to the action of auxin herbicides. It may function as a module in the signaling of a variety of stimuli leading to plant growth regulation [54]. In combination with auxin transport inhibitors, such as the semicarbazone-type phytotropin diflufenzopyr, auxin herbicides abnormally accumulate in meristematic regions, which synergistically enhance their herbicidal effects [11,48]. An additional phenomenon case is caused by the auxin herbicide quinclorac which also controls grass weeds. Here, quinclorac induces the accumulation of phytotoxic levels of cyanide, a co-product of ethylene, which is ultimately derived from herbicide-induced ACC synthase activity in the tissue [7] (Figure 4).

5 Acknowledgements

The author is grateful to Jacek Kwiatkowski, Günter Caspar and Stefan Tresch for technical assistance and Alan Akers (BASF Agricultural Center) for critical reading of the English manuscript.

6 References

[1] K. E. Pallett in Target Sites for Herbicide Action (Eds.: R. C. Kirkwood), Plenum, New York, 1991, pp. 124 – 167.
[2] A. H. Cobb in Herbicides and Plant Physiology (Eds.: A. H. Cobb), Chapman and Hall, London, 1992, pp. 82 – 106.
[3] M. Devine, S. O. Duke, C. Fedtke (Eds.) in Physiology of Herbicide Action, PTR Prentice-Hall, Englewood Cliffs, 1993.
[4] D. Coupland in Herbicide Resistance in Plants. Biology and Biochemistry (Eds.: S. B. Powles, J. A. M. Holtum), Lewis, Boca Raton, 1994, pp. 171 – 214.
[5] T. M. Sterling, J. C. Hall in Herbicide Activity: Toxicology, Biochemistry and Molecular Biology (Eds.: J. D. Roe, J. D. Burton, R. J. Kuhr), IOS Press, Amsterdam, 1997, pp. 111 – 141.
[6] K. Grossmann, Weed Science 1998, 46, 707 – 716.
[7] K. Grossmann in Herbicides and their Mechanisms of Action (Eds.: A. H. Cobb, R. C. Kirkwood), Sheffield Academic Press, Sheffield, 2000, pp. 181 – 214.
[8] Phillips McDougall AgriService 2000.
[9] M. V. Subramanian, P. Bernasconi, B. C. Patel, J. Reagan, Weed Science 1997, 45, 621 – 627.
[10] L. Taiz, E. Zeiger, Plant Physiology, Sinauer, Sunderland, 1998.
[11] S. J. Bowe, M. Landes, J. Best, G. Schmitz, M. Graben, Brighton Conf. – Weeds 1999, pp. 35 – 40.
[12] B. Wuerzer, R. Berghaus, 10th Asian-Pacific Weed Sci. Soc. Conf., 1985, pp. 177 – 184.
[13] P. W. Morgan in Herbicides: Physiology, Biochemistry and Ecology (Ed.: L. J. Audus), Academic Press, New York, 1976, pp. 256 – 280.

[14] F. B. Abeles, P. W. Morgan, M. E. Saltveit (Eds.) in Ethylene in Plant Biology, Academic Press, San Diego, 1992.
[15] H.-G. Zheng, J. C. Hall, Weed Science 2001, 49, 276 – 281.
[16] C. P. Romano, M. L. Cooper, H. J. Klee, Plant Cell 1993, 5, 181 – 189.
[17] F. A. Gilbert, Chem. Rev. 1946, 39, 199 – 218.
[18] J. B. Hanson, ASPP News 1996, 25, 5 – 6.
[19] J. Van Overbeek in The Physiology and Biochemistry of Herbicide Action (Ed.: L. J. Audus), Academic Press, New York, 1964, pp. 387 – 400.
[20] K. Grossmann, Trends Plant Sci. 2000, 5, 506 – 508.
[21] A. Hager, G. Debus, H.-G. Edel, H. Stransky, R. Serrano, Planta 1991, 185, pp. 527 - 537.
[22] D. J. Cosgrove, Plant Physiol. Biochem. 2000, 38, pp. 109 – 124.
[23] H. Macdonald, Physiol. Plant. 1997, 100, pp. 423 – 430.
[24] O. Leyser, Curr. Opin. Plant Biol. 2001, 4, pp. 382 – 386.
[25] T. J. Guilfoyle, G. Hagen, J. Plant Growth Regul. 2001, 20, pp. 281 – 291.
[26] W. M. Gray, S. Kepinski, D. Rouse, O. Leyser, M. Estelle, Nature 2001, 414, pp. 271 – 276.
[27] S. Abel, A. Theologis, Plant Physiol. 1996, 111, pp. 9 – 17.
[28] L. J. Hobbie, Plant Physiol. Biochem. 1998, 36, pp. 91 – 102.
[29] Y. D. Wei, H.-G. Zheng, J. C. Hall, Pest Manag. Sci. 2000, 56, pp. 377 – 387.
[30] C. Luschnig, G. R. Fink, Trends Plant Sci. 1999, 4, pp. 162 – 164.
[31] C. Timpte, Trends Plant Sci. 2001, 6, pp. 586 – 590.
[32] A. M. Jones, K.-H. Im, M. A. Savka, M.-J. Wu, N. G. DeWitt, R. Shillito, A. N. Binns, Science 1998, 282, pp. 1114 – 1117.
[33] P. W. Zimmerman, F. Wilcoxon, Contrib. Boyce Thompson Inst. 1935, 7, pp. 209 – 229.
[34] H. Kende, J. A. D. Zeevaart, Plant Cell 1997, 9, pp. 1197 – 1210.
[35] F. Scheltrup, K. Grossmann, J. Plant Physiol. 1995, 147, pp. 118 – 126.
[36] J. C. Hall, P. K. Bassi, M. S. Spencer, W. H. Vanden Born, Plant Physiol. 1985, 79, pp. 18 – 23.
[37] K. Grossmann, J. Kwiatkowski, J. Plant Physiol. 1993, 14, pp. 457 – 466.
[38] K. Grossmann, T. Schmülling, Plant Growth Regul. 1995, 16, pp. 183 – 188.
[39] H. Hansen, K. Grossmann, Plant Physiol. 2000, 124, pp. 1437 – 1448.
[40] K. Grossmann, F. Scheltrup, J. Kwiatkowski, G. Caspar, J. Plant Physiol. 1996, 149, pp. 475 – 478.
[41] K. Grossmann, J. Kwiatkowski, S. Tresch, J. Experim. Bot. 2001, 362, pp. 1811 - 1816.
[42] J. Dat, S. Vandenabeele, E. Vranova, M. Van Montagu, D. Inze, F. Van Breusegem, Cell. Mol. Life Sci. 2000, 57, pp. 779 – 795.
[43] P. Schopfer, Plant J. 2001, 28, pp. 679 – 688.
[44] K. Grossmann, F. Scheltrup, Pesticide Sci. 1998, 52, pp. 111 – 118.
[45] X. Qin, J. A. D. Zeevaart, Proceed. Nat. Acad. Sci. 1999, 96, pp. 15354 – 15361.
[46] J. Friml, J. Wisniewska, E. Benkova, K. Mendgen, K. Palme, Nature 2002, 415, pp. 806 – 809.
[47] G. K. Muday, A. DeLong, Trends Plant Sci. 2001, 6, 535 – 542.
[48] K. Grossmann, G. Caspar, J. Kwiatkowski, S. J. Bowe, Pest Manag. Sci. 2002, in press.
[49] K. Grossmann, F. Scheltrup, Pest. Biochem. Physiol. 1997, 58, pp. 145 – 153.
[50] K. Grossmann, J. Kwiatkowski, Pest. Biochem. Physiol. 2000, 66, pp. 83 – 91.
[51] K. Grossmann, J. Kwiatkowski, Pest Biochem. Physiol. 1995, 51, pp. 150 – 160.
[52] K. Grossmann, Physiol. Plant. 1996, 97, pp. 772 – 775.
[53] S. J. Koo, J. C. Neal, J. M. DiTomaso, Plant Physiol. 1996, 112, pp. 1383 – 1389.
[54] K. Grossmann, H. Hansen, Physiol. Plant. 2001, 113, pp. 9 – 14.

12 Natural Phytotoxins with Potential for Development in Weed Management Strategies

Stephen O. Duke [(a)], Franck E. Dayan [(a)], Scott R. Baerson [(a)], Joanne G. Romagni [(b)], Ameeta Agarwal [(c)] and Anna Oliva [(d)]

(a) Natural Products Utilization Research, USDA, ARS, P. O. Box 8048, University, Mississippi 38677, USA; E-mails: sduke@ars.usda.gov; fdayan@ars.usda.gov, baerson@ars.usda.gov, (b) Dept. Biology, Univ. St. Thomas, Houston, Texas 77006, USA; E-mail: romagnj@stthom.edu, (c) National Center for Natural Products Research, School of Pharmacy, Univ. Mississippi, University, Mississippi 38677, USA; E-mail: aagarwal@olemiss.edu, (d) Department of Life Science, Second University of Naples, Via Vivaldi 43-81100, Italy; E-mail: anna.oliva@unina2.it

1 Introduction

Evaluation of natural compounds has been a relatively small part of the herbicide discovery effort of most companies involved in pesticide discovery. In some cases, the company has put significant effort into isolation of phytotoxic compounds from various organisms as part of a "bioprospecting" effort. This activity has led to several commercial herbicides, including glufosinate, bialaphos, the triketones, and pelargonic acid.

The need exists to expand the repertoire of tools available for weed management systems. What little we know of the molecular target sites of natural phytotoxins indicates that they are inhibitors of a broad array of enzymes and other molecular targets that have not been the focus of herbicide discovery efforts [1]. At present, commercially available herbicides target a relatively limited number of enzymes and metabolic pathways. The discovery of new target sites is a growing emphasis of pesticide companies, especially since the U.S. Food Quality Protection Act has combined food tolerance levels of pesticides with the same molecular target sites.

The reasons why natural products have not been an area of focus for pesticide company discovery efforts are several. They include the relatively high cost of bioassay-directed isolation, the probability of rediscovery, and the often prohibitively complex structure of the active molecule. Furthermore, the strategies for discovery of natural product-based pesticide discovery are different from those for synthetic compounds, requiring a somewhat different laboratory infrastructure. New methods, including semi-automated bioassays, automated chemical extraction and analysis, and molecular biology methods to probe biosynthesis and mode of action have reduced the time and effort required in natural product-based herbicide discovery efforts. This short review will focus on our own work in this area, but will also mention some of the other literature. Furthermore, we will briefly discuss some of the new approaches to natural products discovery research and for the utilization of natural products for weed management.

2 New Molecular Target Sites

We and others have discovered several new molecular target sites of phytotoxins by studying the mode of action of natural compounds. We will discuss three of these in detail.

2.1 Ceramide Synthase

Ceramides are ubiquitous lipids in plants and animals. In animals, they are particularly important to the integrity of nervous system cells and tissues. Much less is known of their function in plants, although they are significant constituents of the plasma membrane. Abbas and co-workers [2,3] found that the fumonisins are highly phytotoxic, causing rapid cellular leakage and cell death at low concentrations. These compounds are well known mycotoxins that were known to act by inhibiting ceramide synthase of mammals. AAL-toxin, a plant pathogen-produced phytotoxin and structural analogue of the fumonisins, was found to produce physiological effects on plants similar to those of the fumonisins [4]. Both fumonisins and AAL-toxin caused rapid and dramatic increases in phytosphingosine and sphinganine in treated plants [5]. Fumonisins and AAL-toxin are structural analogues of sphinganine. These compounds accumulate when ceramide synthase is inhibited (Figure 1).

Inhibition of ceramide synthase could cause symptoms by depletion of ceramide and ceramide derivatives or by toxic increases in sphinganine and its derivatives. The effects of these phytotoxins are so rapid, that it is unlikely that depletion of ceramides is responsible for cellular death. Others have invoked induction of apoptosis in the mode of action of these compounds [*e.g.*, 6]; however, treatment of plant tissues with phytoshingosine and sphinganine causes symptoms very similar to those caused by inhibition of ceramide synthase [7]. Very low levels of AAL-toxin and other ceramide synthase inhibitiors may cause apoptosis, but the rapidity of plasma membrane damage at one micromolar and higher concentrations makes it unlikely that it is involved in the main phytotoxic effect.

Attempts have been made to find a ceramide synthase inhibitor with good phytotoxicity, but little mammalian toxicity [*e.g.*, 8]. However, even inhibitors with little structural similarity to the fumonisins such as australifungin [9], have relatively little difference between mammalian and plant toxicity. Nevertheless, these studies demonstrate that the ceramide synthase pathway is a viable site for herbicides, provided an inhibitor can be found that is plant specific. This topic is considered in more detail in a recent review [10].

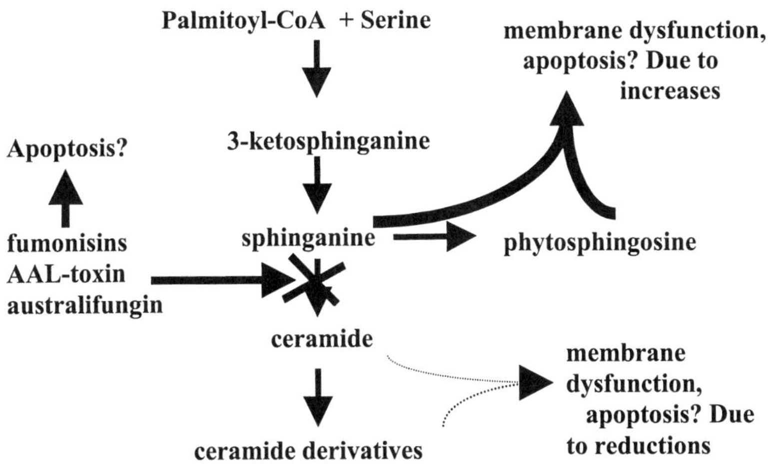

Figure 1. Mechanism of action of ceramide synthase inhibitors in plants. Thick lines indicate increased level effects, whereas thin lines represent decreased level effects.

2.2 Fructose-1,6-Bisphosphate Aldolase

Fructose-1,6-bisphosphate aldolase (FBPase; EC 4.1.2.13) catalyzes the reversible aldol cleavage of Fructose-1,6-bisphosphate (FBP) into the triose phosphates G-3-P and DHAP. Inhibition of this step of glycolysis is known to have long-ranging repercussions on the overall physiology of plants, leading to reduced photosynthetic efficiency, impaired sugar and starch metabolism, and ultimately reduced growth [11]. A significant amount of research has been done to identify inhibitors of animal FBPase to develop novel pharmaceutical compounds. While this enzyme has not been (and probably will not be) developed as a potential herbicide target site because it is too ubiquitous, our studies of the mode of action of phytotoxic natural products showed that some natural phytotoxins may target this enzyme.

The fructose analogue 2,5-anhydro-D-glucitol (AhG) produced by the *Fusarium solani* (Mart.) Sacc. NRRL 18883 is phytotoxic [12], inhibiting lettuce root growth (I_{50} of 1.6 mM). The mechanism of action of this sugar analog is quite complex, involving its sequential bisphosphorylation by hexokinase and phosphofructokinase to yield AhG-1,6-bisphosphate. This phosphorylated sugar analogue is a competitive inhibitor of FBPase (I_{50} of 570 μM and K_i value of 103 μM) [13]. The catalytic mechanism of FBPase involves the formation of a covalent bond between the anomeric hydroxyl group of FPB and the gamma amino functionality of a lysine residue of the enzyme. AhG (and AhG-1,6-bisphosphate) do not have this anomeric hydroxyl group, preventing the normal catalytic function of aldolase.

2.3 Asparagine Synthetase

Asparagine synthetase (AS; EC6.3.5.4), the key enzyme in asparagine synthesis, is a good potential target site for herbicides because asparagine is required for protein biosynthesis and is involved in nitrogen mobilization. Plant AS catalyzes an ATP-dependent reaction where the amine group of glutamine is transferred directly to aspartate [14,15]. The prokaryotic form requires primarily ammonia. An alternative pathway that utilizes hydrogen cyanide through the β-cyanoalanine pathway also exists in some plants.

AS was hypothesized as a potential herbicide target site when it was discovered that the growth inhibition caused by natural phytotoxin 1,4-cineole (or its synthetic analogue cinmethylin) could be reversed with exogenous supply of asparagine [16]. The mode of action of cinmethylin was unknown for several years, despite serious efforts to discover its molecular target site. Further work demonstrated that root uptake of asparagine increased in the presence of the inhibitors, suggesting that the supply of asparagine was low in the treated seedlings. Uptake of aspartate was not affected by the treatments. It was then demonstrated that *in vitro* AS activity was inhibited by 1,4-cineole (Figure 2). Interestingly, the synthetic analogue cinmethylin did not inhibit AS, suggesting that it must be metabolized into a 1,4-cineole analogue, ostensibly 2-hydroxy-1,4-cineole, prior to inhibiting AS activity.

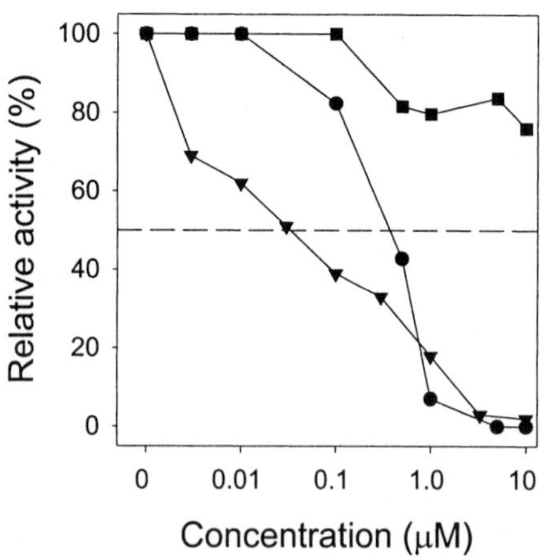

Figure 2. Effect of 1,4-cineole (circles), cis-2-hydroxy-1,4-cineole (triangles) and the commercial herbicide cinmethylin (squares) on the activity of asparagine synthetase from lupin. The dotted line represents 50% inhibition of enzyme activity.

This hypothesis was tested by measuring the inhibitory activity of *cis*- and *trans*- 2-hydroxy-1,4-cineole. The *cis*-form was more than an order of magnitude more effective against AS than 1,4-cineole [16] (Figure 2). The addition of the alcohol to the molecule renders it less volatile, therefore allowing more of the compound to react with the site of inhibition. The trans-form of 2-hydroxy-1,4-cineole was less active than either the cis-diastereomer or 1,4-cineole. Greenhouse studies had demonstrated that the cis-form of cinmethylin was more active than its trans-diastereomer, further supporting the current hypothesis that the orientation of the hydroxyl group may play a key role in the potency of this metabolic intermediate. These studies indicate that AS is a viable site for biorationale herbicide design.

3 New Compounds/Old Target Sites

We have found several natural phytotoxins to have the same molecular target site as a class of commercial, synthetic herbicides. In one case below, the compound targets two herbicide target sites at low doses, something that no commercial herbicide does. Thus, even when natural products target old sites, they might have utility as new herbicides or herbicide classes for these sites or as allelochemicals that might be genetically engineered into crops.

3.1 Sorgoleone

The allelochemical sorgoleone (2-hydroxy-5-methoxy-3-(8'*Z*,-11'*Z*)-pentadeca-8'11'14'-triene-*p*-benzoquinone) exudes as oily droplets from the root hairs of *Sorghum* species. It is relatively stable in soil and accumulates in the rhizosphere. This molecule is highly phytotoxic to broadleaf and grass weeds at concentrations as low as 10 μM [17], and its presence in the soil surrounding sorghum plants represses the growth of weeds.

Sorgoleone was initially found to inhibit mitochondrial respiration, but it was later found to be a more potent inhibitor of PSII [17, 18]. Sorgoleone is structurally similar to plastoquinone (PQ), a benzoquinone involved in photosynthetic electron transport. Sorgoleone competes for the PQ binding site of the D-1 protein in a manner similar to most commercial photosynthetic inhibitors [19]. The *in vitro* PSII inhibiting activity of sorgoleone is similar to some of the commercial herbicides targeting this site (*e.g.*, atrazine and diuron).

A recent study also demonstrated that sorgoleone (and other natural quinones) can inhibit the enzyme *p*-hydroxyphenylpyruvate dioxygenase (HPPD) (Figure 3), the key enzyme in plastoquinone synthesis [20]. In addition to acting as a redox reagent in PSII, plastoquinone is a cofactor for phytoene desaturase, the target of many carotenoid synthesis inhibiting herbicides [21]. Without plastoquinone, phytoene desaturase does not function, resulting in cessation of carotenoid synthesis.

Most commercial HPPD inhibitors (*e.g.*, sulcotrione and isoxaflutole) are competitive time-dependent (tight-binding) inhibitors. As such, these herbicides bind to the enzyme very tightly with T½ of dissociation ranging from a few hours to several days, as opposed to milliseconds for traditional reversible inhibitors. Sorgoleone does not behave as these herbicides and appears to be a reversible inhibitor of HPPD. This quinone is structurally more planar than the traditional HPPD inhibitors, so it may not form a stable reaction intermediate. Instead, its backbone may resemble the conformation of one of the later intermediate step in the reaction mechanism of HPPD.

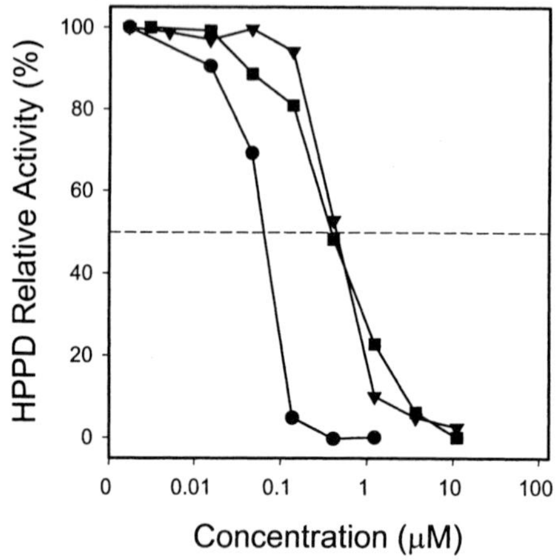

Figure 3. Effect of the β-triketone (-)-usnic acid (circles), the benzoquinone sorgoleone (triangles) and the commercial herbicide sulcotrione (squares) on the activity of *p*-hydroxyphenylpyruvate dioxygenase. The dotted line represents 50% inhibition of enzyme activity.

3.2 Usnic Acid

Usnic acid [2,6-diacetyl-7,9-dihydroxy-8,9b-dimethyl-1,3(2H9βH)-dibenzofurandione] is one of the most common lichen secondary metabolite that is found in large amounts in the genus *Usnea* [22]. Seedlings grown in the presence of usnic exhibit bleaching of the cotyledonary tissues. Loss of chlorophylls in response to phytotoxins can be associated with light-dependent destabilization of cellular and sub-cellular membranes, but usnic acid apparently acts differently since it caused membrane leakage in the absence of light, though the loss of membrane integrity was accentuated in the presence of light. Analysis of the pigment contents showed a strong decrease in β-carotene in samples treated with usnic acid, and total carotenoids decreased with increased usnic concentration. Inhibition of carotenoid synthesis is known to be accompanied with the destruction of chlorophyll (bleaching) due to the destabilization of the photosynthetic apparatus.

Most 'bleaching' herbicides target the enzyme phytoene desaturase that converts phytoene to carotenes. However, bleaching of chlorophyllous tissues can also be associated with the inhibition of the enzyme *p*-hydroxyphenylpyruvate dioxygenase (HPPD) responsible for plastoquinone biosynthesis [23].Usnic acid possesses a 2-keto-cyclohexane-1,3-dione substructure common to many triketone HPPD inhibitors (*e.g.*, sulcotrione and mesotrione) [24]. When tested on this enzyme, usnic acid showed a strong inhibitory activity, with an I_{50} of 70 nM, surpassing the activity obtained with the commercial herbicide sulcotrione (Figure 3) [25].

While the *in vivo* activity of usnic acid is very desirable, a preliminary greenhouse study showed that usnic acid was poorly active on 3-week-old morningglory species, barnyardgrass, velvetleaf, sicklepod, and yellow nutsedge. The difference between the high *in vitro* activity on HPPD and the low *in vivo* activity is most likely due to poor foliar uptake associated with inadequate physicochemical properties of usnic acid.

3.3 Natural Protoporphyrinogen Oxidase Inhibitors

At least two natural products have been reported to inhibit protoporphyrinogen oxidase (Protox). We found the fungal pathogen diphenyl ether compound cyperine to be a weak Protox inhibitor (Figure 4), but this level of activity could not explain its herbicidal properties [26]. It causes chlorophyll loss in darkness that is apparently unrelated to photobleaching. Usnic acid is also a weak Protox inhibitor (Figure 4), but it is much more likely to be herbicidal due to its HPPD activity (Figure 3). Considering the wide structural range of compounds that inhibit Protox [27], other, more potent Protox inhibitors probably exist in nature.

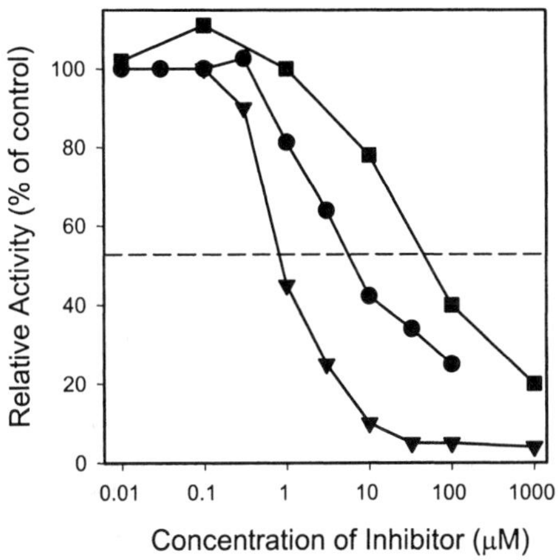

Figure 4. Effect of the synthetic diphenyl ether acifluorfen (triangles), the natural diphenyl ether fungal metabolite cyperine (squares) and the dibenzofurandione lichen metabolite usnic acid (circles) on the activity of protoporphyrinogen oxidase. The dotted line represents 50% inhibition of enzyme activity.

4 New Approaches

Modern molecular biology is transforming the pesticide industry. We will discuss how it can be used in both the search for new molecular target sites and elucidation of modes of action of natural phytotoxins, as well as in modification of crops to fight weeds with allelochemicals.

4.1 Transcriptional Profiling for Mode of Action Clues

Transcriptional profiling represents a potentially effective means for elucidating inhibitor modes of action. Theoretically, specific patterns of up- and down-regulated genes might be associated with particular molecular target sites, allowing the researcher to quickly determine whether a new phytotoxin fits such a pattern from a gene expression library of profiles of known modes of action or might target a new molecular site.
We are in the early stages of building such a gene expression profile library. We began with cinmethylin, since little is known of its mode of action, other than that it is a potent inhibitor of AS (Figure 2). Using a cDNA microarray comprised of 9,200 elements derived from *Arabidopsis thaliana*, we examined global expression pattern changes in 10-day-old *Arabidopsis* seedlings exposed to cinmethylin concentrations sufficient to cause an approximately 50% reduction in root growth.
Preliminary efforts directed at the use of *Arabidopsis thaliana* as a model for studying herbicidal modes of action have thus far been encouraging. A group of cinmethylin up and down-regulated genes is shown below (Figure 5). It will be refined as additional process and biological microarray replicates are completed. Further experiments should allow us to distinguish between gene expression changes that occur in direct response to the presence of inhibitor, versus those that occur indirectly or due to abiotic stresses imposed by the experimental conditions.

4.2 Manipulation of Allelochemical Production by Biotechnology

Phytotoxic compounds are produced by many crop species, and some of these compounds may play a role as allelochemicals in plant interference. This has been hard to prove, much less taken advantage of in weed management. Many crops have been screened for allelopathic potential [28]. However, in all but a few cases, the effects of crop allelopathy are quite subtle compared to the effects of chemical herbicides. Efforts to produce allelopathic crop varieties through conventional breeding have not generated commercial varieties sold with this trait. Molecular biology approaches to this problem might be helpful.
We have proposed strategies for transgene approaches to allelochemical manipulation [*e.g.*, 29]. Fundamental to such an effort is knowing the allelochemical and the genes encoding the enzymes involved in their biosynthesis. Gene isolation strategies which are suitable for most allelopathic species include differential screening approaches such as suppression subtractive hybridization or differential display. These techniques require that the gene(s) of interest be sufficiently differentially expressed under certain conditions. Alternatively, mutants, such as that of *Artemisia annua,* possessing and lacking peltate secretory glands [30] could be used as starting material for differential screens.

Cy 5 (Control)

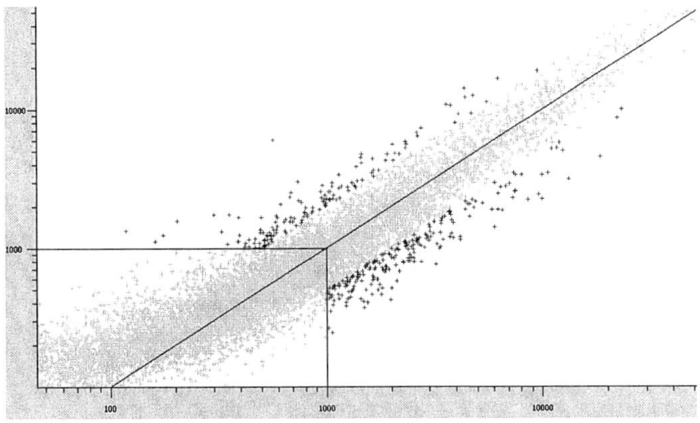

Cy 3 (Treated)

Figure 5. Gene expression of cinmethylin-treated vs. control arabidopsis seedlings. The gray spots in the center of the plot represent genes which remain unchanged in the control and treated samples. Darker spots indicate genes that are at least 2-fold induced or repressed by 0.0165 μM cinmethylin.

Secondary metabolites are often produced by specific cell types. Thus, the genes involved in their biosynthesis may be specifically or preferentially expressed within those cells. If substantial amounts of the compound are produced, the system could be analyzed with Expressed Sequence Tags (EST). Within specialized cells, such as root hairs or glandular trichomes, the genes involved may represent the most highly expressed mRNA species within that cell type, and can thus be identified within a random sampling of a few thousand sequences. This approach was successful in the identification of genes involved in monoterpene biosynthesis in peltate glands of *Mentha piperita* [31].

Sorgoleone is a particularly amenable allelochemical for genetic manipulation. Roots of *S. bicolor* can be propagated under soil-less conditions that either promote or inhibit root hair formation (M. Czarnota, *unpublished*). Thus, an excellent source of starting material is readily available for performing differential screens as described above. Since sorgoleone is the major component of *S. bicolor* root exudates, an EST-based approach is also a viable strategy. Our laboratory has been actively pursuing the isolation of sorgoleone biosynthetic genes using both suppression subtractive hybridization as well as EST analysis. Highly purified root hair preparations can readily be obtained from *S. bicolor*, and from this starting material a cDNA library was prepared for the development of a sequence database comprised of 5,000 randomly chosen sequences.

Preliminary EST data from this study as well as suppression subtractive hybridization results have identified several promising gene candidates. In addition, we have recently developed a secondary screen for determining the expression patterns of candidate sequences based on real-time PCR in the presence of the double stranded DNA-specific fluorophore, SYBR Green [*e.g.*, 32]. Figure 6 illustrates the utility of real time PCR for follow-up analyses of candidate sequences. By comparing PCR amplification signals using cDNAs prepared from six different *S. bicolor* tissues (root hair, whole root, immature leaf, mature leaf, stem, and panicle), the relative expression levels in these tissues can be rapidly determined. In the example shown, a putative O-methyl transferase identified by both EST analysis and suppression subtractive hybridization is shown to exhibit highly root hair-preferential expression. Additional work is being performed to determine if this particular enzyme is capable of methylating sorgoleone precursors and structurally related molecules via heterologous expression in *Escherichia coli*. Ultimately, stably-transformed *S. bicolor* plants or root cultures will be employed to directly demonstrate the involvement of this gene in the sorgoleone biosynthetic pathway.

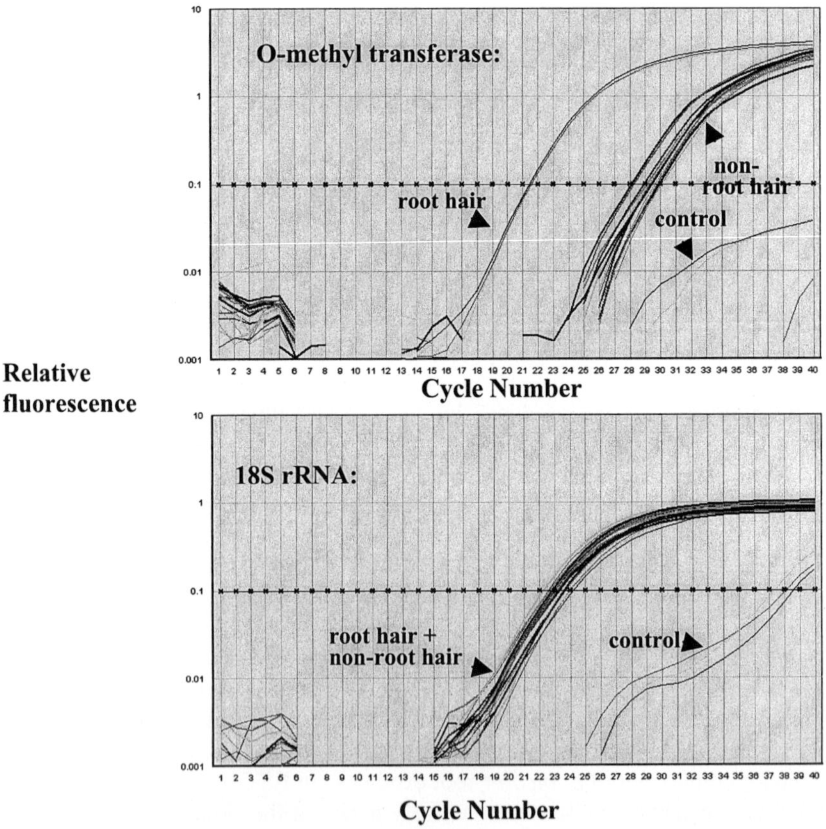

Figure 6. Representative real-time PCR results generated using an ABI 5700 sequence detection system and sequence-specific primers for a putative O-methyl transferase. Upper panel: SYBR Green I signals generated from root hair cDNA vs. non-root hair cDNA. Lower panel: same cDNAs used as in upper panel showing signals generated from an internal control gene (18S ribosomal RNA).

While efforts toward the genetic manipulation of allelochemical pathways are at a very preliminary stage, the tools are now at hand to make significant strides in this area. The likelihood that many, if not most, allelochemicals are synthesized in a cell type or organ-specific manner will greatly facilitate the isolation of genes involved in biosynthesis, as well as possible ancillary proteins involved in regulation, sequestration, or the rhizosecretion of allelochemicals. Moreover, the availability of additional molecular probes for specific allelochemical pathways will ultimately shed further light on the modes of regulation and environmental interactions involved, for which there is currently a paucity of information [33].

5 Outlook

Powerful, new tools and approaches to natural product discovery and utilization are rapidly expanding our knowledge of chemical diversity, the biological activity, and the molecular genetics of naturally occurring compounds. Many of these compounds are phytotoxic and have the potential for use in weed management, either as a herbicide or a genetically engineered allelochemical. But, still, compared to the synthetic herbicide discovery and herbicide-resistant crop development efforts, little has been done with natural products. This knowledge gap and the availability of new capabilities should attract future researchers to focus on discovery and development of natural compounds for weed management.

6 References

[1] S. O. Duke, F. E. Dayan, R.G. Romagni, Crop Protect. 2000, 19, 583-589.
[2] H. K. Abbas, C.D. Boyette, Weed Technol. 1992, 6, 548-543.
[3] H. K. Abbas, R. N. Paul, C. D. Boyette, S. O. Duke, R. F. Vesonder, Can. J. Bot. 1992, 70, 1824-1833.
[4] H. K. Abbas, R. F. Vesonder, C. D. Boyette, S. W. Peterson, Can. J. Bot. 1993, 71, 155-160.
[5] H. K. Abbas, T. Tanaka, S. O. Duke, J. K. Porter, E. M. Wray, L. Hodges, A. E. Sessions, A. H. Merrill, R. T. Riley, Plant Physiol. 1994, 106, 1085-1093.
[6] D. G. Gilchrist, Cell Death Differ. 1997, 4, 689-698.
[7] T. Tanaka, H. K. Abbas, S. O. Duke, Phytochemistry 1993, 33, 779-785.
[8] H. K. Abbas, T. Tanaka, W. T. Shier, Phytochemistry 1995, 35, 1681-1689.
[9] H. K. Abbas, S .O. Duke, A .H. Merrill, E. Want, W. T. Shier, Phytochemistry 1998, 47, 1509-1514.
[10] H. K. Abbas, S. O. Duke, W. T. Shier, M. V. Duke, in Advances in Microbial Toxin Research and its Biotechnological Exploitation (Ed.: R.K. Upadhyay), Kluwer, Amsterdam, 2002, pp. 211-229.
[11] V. Haake, R. Zrenner, U. Sonnewald, M. Stitt, Plant J. 1998, 14, 147-157.
[12] T. Tanaka, K. Hanato, M. Watanabe, H. K. Abbas, J. Nat. Tox. 1996, 5, 317-329.

[13] F. E. Dayan, A. M. Rimando, M. R. Tellez, B. E. Scheffler, T. Roy, H. K.Abbas, S. O. Duke, Z. Naturforsch., C, 2002, In press.
[14] R .J. Ireland, P .J. Lea, in Plant Amino Acids: Biochemistry and Biotechnology (Ed.: B. K. Singh) Marcel Dekker, Inc., New York, 1999, pp. 78-84.
[15] N. G. Richards, S. M. Schuster, Adv. Enzymol. Mol. Biol. 1998, 72, 145-198.
[16] J. G. Romagni, S. O. Duke, F. E. Dayan, Plant Physiol., 2000, 123, 725-732.
[17] C. I. Nimbal, C. N. Yerkes, L. A. Weston, S. C. Weller, Pestic. Biochem. Physiol. 1996, 54, 73-83.
[18] F. A. Einhellig, J. A. Rasmussen, A. M. Hejl, I. F. Souza, J. Chem. Ecol. 1993, 19, 369-375.
[19] V. M. Gonzalez, J. Kazimir, C. I. Nimbal, L. A. Weston, G. M. Cheniae, J. Agric. Food Chem. 1997, 45, 1415-1421.
[20] G. Meazza, B. E. Scheffler, M. R. Tellez, A. M. Rimando, N. P. D. Nanayakkara, I. A. Khan, E. A. Abourashed, J. G. Romagni, S. O. Duke, F. E. Dayan, Phytochemistry, 2002, 59, 281-288.
[21] M. D. Devine, S. O. Duke, C. Fedtke, Physiology of Herbicide Action. Prentice Hall, Englewood Cliffs, NJ, 1993, pp. 395-424.
[22] J. A. Elix, in Lichen Biology (Ed.: T.H. Nash III) Cambridge University Press, New York, 1996, pp. 154-180.
[23] K. E. Pallett, J. P. Little, M. Sheekey, P. Veerasekaran, Pestic. Biochem. Physiol. 1998, 62, 113-124.
[24] D. L. Lee, C. G. Knudsen, W. J. Michaely, H.-L. Chin, N. H. Nguyen, C. G. Carter, T. H. Cromartie, B. H. Lake, J. M. Shribbs, T. Fraser, Pestic. Sci. 1998, 54, 377-384.
[25] J. G. Romagni, G. Meazza, N. P. D. Nanayakkara, F. E. Dayan, FEBS Lett. 2000, 480, 301-305.
[26] P. M. Harrington, B .K. Singh, I. T. Szamosi, J. H. Birk, J. Agric. Food Chem. 1995, 43, 804-808.
[27] F. E. Dayan, J. G. Romagni, S. O. Duke, in Handbook of Pesticide Toxicology, Vol. 2, 2nd Edition, (Ed.: R.I. Krieger) Academic Press, San Diego, 2001, pp. 1529-1541.
[28] H. Wu, J. Pratley, D. Lemerle, T. Haig, M. An, Bot. Rev. 2001, 67, 403-415.
[29] S. O. Duke, B .E. Scheffler, F. E. Dayan, E. Ota, Weed Technol. 2001, 15, 826-834.
[30] M. V. Duke, R.N. Paul, H. K. ElSohly, G. Sturtz, S. O. Duke, Int. J. Plant Sci. 1994, 155, 365-373.
[31] B. M. Lange, M. R. Wildung, E. J. Stauber, C. Sanchez, D. Pouchnik, R. Croteau, Proc. Natl. Acad. Sci. USA 2000, 97, 2934-2939.
[32] M. Hiratsuka, Y. Kishikawa, K. Narahara, T. Inoue, S .I. Hamdy, Y. Agatsuma, Y. Tomioka, M. Mizugaki, Anal. Biochem. 2001, 289, 300-303.
[33] F. A. Einhellig, Amer. Chem. Soc. Symp. Ser. 1995, 582, 1-24.

13 Weed Resistance to Herbicides: Lessons Learned and Future Directions

Malcolm D. Devine

Bayer CropScience N.V., Jozef Plateaustraat 22, B-9000 Gent, Belgium, malcolm.devine@bayercropscience.com

1 Introduction

Herbicide resistance in weeds has been the subject of a huge research effort since the first confirmed occurrence in crops in the late 1960s. Early papers discussed the paucity of cases of herbicide resistance compared to insecticide and fungicide resistance, and often concluded that the relatively slow generation time (one season or year for annual weeds, compared to multiple generations/year for some insects and pathogens), low selection pressure, and other factors would not lead to resistance development on the same scale as with other pesticides. Time has shown otherwise. According to recent figures [1], there are now over 250 distinct cases of herbicide resistance in weeds, covering most regions of the globe. Indeed, in almost all cropping systems in which herbicides are used intensively, herbicide resistance has developed in one or more weed species. A summary of herbicide resistance in weeds broken down by herbicide class/mode of action is shown in Table 1. Taking another perspective, for only a very small group of herbicides families (e.g. HPPD inhibitors, glufosinate, and a few others) is there no evidence of resistance development.

Although there were sporadic reports of weed resistance to herbicides earlier than 1970, the evolution of triazine resistance in many parts of North America and Europe in the 1970s marked the beginning of herbicide resistance as the subject of widespread scientific investigation. Since then, the scientific community has developed a much better understanding of the evolution, genetics, biochemistry and agronomy of herbicide resistance. Indeed, the determination of the molecular basis of triazine resistance marked a milestone in the discipline and provided the first detailed account of herbicide-target interactions. On the agronomic front, various management strategies to deal with herbicide-resistant weeds have been proposed, both proactive and retroactive. In addition, a variety of predictions has been put forward concerning likely future developments, in part based on some of the early observations on resistance.

This paper reviews the major lessons learned concerning herbicide resistance in weeds, the current state of knowledge, and presents some possibile future scenarios for consideration.

Table 1. Summary of confirmed cases of herbicide resistance in weeds, arranged by herbicide class and target site. From [1].

Herbicide class	Target site	Number of confirmed resistant biotypes
Sulfonylureas, imidazolinones, triazolopyrimidine sulfonamides	Acetolactate synthase	73
s-Triazines	Photosystem II electron transport	63
Aryloxyphenoxypropanoates, cyclohexanediones	Acetyl CoA carboxylase	28
Bipyridiliums	Photosystem I electron transport	22
Synthetic auxins	Auxin-binding protein(s)	21
Substituted ureas, amides	Photosystem II electron transport	20
Dinitroanilines	Microtubule assembly	10
Thiocarbamates	Fatty acid biosynthesis	6
Triazoles, isoxazolidiones	Caroteinoid biosynthesis	4
Glyphosate	*enol*-Pyruvylshikimate phosphate synthase	4
Chloroacetamides, others	Unknown	2
Nitriles	Photosystem II electron transport	1
Nitro-diphenylethers, others	Protoporphyrinogen oxidase	1
Diflufenican, others	Phytoene desaturase	1
Carbamates	Microtubule assembly	1
Organoarsenicals	Unknown	1
Flamprop methyl	Unknown	1
Pyrazoliums	Unknown	1
TOTAL		**260**

2 Lessons Learned

Much has been learned over the past 30 years concerning weed resistance to herbicides. Some of the events that have occurred have been unexpected or have gone against conventional wisdom; it is perhaps from these that we learn the most.

The evolution of resistance is not necessarily a slow process, as was first believed. Experience has shown, for example with sulfonylurea herbicides, that resistance can develop very quickly, in some cases fully evident after only 3 years of use of a particular herbicide family. That the genetic composition of a weed population can change so quickly is a sobering thought.

One early assumption was that herbicide resistance was linked to deleterious pleiotropic effects. The well-characterized serine$_{264}$ to glycine Q_B mutation conferring *s*-triazine resistance alters photosynthetic electron transport and reduces the ecological fitness of plants carrying this mutation (e.g. [2]). Reduced fitness of resistant biotypes provided a convenient explanation for the slow development of resistance, and for some time there was a general belief that weeds resistant to other chemical classes would be at a similar disadvantage. It now appears that many (if not most) mutations for herbicide resistance do not confer a fitness penalty on the resistant individuals [3, 4], although other pleiotropic effects may be present.

Surprises or unpredicted events can happen. The early dogma, also based largely on our experience of triazine resistance, suggested that soil residual activity of herbicides was an important, if not essential, contributor to the selection of resistant biotypes. The subsequent development of multiple instances of paraquat resistance, followed by resistance to acetyl CoA carboxylase (ACCase) inhibitors, demonstrated that soil activity was not necessary for resistance to evolve.

Similarly, a variety of arguments was put forward as to why resistance to glyphosate was very unlikely to occur. Apparently the weeds were not listening, as glyphosate resistance has now developed in at least four weed species [1]. The process has been slow and the frequency very low, but it took resistant weeds in farmers' fields to convince us it could happen.

Despite the increased awareness of herbicide resistance and the factors that cause it to evolve, it continues to increase. The research community has a much more complete understanding of the factors influencing resistance evolution, farmers are aware of the consequences of repeated use of herbicides in the same families or with the same mechanism of action, yet we continue to see new cases of resistance every year. Indeed, the number of confirmed cases of herbicide resistance in weeds has been increasing linearly since the early 1980s and shows no sign of decreasing. This is not to discount the valuable effort of extension and advisory personnel, who have developed various programs to alert farmers to the consequences and management of resistance. However, awareness of the *possibility* of resistance developing is often not enough to alter the short-term decisions that farmers have to make on a yearly basis [5]. Unfortunately, the difficulty in screening for resistance on a large scale means that the farmer usually becomes aware of the problem after it has developed.

The reason for the continuing increase in resistance is that despite the development of resistance to most classes of herbicides, in many weed species and in many cropping systems, herbicides remain an important component of weed control programs for many farmers. Indeed, some herbicides to which resistance has developed are still used in fields containing the resistant weeds, as a sequential or tank-mix partner to control susceptible weeds. For example, despite the presence of triazine-resistant weeds throughout much of the U.S. corn-growing area, 2001 data indicate that atrazine was used on 75% of the total treated area in the 19 States surveyed [6]. Much of this area likely contained triazine-resistant weeds. Presumably, atrazine was used to control susceptible members of the relevant species, in addition to other susceptible weeds present; a second herbicide may have been used specifically to control the atrazine-resistant weeds and other species not susceptible to atrazine. In other words, triazine resistance did not signal the end for atrazine, and the same applies to other herbicide classes to which resistance has developed. Although specific herbicide use practices may change after resistance develops, no individual herbicide or family of herbicides has been withdrawn solely as a result of weed resistance.

3 Strategies to Deal with Herbicide-Resistant Weeds

One might imagine the extensive development of herbicide resistance in weeds would have led to the development of new tools or agronomic practices to deal with the problem. Unfortunately, this is not the case. A variety of new approaches to weed control can be imagined, but in only a few cases has this translated into practical solutions that can be used to manage the problem in farmers' fields. Some strategies proposed many years ago remain theoretical, with no clear examples to demonstrate practical utility.

3.1 Herbicide Rotation

Perhaps the most obvious strategy to delay or prevent the build up of resistance is to reduce selection pressure by using particular herbicides or groups of herbicides less often. In particular, the evolution of target site-based resistance can be slowed using this approach. Both modeling [7] and practical experience [8] indicate that greater frequency of use of herbicides with the same mechanism of action promotes the development of resistance in weed populations. Therefore, using a given mode-of-action group sparingly can result in at least a delay in the development of resistance.

In practice, herbicide selection is limited by the number of different products available for use in a particular crop, the weed spectrum, the possible rotational consequences of using residual products, and the treatment cost. In some crop rotations choosing a herbicide from a different mode-of-action group each year may be a viable option, whereas in others the choice of registered products may be limited, and the farmer may have to rely on similar products more frequently. Again, short-term consideration of the important weed species present, their likely economic effect, and cost of control are likely to drive the decision-making process.

The use of herbicide rotation to avoid resistance will not be successful in cases where resistance is conferred by non-specific detoxification mechanisms that act on herbicides with different modes of action. In this case, selection for weeds resistant to members of two or more mode-of-action groups can and does occur [9]. Therefore, alternating or rotating amongst herbicides from different mode-of-action groups does necessarily delay resistance development. Clearly, there is no simple herbicide rotation solution to resistance avoidance. The tremendous genetic diversity in some seed populations allows the evolution of resistance, with the resistance mechanism simply reflecting the nature of the selection pressure that was applied.

3.2 Manipulation of Herbicide Use Rates

Two conflicting approaches have been proposed concerning herbicide use rates. The first is a high-dose strategy, aimed at ensuring that biotypes carrying minor genes conferring a low level of resistance do not survive, thus preventing the accumulation of such genes in the population. This strategy was borrowed, in part, from the insecticide and fungicide fields, where evidence showed that such minor gene accumulation did occur and contributed to the evolution of resistant populations. In contrast, in the great majority of cases, herbicide resistance is conferred by single major genes conferring a relatively high level of resistance, so this strategy does not apply. There are some cases of multi-gene herbicide resistance in weeds in which such an approach may be valid, but these are so few as to make the approach of limited value.

The second strategy is to use low herbicide doses that will result in reduced kill rates and lower selection pressure for resistant weeds. This - in theory - will allow more susceptible plants to survive, thereby increasing competition and decreasing the rate at which resistance accumulates in the population. The trick in adopting such an approach is to choose a dose at which mortality is reduced slightly but still ensures that weed density is reduced below the economic threshold. However, it is notoriously difficult to choose such a dose accurately and farmers, given the choice, will typically use a sufficiently high dose to ensure good control of the major weeds in the field. If s/he judges wrongly, one outcome can be a rapid (and economically important) increase in populations of susceptible weeds in the field [10]. In addition, a recent model suggests that altering herbicide effectiveness (kill rate) does not dramatically change the time for resistance evolution [7]; decreasing the frequency of use of particular herbicide classes was the only practice that substantially lengthened the time for resistance to develop.

3.3 Appropriate Tank Mixes of Herbicides

One elegant approach to reducing the rate of resistance evolution could be to apply herbicides as mixtures of active partners which cannot be rendered inactive by the same resistance mechanism. This is analogous to the use of drug combinations to ensure broad-sectrum control of susceptible and potentially resistant pathogen strains. In such a scenario a plant carrying a mutation conferring resistance to one partner would be controlled by the other, and the resistace mutation would not build up in the population. In practice, however, there are very few examples of "ideal" mixing partners. ALS inhibitors and auxin-type herbicides could be suitable partners in such a strategy for control of broadleaf weeds. However, individual members of the two groups of herbicides do not have the same activity against all target weeds, and individuals resistant to the ALS inhibitor may not be controlled by the auxin-type herbicide, or *vice versa*. In addition, cost often precludes the addition of a second herbicide with essentially the same weed control spectrum. Using 50% of the normal dose of each herbicide would only be effective if the activity of the two components were equal and additive, which is unlikely to be the case with many weed species. If individuals resistant to one component were not controlled by the half dose of the second component, their population would increase.

3.4 Herbicides with New Modes of Action

An obvious strategy would be to develop herbicides with new modes of action, to which the existing weeds are not resistant. However, the introduction of new modes of action has been slow, and only one new target site, hydroxyphenylpyruvate dioxygenase, has been added to the list for registered herbicides in recent years. Increased research, development and regulatory costs associated with finding the highly effective, selective, safe and environmentally benign products required in today's marketplace have not helped the introduction of new herbicides. In addition, despite the potential of many lethal herbicide target sites in plants, the known herbicides only target around 20 of these sites; finding potent *in vivo* inhibitors for the other sites remains a challenge.

Ironically, the introduction of herbicide-resistant crops, in particular to glufosinate or glyphosate, now allows the use of a "new" herbicide in terms of selective in-crop weed control. Glufosinate or glyphosate can be used to control weeds resistant to other commonly used selective herbicides. For example, weeds resistant to ACCase inhibitors and/or dinitroanilines can be controlled in glufosinate- or glyphosate-resistant oilseed rape (*Brassica napus*) using these herbicides. In growing such crops, the question of control of volunteer crop plants in subsequent years must be included in the overall weed management plan.

3.5 Other Possibilities

3.5.1 Additives to Overcome Resistance

In certain situations it is possible to overcome herbicide metabolism-based resistance by adding an ingredient that will block detoxification of the herbicide in the resistant weed. One example is with propanil-resistant *Echinochloa colona* in rice in Latin America. The addition of piperophos, an organophosphate insecticide that inhibits the aryl acylamidase activity that confers resistance on the weed biotype [11]. This combination, based on an understanding of the resistance mechanism, has been approved for use on resistant populations in the field. Several other examples exist which draw attention to the possibility of using inhibitors to overcome resistance, in particular where resistance is conferred by elevated activity of specific cytochrome P450 monooxygenases [9]. The ideal outcome is that the rate of herbicide metabolism is reduced in the resistant weed biotype but not in other weeds or in the crop, if selectivity is based on rapid detoxification by the crop. In practice this is difficult to achieve, and the extensive screening efforts for appropriate inhibitors and subsequent re-registration of the new mixture are seldom justified.

3.5.2 New Herbicides that Target both Resistant and Susceptible Weeds

One possible approach to developing new herbicides is to screen candidate molecules against both resistant and susceptible weeds, or the resistant and susceptible target enzymes from these weeds. This approach was used to identify novel ACCase inhibitors that inhibited the target enzyme from a collection of resistant and susceptible biotypes [12]. The challenge here is to find suitable inhibitors at the enzyme level that retain this activity at the whole-plant level. In addition, this approach is only feasible when resistance is conferred by target site mutations; if resistance were conferred also by metabolic detoxification an additional set of screens against detoxifying enzymes would have to be included.

3.5.3 *A Priori* Resistance Management

It is tempting to believe that a better understanding of the potential for resistance evolution could be useful in managing the use of a new herbicide after its introduction. The ability to anticipate or predict the development of herbicide resistance could lead to use patterns through which resistance could be avoided, or at least delayed. This requires an understanding of possible resistance mechanisms, the frequency of genes conferring resistance, and other aspects of weed biology.

Many different bioassays and biochemical or genetic tests have been developed to identify resistant weeds. However, these are normally conducted after the suspected development of resistance, not in a proactive or preventive manner. The potential for evolution of resistance to a new herbicide can be examined in several ways: wild-type populations can be screened for resistant individuals, model plant populations can be mutagenized and screened for resistance, resistant cells can be selected in culture, with or without prior exposure to the herbicide, or biochemical or genetic assays can be used to identify known resistance mechanisms. However, more complex or obscure resistance mechanisms may exist, and certain mechanisms may only be expressed in whole plants, not in cell cultures. More recent techniques focused on rapid genetic evolution can also provide a clue to the relative ease with which resistance can be generated, but still require a large investment. However, as in many predictive studies, it is often difficult to relate the results of such experiments to resistance evolution in the field.

The frequency of resistance genes in an unselected population is generally unknown, and there is no efficient way of screening large populations (at least 1 000 000 plants, perhaps much higher) to determine such gene frequencies. Other important information on weed biology, such as seed longevity in the soil seedbank and gene flow behaviour, is often not available. The relative importance of these parameters is unclear, and some may have little bearing on the rate of resistance evolution [7]. In the end, a farmer might ask "Why should I do anything different?" if s/he is not certain resistance will develop in a particular field. Even if s/he is certain that it will develop, it is difficult to imagine that s/he will opt for a more expensive or a less effective weed control program only to delay resistance development in a particular field.

3.5.4 Restrictions on Herbicide Use

Herbicide resistance evolves under selection pressure, which is largely related to the frequency of herbicide use [8]. One strategy to avoid resistance, therefore, is less frequent herbicide use, either mandated or voluntary. Mandated restrictions on herbicide use would be unpopular in most places, and would be seen by many farmers as a move to restrict choice and potential profitability. Voluntary restrictions are perhaps more viable, but in the end depend on the nature of the (potential) problem and the short-term agronomic and economic consequences of not using the herbicide in question or of using reduced rates [13]. This approach may have merit in situations where alternative herbicides are available that will control the problem weeds without exacerbating the potential resistance problem. However, in many cropping situations few or no alternative mode of action groups exist to control the problem species, and the default is often a herbicide that will further select for resistance. Again, the short-term nature of the decision-making process often results in the farmer making the pragmatic herbicide choice for that season, at the possible expense of long-term consequences [5].

4 Future Directions

Given our extensive understanding of resistance, what is likely to change in the foreseeable future? Several possible futures present themselves. These include:

- Farmers continue to manage resistance retroactively, responding to the presence of resistant weeds by using different herbicides or alternative control practices, perhaps accompanied by changes in cropping practices.
- Farmers realize the long-term consequences of applying similar herbicides frequently and make proactive decisions to change herbicide use and other weed control practices to delay resistance development.
- Our understanding of the biology and biochemistry of resistance allows the development of new concepts in herbicide discovery and development that reduce or eliminate resistance concerns.

The realistic view is that the first option will be most prevalent, the second option will be practiced by a minority of farmers (but may increase?), and the third will remain a very long-term hope. Regardless of the practicalities of these different scenarios, it must be recognised that weed resistance to herbicides has become firmly established in many

cropping systems and is here to stay. In the short term, the need to produce food on a limited land base requires the use of intensive agricultural practices, including judicious herbicide use. Also in the short term, the frequency and area affected by herbicide-resistant weeds will continue to increase. The scientific community will continue to use this as a source of interesting biological and genetic material for investigation. Our hope must be that in time this leads to sustainable management practices that allow weed control measures to be implemented with a balance between short- and long-term goals, meeting agronomic, environmental and economic needs.
In parallel, efforts should continue to develop competitive crops that can withstand intense weed competition. This is an on-going goal of many crop breeding programs, and new approaches to genetic enhancement of crops may allow further advances. More vigorous crop varieties, grown with improved agronomic practices (more precise seed and fertilzer placement, etc.), will be an important component of weed management in the future.

5 References

[1] I. M. Heap, International Survey of Herbicide Resistant Weeds, can be found under http://www.weedscience.org/in.asp, 2002.
[2] J. J. Hart, A. Stemler, Plant Physiol. 1990, 94, 1301-1307.
[3] S. I. Warwick, L. D. Black, Phytoprotection 1994, 75, 37-49.
[4] J. S. Holt, T. C. Thill in Herbicide Resistance in Plants: Biology and Biochemistry (Eds.: S. B. Powles, J. A. M. Holtum), Lewis Publishers, Boca Raton, FL, USA 1993, 299-316.
[5] D. J. Pannell, D. Zilberman in Herbicide Resistance and World Grains (Eds.: S. B. Powles, D. L. Shaner), CRC Press, Boca Raton, FL, USA, 2000, 251-277.
[6] Anonymous, AGROW, PJB Publications, UK, No. 401, May 31, 2002.
[7] G. Cavan, J. Cussans, S. Moss, Weed Sci. 2001, 49, 236-240.
[8] A. Légère, H. J. Beckie, F. C. Stevenson, A. G. Thomas, Weed Technol. 2000, 14, 366-376.
[9] M. D. Devine, C. Preston in Herbicides and their Mechanisms of Action (Eds.: A. H Cobb, R. C. Kirkwood), Sheffield Academic Press, Sheffield, UK, 2000, 72-104.
[10] I. N. Morrison, L. F. Friesen in Proc. 2nd Int. Weed Control Congr., Copenhagen, Denmark, 1996, 377-385.
[11] B. E. Valverde in Proc. 2nd Int. Weed Control Congr., Copenhagen, Denmark, 1996, 415-420.
[12] A. Shukla, C. Nycholat, M. V. Subramanian, R. J. Anderson, M. D. Devine, unpublished data.
[13] J. E. Jensen, P. H. Petersen in Proc. Brighton Crop Prot. Conf. – Weeds, 2001, 449-454.

Insect Control

Chemistry, Biochemistry, Molecular Biology

14 Voltage-Gated Sodium Channels as Insecticide Targets

Ke Dong

Department of Entomology and Neuroscience Program, Michigan State University, East Lansing, MI 48824, U.S.A., dongk@msu.edu

1 Introduction

Voltage-gated sodium channels are large transmembrane proteins present in excitable cells. Sodium channels open and inactivate in response to a depolarization of the resting membrane potential. When these channels open, an influx of Na^+ occurs, depolarizing the membrane potential, and generating the upstroke phase of action potential. Inactivation of sodium channels stops Na^+ in flux, preventing a permanent depolarization of the resting membrane potential. Recovery from inactivation allows the cell membrane to regain its resting excitable properties. Therefore, the voltage-gated sodium channels are essential for action potential initiation and propagation in the nervous system and other excitable cells.

Sodium channels are targets for the actions of many neurotoxins and drugs. These neurotoxins affect various aspects of sodium channel functional properties and can be classified according to their unique binding sites on the sodium channel [1,2]. DDT and pyrethroid insecticides are among the earliest pharmacological agents found to act on sodium channels [3]. These compounds cause repetitive firing and membrane depolarization by slowing the gating kinetics of activation and inactivation of sodium channels [3]. Sodium channel modification by pyrethroids is evident in the induction of a large, slow decaying tail current associated with membrane repolarization and in the delayed and prolonged opening of single sodium channels [3].

Pyrethroids are widely used to control many agriculturally and medically important insect pests. Due to intensive use of pyrethroids in pest control, many pest populations have developed resistance to these compounds. One major mechanism of pyrethroid resistance, conferred by the knock down resistance gene (*kdr*), is reduced target site (sodium channel) sensitivity to DDT and pyrethroids. Studies on the molecular basis of *kdr* and *kdr*-type resistance in various insects are enhancing our understanding of the structure and function of insect sodium channels and the molecular interaction between insect sodium channels and pyrethroids. In this chapter, I will review recent advances in the study of insect sodium channels, with emphasis on the molecular and functional diversity of insect sodium channels and the involvement of sodium channel mutations in insect resistance to pyrethroid insecticides.

2 At least Nine Voltage-Gated Sodium Channel Genes in Mammals versus One in Insects

Voltage-gated sodium channels are integral transmembrane proteins responsible for the rapidly rising phase of action potentials in most excitable cells. The sodium channel α-subunit contains four repeated homologous domains (I-IV), each having six membrane spanning segments (S1-6) connected by intracellular or extracellular loops of amino acid sequences (Figure 1). In the past two decades, ten different mammalian sodium channel α–subunit genes were isolated and a new nomenclature, $Na_v1.1$ to $Na_v1.9$, and Na_x, was proposed [4]. Many of the mammalian sodium channel genes have specific developmental, tissue, or cellular expression patterns [5]. Different sodium channel isoforms also exhibited varied gating properties and pharmacological profiles [5]. It appears that the functional diversity in mammals is achieved primarily through expression of distinct sodium channel genes.

In *Drosophila melanogaster*, however, *para* is the only gene that has been demonstrated to encode a functional sodium channel [6,7]. *para* was identified by Barry Ganetzky's laboratory from a *Drosophila* mutant that exhibited a temperature-sensitive paralytic phenotype [8]. The overall structure of and the amino acid sequence of the Para sodium channel share a high similarity with those of mammalian sodium channel α–subunits. *In situ* hybridization revealed the expression of *para* transcripts in both the central and peripheral nervous systems [9]. Another putative sodium channel gene, *DSC1*, was identified from *D. melanogaster* genomic DNA library using a cDNA probe encoding the eel sodium channel [10]; however, the functional identity of the DSC1 channel was not determined. An ortholog of DSC1, BSC1, isolated from *Blattella germanica* [11], shares a high sequence similarity with *para* and *para*-homologous sodium channel proteins. Functional expression of BSC1 in *Xenopus* oocytes, however, revealed that BSC1 channels are Ca^{2+}-permeable cation channels with gating properties distinct from those of classic sodium or calcium channels (Chung I., Liu Z., Zhou W., Goldin A.L., Dong K., unpublished results.) Whether the BSC1 channel is another target-site of various sodium channel neurotoxins remains to be investigated.

3 Cloning and Functional Expression of Insect Sodium Channels

Because of the intimate involvement of sodium channels in pyrethroid insecticide resistance (see below), partial cDNA clones of *para*-orthologs were isolated from other insects by polymerase chain reaction (PCR) using degenerate primers based on the *Drosophila para* and mammalian sodium channel sequences. The entire coding sequence (*ca.* 6 kb), however, has been determined only for the *para*-orthologs from the housefly and the German cockroach [12-14]. Furthermore, cloning of functional full length cDNAs proved to be difficult for two reasons. First, the *para* transcripts are large (>12 kb in size). The open reading frames are at least 6 kb in length. Second, for largely unknown reasons, the full length *para* cDNA inserts, which contain many repeats of similar sequences, were often rearranged when cloned in common laboratory *E. coli* strains. The first full length cDNA clone was constructed by ligating four overlapping PCR products of the *para* cDNA [7]. Co-expression of this *para* clone with *tipE*, a *Drosophila* gene encoding a 50-kDa transmembrane protein, produced sodium currents in *Xenopus* oocytes [6]. TipE likely serves as an auxiliary subunit, functionally similar to β-subunits of mammalian sodium channels. Subsequently, a full length house fly sodium channel clone (Vssc1) was constructed using two overlapping partial cDNA clones and functionally expressed in *Xenopus* oocytes [15]. Three full length German cockroach sodium channel clones (*BgNa$_v$* hereinafter) were isolated by direct RT (reverse transcription)-PCR using RNA from heads and thoraces [16,17]. This one step isolation protocol made it possible to recover naturally occurring sodium channel gene variants without in vitro recombination. To this end, we have been able to isolate a total of 69 full-length *BgNa$_v$* clones (Nomura Y., Liu Z., Dong K., unpublished results.) Three of these clones have been expressed in *Xenopus* oocytes and functionally and pharmacologically characterized [17, also see below].

4 Distinct Insect Sodium Channel Isoforms Generated by Alternative Splicing

Structural analysis of the *D. melanogaster para* gene revealed extensive alternative splicing of the *para* transcript [18]. A total of nine alternatively spliced exons have been identified in a region encoding domain II and part of the linkers connecting domains I and II, and domains II and III (Figure 1). Interestingly, these alternative splice sites are conserved in *D. virilis* [19] and also in housefly [20]. The actual number of *para* splice sites may well be much greater than nine, because the region examined in detail in these studies represents only about 30% of the complete *para* open reading frame. Most splicing sites identified in the *Drosophila para* genes were conserved also in the cockroach *BgNa$_{vv}$* gene (Figure 1). Alternative exons were also identified in *Heliothis virescens* [21], and *Haematobia irritans* [22].

In addition to the sites homologous to those in the *Drosophila para* gene, one novel splicing site was identified in *BgNa$_v$* producing exons G1, G2 and G3, which are mutually exclusive and encode portions of IIIS3-4 (in Figure 1) [17]. The identification of exons G1/G2/G3 is especially significant because the exon/intron arrangement in this region bears a striking resemblance with that of the genomic region coding for IIIS3-4 of the mouse sodium channel protein Na$_v$1.6 (Scn8a) and its human and fish counterparts [23]. Like the mammalian counterparts, exon G3 possesses a stop codon that would

generate a truncated protein having only the first two domains. Therefore, splicing in this region appears to be of a very ancient origin and is preserved in organisms ranging from insects to humans. Alternative exons (named k/l) homologous to exons G1/G2 were detected in the housefly Vssc1 sodium channel gene and also in *para* [20]. The functionally expressed *para* cDNA clone apparently contains exon G1 [7,8], whereas the functionally expressed house fly Vssc1 sodium channel contains exon G2 [12].
Two *BgNa*$_v$ splice variants, one containing exon G1 and the other containing exon G2 (BgNa$_v$1 and BgNa$_v$2, previously called KD1 and KD2), exhibited distinct gating properties [17]. The BgNa$_v$1 channel activates at a more positive membrane potential and displays a hyperpolarizing shift in the voltage-dependence of steady state inactivation compared to the BgNa$_v$2 channel [17]. Interestingly, the gating differences between the two cockroach sodium channel splice variants are strikingly reminiscent of those between the mammalian sodium channel isoforms produced by different genes. For example, the Nav1.6 channel activates at a more positive membrane potential and displays a hyperpolarizing shift in the voltage-dependence of steady-state inactivation compared to the Na$_v$1.1 and Na$_v$1.2 channels [24]. Therefore, alternative splicing of a single transcript appears to be a major mechanism by which insects generate sodium channel functional diversity, whereas a distinct mechanism involving selective expression of multiple sodium channel genes has evolved among vertebrates. Differences in gating properties have been observed among oocyte expressed sodium channels from *Drosophila*, housefly, and cockroach. In light of our findings, it is possible that some of the differences result from alternative exon usage in different cDNA clones.

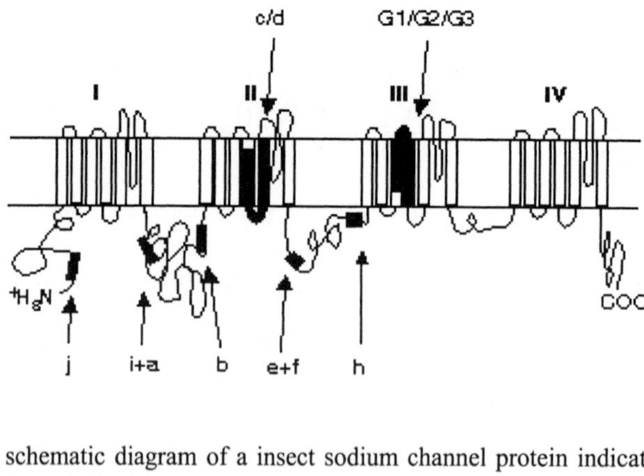

Figure 1. A schematic diagram of a insect sodium channel protein indicating four homologous domains (I-IV), each having six transmembrane segments. Locations of segments encoded by alternative exons are indicated with solid blocks. Optional exons a, b, e, f, h, i, and j are present in *D. melanogaster para* [18], the housefly sodium channel gene *Vssc1* [20] and *BgNa*$_{vv}$ (Nomura Y., Liu Z., Dong K., unpublished). Mutually exclusive exons c/d are identified in *para* and *Vssc1*. Mutually exclusive exons G1/G2/G3 were identified in *BgNa*$_v$ [17]. Exons corresponding to G1 and G2 were also found in *para* and *Vssc1* [20].

5 Differential Sensitivity of Sodium Channel Isoforms to Pyrethroid Insecticides

Differential sensitivities to pyrethroid insecticides were found for mammalian sodium channel isoforms encoded by distinct genes. For example, tetrodotoxin (TTX) sensitive sodium channels in the rat dorsal root ganglion neurons are less sensitive to pyrethroids than TTX resistant sodium channels in the same neurons [25-28]. Rat Nav1.2 and Na1.4 channels are very insensitive to pyrethroids [7,29-31]. However, rat Nav1.8 sodium channel is highly sensitive to pyrethroids [32]. It is not known, however, how each sodium channel isoform contributes to the overall mammalian sensitivity to pyrethroids.
The existence of pharmacologically distinct sodium channel isoforms in insects has been revealed recently. It was found that the BgNa$_v$2 channel (containing exon G2) was 100-fold less sensitive to deltamethrin than BgNa$_v$1 (containing exon G1) [17]. The BgNa$_v$2 variant appears to be a minor isoform in a pyrethroid susceptible cockroach strain CSMA. Only three (including BgNa$_v$2) of the 69 full length BgNa$_v$ cDNA clones possess the G2 exon, whereas 64 clones have the deltamethrin-sensitive G1 exon (Nomura Y., Liu Z., Dong K., unpublished results.) The frequency of this pyrethroid resistant isoform remains low in a pyrethroid resistant German cockroach strain Ectiban-R (Dong unpublished result). Therefore, the BgNa$_v$2 channel is not associated with the increased pyrethroid resistance in the *kdr*-type strain Ectiban-R (see below for further discussion). The BgNa$_v$2 variant is also expressed mainly gut and ovary, whereas BgNa$_v$1 is exclusively expressed in the nerve cord and leg [17]. Therefore, the distinct gating and pharmacological properties of the BgNa2 variant likely reflect adaptations of the BgNa$_v$2 sodium channel to fulfill its unique function in specific tissues or cell types, such as gut and ovary. These findings also suggested caution in the interpretation of electrophysiological and pharmacological results when different tissues, cells, or sodium channel clones are used.

6 Naturally Occurring Point Mutations in the Sodium Channel Gene Responsible for Knockdown Resistance to Pyrethroids.

Several point mutations in the sodium channel gene are associated with *kdr* and *kdr*-type resistance to pyrethroids in agriculturally or medically important arthropod species. The most common mutation is the leucine (L) to phenylalanine (F) or histidine (H) or serine (S) substitution in domain II segment 6 (IIS6). Mutation at this position was present in *kdr* houseflies [13]; German cockroach *B. germanica* [33,14,34]; tobacco budworm *Heliothis virescens* [35]; diamondback moth, *Plutella xylostella* [36,37]; peach-potato aphid, *Myzus persicae* [35,38]; mosquito, *Anopheles gambiae* [39,40]; *Cluex pipiens* [41]; horn fly, *Haematobia irritans* [22]; and Colorado potato beetle, *Leptinotarsa decemlineata* [42].

Additional sodium channel mutations were found to co-exist with the L to F mutation in IIS6 in highly resistant strains. For example, in super-*kdr* housefly strains, which exhibit greater levels of resistance to DDT and pyrethroids than *kdr*, a second mutation [methionine (M) to threonine (T) in the linker connecting IIS4 and IIS5] was found in addition to the L to F mutation [13]. The M to T mutation together with the L to F mutation was also found in a horn fly strain that exhibited high-level resistance to pyrethroids [22]. In a highly resistant diamondback moth strain, a T to I (isoleucine) mutation was observed along with the L to F mutation [37]. The same mutation (T to I) was also found in head louse populations [43]. Finally, in the German cockroach, four unique point mutations, aspartic acid to glycine (D58G) at the N-terminus, glutamic acid to lysine (E434K) and cysteine to arginine (C764R) in the linker between IS6 and IIS1, and proline to leucine (P1888L) at the C-terminus, together with the L to F mutation in IIS6, were associated with high levels of resistance [44].

Some *kdr*-type strains lack the L to F (H or S) mutation in IIS6. Instead, a valine (V) to methionine (M) change in IS6 was found in several pyrethroid-resistant *H. virescens* populations [35]. An F to I mutation in IIIS6 was identified in several pyrethroid resistant southern cattle tick (*Boophilus microplus*) strains [45]. Two mutations, D (aspartic acid) to V and E (glutamic acid) to G (glycine), in the linker region connecting domains III and IV, were associated with pyrethroid resistance in the tobacco budworm (*H. virescens*) and the cotton bollworm (*Helicoverpa armigera*) [46].

To confirm the involvement of *kdr* associated mutations in pyrethroid resistance, a number of laboratories developed the *Xenopus* oocyte sodium channel expression system to determine the sensitivity of mutant sodium channels to pyrethroids. Several mutations have already been confirmed to reduce the sodium channel sensitivity to pyrethroids. The L to F (or H) mutation in IIS6 reduced the insect sodium channel sensitivity to deltamethrin by 5 to 10-fold [15-17,48]. The M 918T mutation in the linker region connecting IIS4-S5, in combination with the L 1014F mutation further reduced the sensitivity of the house fly sodium channel to cypermethrin [49]. Furthermore, the M918T mutation alone reduced the channel sensitivity to pyrethroids [49,50]. In contrast, two cockroach mutations, E434K and C764R, by themselves did not reduce the sensitivity of the cockroach sodium channel to deltamethrin. However, when either the E434K or C764R mutation was combined with the L993F mutation (i.e., KF or RF), the channel sensitivity was reduced by 100-fold [16]. Concomitant presence of all three mutations (KRF) reduced channel sensitivity to deltamethrin by 500-fold [16]. The two mutations (D58G and P1888L) at the N- and C-termini did not reduce the channel sensitivity to deltamethrin (Tan J., Liu Z., Dong K., unpublished result). The *H. virescens* V421M mutation in IS6 reduced the pyrethroid sensitivity when introduced into *Drosophila*, housefly and cockroach sodium channels [47,51,52]. Finally, the T929I mutation in IIS5, identified in diamondback moth and also in human head louse, reduced the Para channel sensitivity to deltamethrin [50]. It appears that these heterologous sodium channels expressed in *Xenopus* oocytes can be used to determine the effects of pyrethroid resistance associated mutations identified in other insect species for which an in vitro sodium channel expression assay is not yet available.

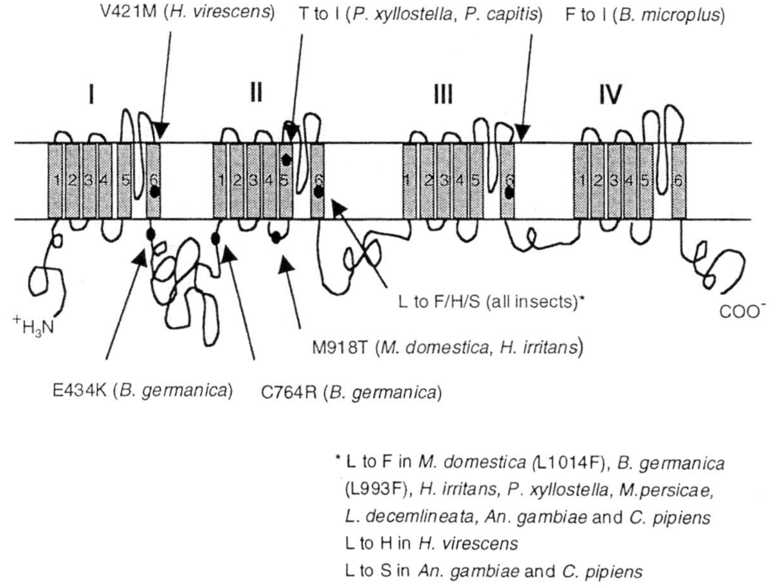

Figure 2. Sodium channel protein topology indicating the naturally occurring mutations that are confirmed to reduce insect sodium channel sensitivity to pyrethroids. The sodium channel contains four repeated homologous domains (I-IV), each having six membrane-spanning segments (S1-6). See the text for the details.

7 Evolution of *kdr* and *kdr*-Type Mutations

None of the cockroach *kdr*-associated mutations significantly affected sodium channel gating properties [16]. However, the peak current amplitude of the mutant channel carrying either E434K or C764R alone was reduced by 5-fold compared to those of the wild-type channel or the channel carrying the double or triple mutations (KF, RF and KRF) [16]. The reduction of sodium peak current by the E434K or C764R mutation could be detrimental to insects carrying these single mutations. Consistent with this prediction, we did not identify any cockroaches in natural populations carrying only the E434K or C764R single mutation, suggesting a possible fitness problem associated with the E434K or C764R single mutation. We therefore proposed a sequential selection mechanism for the evolution of high-level pyrethroid resistance in the German cockroach: an initial selection of a primary mutation at L993 (L993F) provides a platform for the subsequent selection of two secondary mutations, E434K and C764R, which by themselves cannot be selected because of possible fitness costs [16]. A similar mechanism was proposed earlier for the evolution of the M918T mutation in housefly by Soderlund and associates because the M918T mutation dramatically reduced the amplitude of peak current [49]. Different from double mutants in the German cockroach, however, a combination of the L1014F in IIS6 and M918T still reduced the amplitude of peak current [49].

The molecular basis for the evolution of distinct *kdr* mutations in different insects and arachnids remains unclear. Assuming that the pyrethroid binding site(s) (and/or the pyrethroid response domain) is composed of multiple amino acid residues, there are two ways by which different mutations can be selected in different insects and arachnids. First, the random mutation hypothesis: mutation in any pyrethroid binding site/response domain affects pyrethroid toxicity without impacting normal sodium channel functional properties. Thus, selection of different mutations in different insects and arachnids is purely random. Second, the nonrandom mutation hypothesis: mutation in any pyrethroid binding site/response domain affects pyrethroid toxicity, but some mutations also drastically alter normal sodium channel functional properties in one species, but not in another, presumably because of different sodium channel backbone sequences. That is, there may be severe fitness costs for some mutations, if placed out of their native protein context.

When the *H. virescens kdr*-type mutations V421M and L1029H were introduced into the *Drosophila* Para channel protein, they modify the voltage-dependence of activation and inactivation of Para channels. However, the modification by the L1029H mutation was less than by V421M: a 3-mV versus a 6-mV positive shift in the voltage-dependence of activation and a 2-mV versus a 5-mV positive shift in voltage-dependence of inactivation [47]. Insects carrying the V421M mutation were found to exhibit decreased excitability and sluggish behavior, suggesting a fitness cost [47]. The effect of these mutations on the native *Heliothis* Para-homologous channel has not been examined. However, the neurons isolated from the resistant *Heliothis* individuals carrying the V421M mutation also exhibited altered sodium channel gating properties [53].

We examined the effects of three heterologous *kdr* mutations, the housefly super-*kdr* mutation M918T, the *H. virescens* V421M mutation in IS6, and the *B. microplus* F to I mutation in IIIS6, on the gating properties of the $BgNa_v1$ channel. All three mutations reduced the channel sensitivity to deltamethrin. However, their effects on the gating properties varied. The M918T mutation did not alter the channel gating properties (Tan and Dong, unpublished result). The V421M mutation caused a 4-5 mV positive shift in both voltage-dependence of activation and inactivation of both the $BgNa_v1$ channel [52], and the Vssc1 housefly channel [51]. The F to I mutation caused an 8-mV positive shift in the voltage-dependence of activation (Tan J., Wang R., Liu Z., Huang Z., Chen A., Gurevitz M., Dong K., unpublished result). More strikingly, the sodium channel carrying the F to I mutation revealed an additional sustained sodium current at the end of a 20-ms depolarization, whereas the wild-type sodium channel inactivates completely (Tan J., Wang R., Liu Z., Huang Z., Chen A., Gurevitz M., Dong K., unpublished result). The modification of channel properties by the F to I mutation would have a negative impact on the physiology and behavior of the individuals carrying this mutation, which could be the reason why this mutation is not found in cockroaches in nature. Therefore, these results provided supporting evidence for both the random and nonrandom mutation hypotheses.

8 Acknowledgements

I thank Michelle Carlson and Tom Brown (GeneCTAr at www.genectar.com) for proof-reading, editing and valuable comments. The research in the author's laboratory is supported by NSF grants (IBN 9696092 and IBN 9808156), NIH (08-1GM57440A), and the Michigan State University Rackham Endowment Fund.

9 References

[1] W. A. Catterall, ISI Atlas Sci. Pharmacol. 1988, 190-195.
[2] E. Zlotkin, Ann. Rev. Entomol. 1999, 44, 429-455.
[3] T. Narahashi, J. Pharmacol. Exp. Ther. 2000, 294, 1-26.
[4] A. L. Goldin, R. L. Barchi, J. H. Caldwell, F. Hofmann, J. R. Howe, J. C. Hunter, R. G. Kallen, G. Mandel, M. H. Meisler, Y. B. Netter, M. Noda, M. M. Tamkun, S. G. Waxman, J. N. Wood , W. A. Catterall, Neuron 2000, 28, 365-368.
[5] A. L. Goldin, Ann. NY Acad. Sci. 1999, 868, 38-50.
[6] G. Feng, P. Deak, M. Chopra, L. M. Hall, Cell 1995, 82, 1001-1011.
[7] J. W. Warmke, R. A. G. Reenan, P.Wang, S. Qian, J. P. Arena, J. Wang, D. Wunderler, K. Liu, G. J. Kaczorowski, L. H. T. Van der Ploeg, B. Ganetzky, C. J. Cohen, J. Gen. Physiol. 1997, 110, 119-133.
[8] K. Loughney R. Kreber, B. Ganetzky B, Cell 1989, 58, 1143–1154.
[9] C.-S. Hong, B. Ganetzky, J. Neurosci. 1994, 14, 5160-5169.
[10] L. Salkoff, A. Butler, A. Wei, N. Scavarda, K. Giffen, C. Ifune, R. Goodman, G. Mandel, Science 1987, 237, 744-749.
[11] Z. Liu, I. Chung, K. Dong, Insect Biochem. Molec. Biol. 2001, 31, 703-713
[12] P. J. Ingles, P. M. Adams, D. C. Knipple, D. M. Soderlund, Insect Biochem. Molec. Biol. 1996, 26; 319-326.
[13] M. S. Williamson, D. Martinez-Torres, C. A. Hick, A. L. Devonshire, Mol. Gen. Genet. 1996, 252, 51-60.
[14] K. Dong, Insect Biochem. Molec. Biol. 1997, 27, 93-100.
[15] T. J. Smith, S. H. Lee, P. J. Ingles, D. C. Knipple, D. M. Soderlund, Insect Biochem. Molec. Biol. 1997, 27, 807-812.
[16] J. Tan, Z. Liu, T. D. Tsai, S. M. Valles, A. L. Goldin, K. Dong, Insect Biochem. Molec. Biol. 2002a, 32, 445-454.
[17] J. Tan, Z. Liu, Y. Nomura, A. L. Goldin, K. Dong, J. Neurosci. 2002b, 22, 5300-5309.
[18] J. R. Thackeray, B. Ganetzky, J Neurosci. 1994, 14, 2569-2578.
[19] J. R. Thackeray, B. Ganetzky, Genetics 1995, 141, 203-214.
[20] S. H. Lee, P. J. Ingles, D. C. Knipple, D. M. Soderlund, Invert. Neurosci. 2002, 4, 125-133.
[22] Y. Park, M. F. Taylor, R. Feyereisen, Insect Mol. Biol. 1999, 8, 161-70.
[22] F. D. Guerrero, R. C. Jamroz, D. Kammlah, S. E. Kunz, Insect Biochem. Molec. Biol. 1997, 27, 745-755.
[23] N. W. Plummer, M. W. McBurney, M. H. Meisler, J. Biol. Chem. 1997, 272, 24008-24015.
[24] M. R. Smith, R. D. Smith, N. W. Plummer, M. H. Meisler, A. L. Goldin, J. Neurosci. 1998, 18, 6093-6102.
[25] K. S. Ginsburg, T. Narahashi, Brain Res. 1993, 627, 239-248.
[26] H. Tatebayashi, T. Narahashi, J. Pharmacol. Exp. Ther. 1994, 270, 595-603.
[27] J. H. Song, T. Narahashi, Brain Res. 1996, 712, 258-264.

[28] I. V. Tabarean, T. Narahashi, J. Pharmacol. Exp. Ther. 1998, 284:958-965.
[29] H. Vais, M. S. Williamson, C. A. Hick, N. Eldursi, A. L. Devonshire, P. N. R. Usherwood, FEBS Lett. 1997, 413, 327-332.
[30] T. J. Smith, D. M. Soderlund, Neurotox 1998, 19, 823-832.
[31] S.-Y. Wang, M. Barile, G. K. Wang, Mol. Pharmacol. 2001, 68, 620-628.
[32] D. M. Soderlund, T. J. Smith, S. H. Lee, Neurotox. 2000, 21, 127-138.
[33] M. Miyazaki, K. Ohyama, D. Y. Dunlap, F. Matsumura, Mol. Gen. Genet. 1996, 252, 61-68.
[34] K. Dong, S. M. Valles, M. E. Scharf, B. Zeichner, G. W. Bennett, Pestic. Biochem. Physiol.1998, 60, 195-204.
[35] Y. Park, M. F. Taylor, Insect. Biochem. Molec. Biol. 1997, 27, 9-13.
[36] D. Martinez-Torres, A. L. Devonshire, M. S. Williamson, Pestic. Sci. 1997, 51, 265-270.
[37] T. H. Schuler, D. Martinez-Torres, A. J. Thompson, I. Denholm, S. P. Foster, A. L. Devonshire, I. R. Duce, M. S. Williamson, Pestic. Biochem.. Physiol. 1998, 59, 169-182.
[38] D. Martinez-Torres, S. P. Foster, L. M. Field, A. L. Devonshire, M. S. Williamson, Insect Mol. Biol. 1999a, 8, 339-46.
[39] D. Martinez-Torres, F. Chandre, M. S. Williamson, F. Darriet, J. B. Berge, A. L. Devonshire, P. Guillet, N. Pasteur, D. Pauron, Insect Mol. Biol. 1998, 7, 179-184.
[40] H. Ranson, B. Jensen, J. M. Vulule, X. Wang, J. Hemingway, F. H. Collins, Insect Molec. Biol. 2000, 9, 491-497.
[41] D. Martinez-Torres, C. Chevillon, A. Brun-Barale, J. B. Berge, N. Pasteur, D. Pauron, Pestic. Sci..1999b, 55, 1012-1020.
[42] S. H. Lee, J. B. Dunn, J. M. Clark, D. M. Soderlund, Pestic. Biochem. Physiol. 1999a, 63, 63-75.
[43] S. H. Lee, K. S. Yoon, M. S. Williamson, S. J. Goodson, M. TakanoLee, J. D. Edman, A. L. Devonshire, J. M.. Clark, Pestic. Biochem. Physiol. 2000, 66, 130-143.
[44] Z. Liu, S. M. Valles, K. Dong, Insect Biochem. Molec. Biol. 2000, 30, 991-997.
[45] H. He, A. C. Chen, R. B. Davey, G. W. Ivie, J. E. George, Biochem. Biophys. Res. Comm. 1999, 261, 558-561.
[46] D. J. Head, A. R. McCaffery, A. Callagham, Insect Molecular Biology 1998, 7, 191-196.
[47] Y. Zhao, Y. Park, M. E. Adams, Biochem. Biophys. Res. Commun. 2000, 278, 516-521.
[48] H. Vais, M. S. Williamson, S. J. Goodson, A. L. Devonshire, J. W. Warmke, P. N. R. Usherwood, C. J. Cohen, J. Gen. Physiol. 2000, 115, 305-318.
[49] S. H. Lee, T. J. Smith, D. C. Knipple, D. M. Soderlund, Insect Biochem. Molec. Biol. 1999b, 29, 185-194.
[50] H. Vais, M. S. Williamson, A. L. Devonshire, P. N. R. Usherwood, Pest Manag. Sci. 2001, 57, 877-888.
[51] S. H. Lee, D. M. Soderlund, Insect Biochem. Mol. Biol. 2001, 31, 19-29.
[52] Z. Liu, J. Tan, S. M. Valles, K. Dong, Insect Biochem. Molec. Biol. 2002, 32, 397-404.
[53] D. Lee, Y. Park, T. M. Brown, M. E. Adams, Mol. Pharmacol. 1999c, 55, 584-593.

15 Structural Studies of Insect Nuclear Receptors

Isabelle M.L. Billas and Dino Moras

Département de Génomique et de Biologie Structurales, Institut de Génétique et de Biologie Moléculaire et Cellulaire, 1, rue Laurent Fries, 67404 Illkirch (France); email: billas@igbmc.u-strasbg.fr; moras@igbmc.u-strasbg.fr

1 Introduction

Insect development, molting and metamorphosis are orchestrated by ecdysteroids, the insect steroid related hormones [1]. The main hormones involved are α-ecdysone and its biologically active metabolite, 20-hydroxyecdysone (20E) (Fig. 1). The action of ecdysteroids on insect molting and metamorphosis is modulated by esterified sequiterpenoid hormones, the juvenile hormones (JHs) [1,2,2,3,3,4]. In the absence of JH, metamorphosis occurs, while in its presence, the molt is to a next larval stage (for a review, see [5]). Pulses of ecdysone trigger a genetic regulatory cascade that coordinates the developmental pathways associated with metamorphosis [6]. The ecdysone signal is mediated through the ecdysone receptor (EcR) that belongs to the large superfamily of nuclear receptors (NRs). NRs are intracellular receptors that regulate target gene expression upon the binding of small, hydrophobic molecules like steroids, retinoids, thyroid hormones and vitamin D_3 [7,8]. The EcR gene, first isolated and characterized in *Drosophila melanogaster*, was shown to encode for the ecdysone receptor which functions as a ligand-dependent transcription factor [9]. However, continued investigation showed that while EcR is an essential component of the receptor, it is not sufficient for rendering mammalian cells ecdysone responsive. In fact, the functional active receptor for ecdysone contains both EcR and the NR family member Ultraspiracle (USP), the insect ortholog of the vertebrate retinoid X receptor (RXR), as demonstrated *in vitro* by DNA binding, ligand binding and transactivation assays and *in vivo* on fly *usp* mutants [10-12].

USP and its mammalian homolog RXR shares the ability to heterodimerize with NRs of type II, such as the vertebrate TR, RAR, VDR, PPAR [11]and the insect EcR. For insects, in addition to EcR, USP was shown to form an heterodimeric complex with DHR38 [13]. However unlike its vertebrate homolog for which ligands are known, such as the terpene 9-*cis* retinoic acid (9-*cis* RA) or the docosahexaenoic acid (DHA) [14], no hormone ligand has been identified for USP up to now. Based on the chemical analogy with the vertebrate terpenes, JHs have been put forward as the putative USP ligands. The idea that USP could be the receptor of JH or any of its derivatives raises the attractive possibility that JH might directly modulate the activity of the EcR/USP complex [15]. In addition, some evidence was given that JH can bind to USP and stimulate oligomerisation of USP in vitro and in yeast cells [16,17]. However, this is largely debated and still awaits further substantiation. In particular, the dissociation constant for binding of JH to USP was measured to be about 0.5 μM [16]. Compared to the typical affinity of hormones for their nuclear receptor (in the nM range), this low affinity questions whether such concentrations might be meaningful at a physiological level, where other cellular molecules might compete with JH for USP binding.

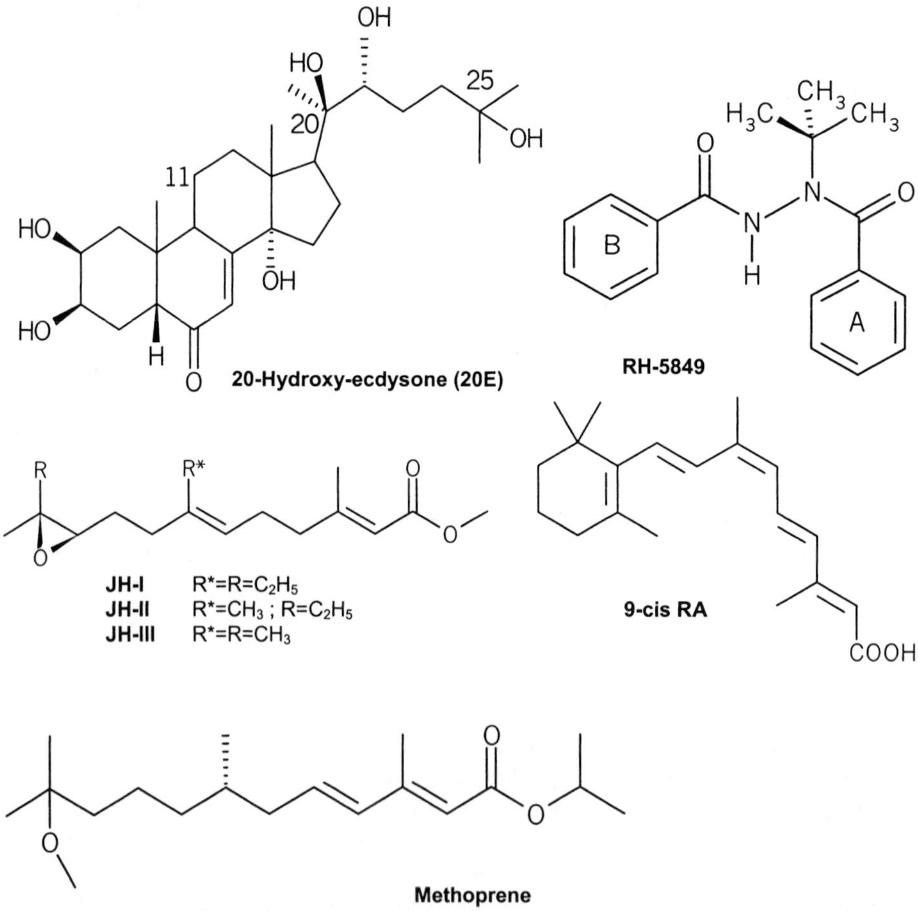

Figure 1. Structures of the different ligands discussed in the text.

Both EcR and USP share the features of typical NR family members with a modular organization encompassing a highly conserved domain for DNA-binding (DBD) and a moderately conserved ligand binding domain (LBD) [18]. In Drosophila, EcR occurs in three isoforms [19]. The various EcR isoforms have different amino-terminal A/B domains and identical DNA and ligand binding domains, all able to interact with USP and bind ecdysteroids with similar affinity. For USP, one single isoform is known in *D. melanogaster* and in the lepidopteran *B. mori and C. fumiferana*, while two isoforms exist in the lepidopteran *M. sexta* and in the dipteran *C. tentans* and *A. aegypti* [5].

The ligand binding domain does not only confer specificity to ligand-binding, but it is also responsible for ligand-dependent transactivation [20-22]. Crystallographic investigations of the LBD structures of several NRs indicate a conserved fold described as an antiparallel α-helical sandwich [23] composed of 11 α-helices (H1, H3-H12) and two short β-strands (s1-s2). Most importantly, the crystal structures of the LBDs of unliganded (apo) [23] and liganded RXR [24] and of other liganded NRs (for a review see [22,25]) have allowed us to gain insight into the molecular mechanisms that underlie the dramatic structural reorganization which accompanies the ligand binding to the LBD. This conformational rearrangement mostly affects the N-terminal part of H3, H11 and H12 which carries the autonomous activation function (AF-2) [23,24,26]. Upon binding of an agonist ligand, the corepressors are released and the LBD adopts a unique conformation that generates an interaction surface for the coactivators which then recruit multiprotein complexes and lead to the activation of responsive genes [27]. In contrast, antagonist ligands induce a transconformation of the LBD which does not allow the binding of coactivators. Several antagonist conformations have been observed that can be considered variations around a common theme [28-31]. In these cases, the activation helix H12 lies in a groove similar to the binding site of the helical NR-box module of nuclear coactivators (the so called antagonist groove) [29,32,33].

The major effort in the agrochemical research against the main larval lepidopteran, coleopteran and dipteran pests species concerns the development of insecticides that mimics the action of the two insect growth and developmental hormones, 20E and JH [34]. The most widely used insecticides belong to the bisacylhydrazine class of compounds which are non-steroidal ecdysone agonists that exert their action through binding to EcR and permanent activation [35]. Some of these bisacylhydrazines are very selectively toxic to lepidopteran insect species (for e.g. tebufenozide RH-5992 and methoxyfenozide RH-2485), and generally devoid of toxicity to non-lepidopteran species. Other compounds, like halofenozide RH-0345, have a broader spectrum of action being used as agents controlling scarabid beetle larva as well as caterpillars (for a review see [34] and references therein). The understanding of the binding modes of ecdysteroids and bisacylhydrazines to EcR has mainly relied on three-dimensional quantitative structure-activity relationship (3-D QSAR) analyses [36-38]. In addition, in the absence of crystal structure, models for the ecdysone receptor have been proposed [39]. These homology models will be discussed below.

Compounds that mimic the action of JHs could also be used as insecticides. In fact, numerous JH analogs were derived among which methoprene. This ligand, used commercially as an insect growth regulator, closely resembles JHs, but lacks the epoxide function. Other analogs which differ significantly from JHs, like fenoxycarb, pyriproxyfen and diofenolan, have been used for insecticidal activity [34]. However, the JH analogs have ecotoxicologial effects on a number of non-target invertebrates and vertebrates [40]. Furthermore, some of the target insect species develop insecticide resistance towards JH analogs [34]. Although the biological role of JH has been extensively studied [41], its mode of action at the molecular level is not known. In particular, it is not clear-whether USP is the JH receptor. Recently, the crystal structure of USPs from the lepidopteran *Heliothis virescens* (hvUSP) [42] and the dipteran *Drosophila melanogaster* (dmUSP) [43] were resolved. These structures discussed below suggest that USP is most likely a ligand-regulated hormone receptor.

2 Results

2.1 Sequence Comparisons of USPs and RXRs

The ability to bind 9-cis RA is shared by all RXRs, from cnidarians (*Tripedalia*) to vertebrates [8]. In arthropods, there is no requirement for retinoids and no activation with retinoids was observed for the lepidopteran and the dipteran insects [5] or for the tick RXR [44]. The ultraspiracle protein, first cloned for *Drosophila*, was considered to be the insect homolog of RXRα based on the sequence homology of their DBDs [45]. However, while the DBDs of arthropod RXRs and USPs are evolutionary very well conserved, striking differences are observed in the LBDs. The major divergences are observed for the Diptera and the Lepidoptera insect orders. Sequence alignments of USP and RXR LBDs can be found in [42,46,47]. Because of these divergences, the term USP is commonly used in the literature to specify the RXR orthologs of the highly derived insects species (Lepidoptera and Diptera), whereas RXR is used for other arthropods. In fact, the RXR LBD sequences of vertebrates and arthropods are highly conserved (more than 60 % identity). The conservation between USP and RXR LBD sequences is not so high, but still quite good (between 42 and 51 % identity).

The crystal structures of hsRXRα and hvUSP, 11 helices and a β-sheet confirm the secondary structure prediction using the canonical structure of nuclear receptor LBDs. Structurally, the major difference between the RXR and USP sequences is a long insertion between helix 5 and the β-sheet observed for all USPs, but absent in all vertebrate and arthropod RXRs. Most of the conserved residues between USP and RXR sequences are located in the helices, in particular those forming the core of the LBD and the dimerization region as well as the signature region which encompasses the C-terminal region of H3, the loop between helices 3 and 4 and most of helix 4 [21]. In addition, the loop connecting the helices 1 and 3 (L1-3) is rather well conserved for USPs, while it is poorly conserved for RXRs. This is remarkable, because within the whole NR superfamily, L1-3 is usually found to be extremely variable in length and in sequence, consistent with its nature of a rather flexible and loosely structured region. The loop contains a highly conserved sequence motif that is conserved in dipterans and lepidopterans, but not for other arthropods. As discussed in section 2.4, the crystal structures of USP indicate that L1-3 plays a crucial role by locking the activation helix H12 in an antagonist conformation.

If we consider only the residues lining the ligand binding pocket (LBP) according to the RXRα/9-cis RA complex (22 residues at 4.5 Å as indicated in [24]), their sequence identity is much higher compared to the whole LBD, around 60 % (12 residues for ctUSP) for dipterans and above 95 % for lepidopterans. In particular, the residues Gln 275 and Arg 316 of hsRXRα anchoring the carboxylate moiety of the 9-cis RA are conserved for all vertebrate and arthropod RXRs except for a few dipteran insects, like *Drosophila melanogaster* and *Chironomus tentans*. This observation suggests that most of the USP LBDs could form a similar electrostatic interaction with a ligand.

2.2 Analysis of the EcR Protein Sequences

The ecdysone receptor belongs to the same subfamily of nuclear receptors as the liver X receptor LXR, the farnesoid X receptor FXR and the vitamin D receptor VDR [8,[48]. All these receptors bind steroid-related compounds (oxysterol, bile acids and vitamin D). A detailed sequence alignment and a phylogenetic analysis encompassing 19 EcR protein sequences of various arthropods has been performed by Bonneton *et al* [46]. We will summarize here the main issues of this study.

The EcR LBDs show a strong conservation both in length and sequence (54 % residue identity) that is even higher within the dipteran and lepidopteran lineages (73 % and 84 %, respectively). Structurally, the 11 helices (H1 to H12) of the antiparallel helical sandwich fold of NRs are well identified [39]. In particular, the NR signature region encompassing helices 3 and 4 exhibits a strong amino acid conservation.

The similarity between the different EcR sequences contrasts with the results obtained for USPs-RXRs, where the sequence comparison indicates a clear-cut divergence between the highly derived insects and the other arthropods. The differences between the lepidopterans and dipterans and the other arthropods lie in the DBD and in the F domain at the carboxy-terminal end of the receptor. The difference in the DBD lies in the region located just upstream the second zinc finger. However, it is not possible to draw conclusion on the functional consequences of these conservative changes specific to the highly derived insects. More remarkable is the long and variable F domain specific to dipterans and lepidopterans. It varies in length between 18 (for *D. melanogaster*) and 226 residues (*C. tentans*), while other arthropod EcRs and the vertebrate homologs LXR or FXR either lack a F domain or have a very short one (3-5 residues for the locust, tick and crab EcRs). The F domain may contribute to the stabilization of helix 12 and modulate transcriptional activity of the ligands. In fact, *in vitro* studies showed that *D. melanogaster* EcR harbours an autonomous silencing function in its carboxy-terminus [49]. Furthermore, for the vertebrate estrogen receptor α, the F domain was shown to modulate transcriptional activity and dimerization signal, probably through interaction with the ligand-dependent AF-2 domain [50-52].

Both nuclear receptors EcR and USP that form the functional receptor for ecdysone seem to have evolved in a concerted way during insect evolution. The evolutionary conserved divergences in the protein sequences of arthropod USPs-RXRs and EcRs are essentially confined to the highly derived holometabolous insects, Diptera and Lepidoptera. The analysis of these differences suggests that the ligand-dependent transactivation properties of the EcR/USP heterodimeric complex of dipterans and lepidopterans might be different from those of other arthropod EcR/USP heterodimers.

2.3 Modeling of the EcR LBD and Ligand Docking

In the absence of crystallographic structure for the EcR LBD, two homology models have been constructed for EcR [39] by taking as templates the known LBD crystal structures of the retinoic acid and vitamin D receptors [26]. This choice of crystal structures is based on the belonging of EcR to the same subfamily as VDR and RAR, although the sequence conservation is not so high (only 25 %). The goal of the study was to structurally rationalize the selectivity and binding mode of bisacylhydrazines versus ecdysteroids. The study focused on the docking of the natural ligand 20E and of the synthetic agonist RH-5849 (1,2-dibenzoyl-1-*tert*-butylhydrazine) in the two homology models (EcRvd and EcRra for the models based on hVDR and hRARγ, respectively) of the *C. tentans* ecdysone receptor (ctEcR).
In both models, the ligand binding pockets are delineated by the helices 5, 7, 11 and 12, the β-turn and the loops 6-7 and 11-12. Their shape is similar, a bulky envelope and a shallow tube, but their orientation is reversed, the bulky envelope being oriented towards the β-sheet for EcRvd and towards helices 11 and 12 for EcRra. The cavity of EcRra (764 $Å^3$) is slightly smaller than that of EcRvd (840 $Å^3$), but large enough to accommodate a ligand such as 20E (473 $Å^3$). The resulting percentages of LBP occupancy (56 % and 62 %, respectively) compare well with the values found for RAR/all-*trans* RA (66 %) and VDR/vitamin D (56 %). The residues that line the cavity (23-24 residues) are mostly hydrophobic. Polar residues are found for both models in helices 3 (Thr-324, Thr-327), 5 (Arg-376, Arg-371), 6 (Tyr-392) and 11 (Asn-488, Cys-492 and Asn-499). In addition, Glu-294 (in loop 1-3) is only present in the LBP of EcRra, while Lys-501 (in loop 11-12) is only found in the cavity of EcRvd.
The natural ligand 20E docked in the LBP of ctEcR orients its A-ring in the bulky envelope and its long hydroxylated side chain in the shallow tube. Since the orientation of the LBPs is opposite, the orientation of 20E is inverted in EcRra with respect to EcRvd. Nevertheless, both models can accommodate 20E and its polar groups can form hydrogen-bonds with the polar residues of the cavity. The pattern of H-bonds anchoring 20E is then also inverted, except for the C14-hydroxyl group of 20E that is H-bonded to Thr-327 in both models. As an example, the C6-ketone moiety of 20E is H-bonded to Tyr-392 and the C20-hydroxyl group forms a H-bond with Asn-488 in the EcRvd model (Fig. 2a), while the inverse is observed in the EcRra model.
The synthetic non-steroidal agonist RH-5849 occupies a smaller fraction of the LBP than 20E. For both models, a good correspondence is observed between the hydroxylated side chain of 20E at the C17 position and the B-ring of RH-5849 within the complex (Fig. 2b). Therefore, again a head-to-tail orientation of the ligand is observed when the EcRra and EcRvd models are compared. However, interestingly, in both models, the *tert*-butyl group characteristic of bisacylhydrazines is located in the same hydrophobic groove,

(lined by helices 5, 6, 7, 11) that is left unoccupied by 20E. This bulky group fits nicely in this groove and locks the ligand in the pocket preventing any displacement.

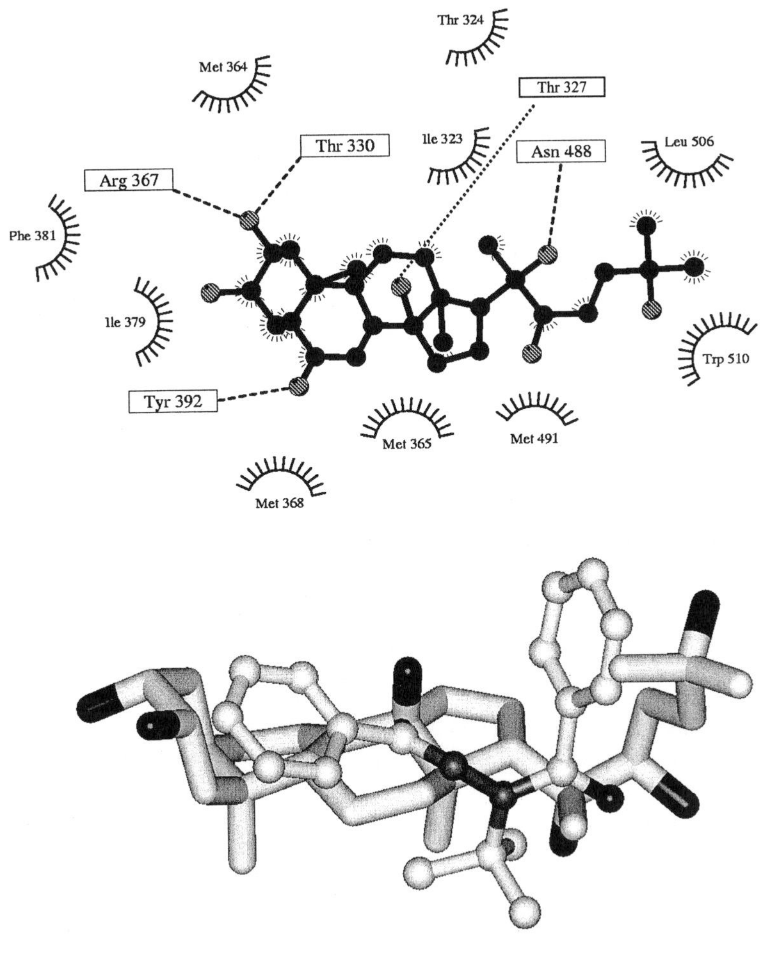

Figure 2. a) Ecdysone agonists in the EcR model based on the structure of the vitamin D receptor. Two-dimensional projection of 20-hydroxy-ecdysone (20E) in the ligand-binding pocket of ctEcRvd. Residues that make hydrophobic contacts with the ligand are shown by hatched half-circles, those that are hydrogen-bonded to the ligand are surrounded by a rectangular box. The hydrogen bonds between these residues and hydroxyl groups of 20E are shown by dashed lines.
b) Superimposition of 20E and RH-5842 as observed in the ligand-binding pocket of EcRvd. 20E is depicted by sticks, RH-5842 by balls and sticks. The carbon atoms are colored in light grey, the nitrogen atoms in grey and the oxygen atoms in black.

In order to evaluate which of the two EcR models is best, the few available experimental data on receptor ligand binding activity were exploited (see references in [39]). To summarize, the activity data on 20E derivatives do not lead to a clear preference for one or the other model. For both models, only partial agreement is found with the experimental activity data, that suggests that EcR might adopt a different structural arrangement as both RAR and VDR.
The situation is somewhat easier to rationalize for bisacylhydrazines, where the results of numerous three-dimensional quantitative structure-activity relationship (3-D QSAR) studies are available (see references in [39] and more recently in [37,54,55]. The data available for A- and B-ring substitutions would favor an orientation of RH-5849 as observed for EcRvd, with the A-ring oriented towards helix 5 and the β-sheet. For the *tert*-butyl group which is known to confer a high ecdysone-like activity, substituents with large groups at this position do not affect the binding activity. In terms of modeling and docking, this would then require a slight structural rearrangement of the receptor where the loop 6-7 would be displaced to allow the ligand to fit.

2.4 Crystal Structure of the USP LBD

Two crystal structures of the LBD of USP were published, that of the lepidopteran *Heliothis virescens* (hvUSP) at 1.65 Å resolution [42] and that of the dipteran *Drosophila melanogaster* (dmUSP) at 2.4 Å resolution [43]. Both structures exhibit the same general features that we will discuss here mostly in light of the hvUSP structure.

2.4.1 Overall Structure

The overall architecture of USP LBD exhibits the canonical fold of the nuclear hormone receptor family [23] with 11 α-helices (H1, H3-H12) and two short β-strands (s1-s2) (Fig. 3a). To facilitate comparison with other NR structures, the structures of RXRα in the agonist and in the antagonist conformation will be used, as reported for hsRXRα complexed to 9-*cis* RA [24] and for msRXRα complexed to oleic acid [30], respectively. Superimposition of the hvUSP LBD to holo RXRα LBD gives a root mean square deviation (r.m.s.d.) of 1.22 Å for 183 matched Cα's atoms out of 246. Seven helices match rather well, while helices 1, 3 and 6 and the β-sheet show larger deviations (Fig. 3b). In fact, H1 and H3 which contribute to the outermost shell of the LBD are less coplanar for hvUSP than for RXRα, the angle between their helical axes being 12.1° larger than the corresponding value in RXRα. This correlates with the path adopted by the loop L1-3 connecting H1 to H3. It induces a considerable difference in the positioning of the N- and C-terminal parts of H3 compared to RXRα. The N-terminal region of H3 (Pro-240 to Cys-250) is displaced outwards the protein core in a substantial manner, being tilted by about 24° with respect to the same region in holo RXRα. This position is intermediary between the positions of the N-terminal region seen in the apo RXRα and the holo RXRα LBD structures.

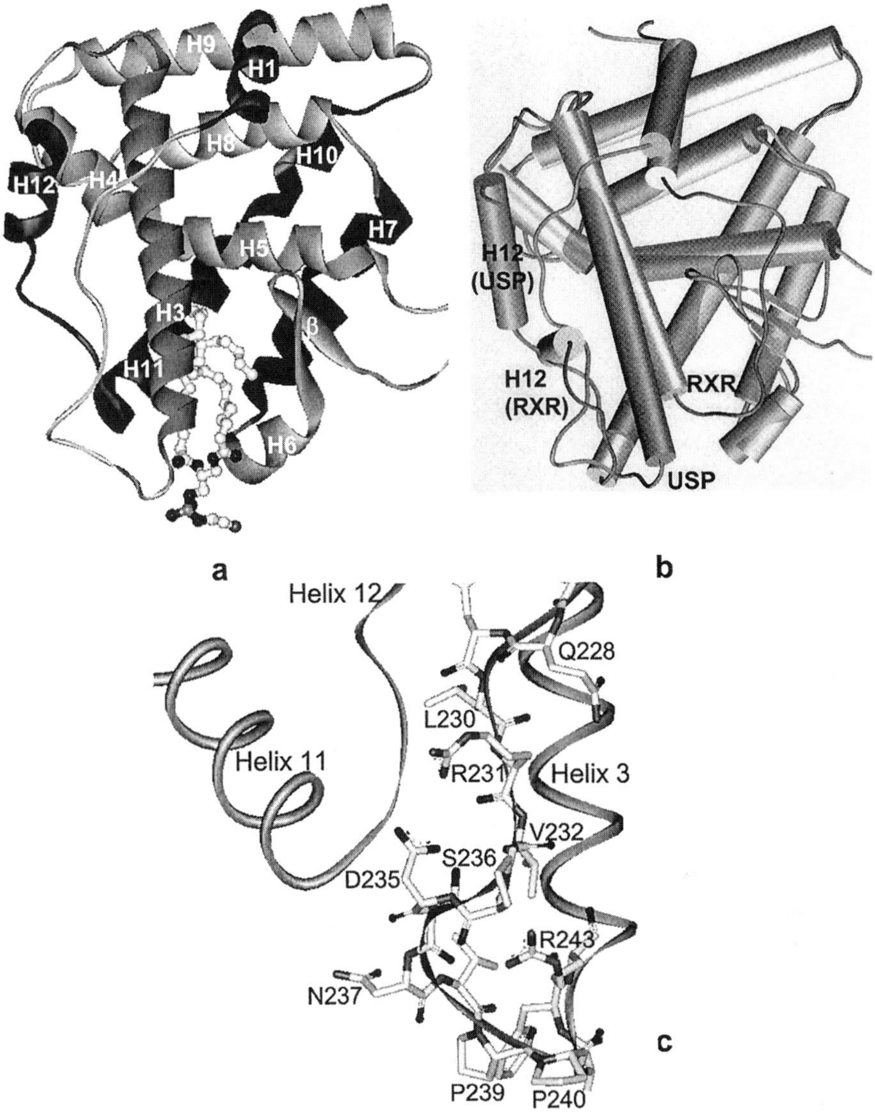

Figure 3. The crystal structure of USP.
a) Overall structure of hvUSP LBD showing the canonical fold of NR LBDs with 11 helices and two short β-strands. Helices H1 to H12 are labeled accordingly. The fortuitous phospholipid co-crystallized with the USP LBD is represented by balls and sticks. For the ligand, the carbon atoms are colored in light grey, the phosphorous atom in grey and the oxygen atoms in black.
b) Superimposition of the crystal structure of hvUSP LBD (grey) with that of agonist-bound hsRXRα LBD (light grey). Notice that the position of the loop between helices 1 and 3 sterically prevents helix 12 of hvUSP from adopting the canonical agonist conformation.
c) Detailed view of the residues of the loop connecting helices 1 and 3 interacting with helices 3 and 11. The protein is depicted as a backbone C_α-trace. Protein atoms are colored in light grey for carbon, in grey for nitrogen and in black for oxygen. Residues are labeled accordingly.

2.4.2 The Activation Helix Locked into an Antagonist Conformation

Two remarkable regions differ considerably from all other known NR LBD structures, and in particular from RXRα: the loop connecting helices 1 and 3 and the activation helix H12. The connecting loop L1-3 adopts an unusual conformation for NRs, it wraps over the top of helix 3 and lies between helices 3 and 11. In most of the NRs, it passes outside the β-strand (for example, in holo hsRXRα [24], hPPARγ [32], TRβ [56]) or between H3 and the tip of the β-sheet, as in ER [28,29,31] and VDR [53].
The rather tensed conformation adopted by this loop in USP allows it to establish direct contacts with residues in H3, H11 and H12 and stabilize them in their actual position. Notice that these helices are the structural elements shown to undergo the largest conformational changes upon ligand-binding [24]. The loop L1-3 plays a crucial role in USP by sterically interfering with the loop L11-12 and precluding the positioning of helix 12 in the agonist conformation. As a consequence, H12 adopts a conformation observed in the case of several NRs such as RXRα /oleic acid (Fig.3b), RARα/BMS614 [30] and antagonist-bound ER LBDs [28,29,31]. The groove in which H12 lies is the binding site for the helical NR-box module of nuclear coactivators with a consensus sequence LXXLL, as shown for the LBDs of PPARγ [32], TRβ [33] and ERα [29]. However, in contrast to these structures, the helix 12 of hvUSP and dmUSP is locked firmly in the antagonist position by making extensive contacts with the loop L1-3. As mentioned in section 2.1, most of the residues belonging to the loop L1-3 are highly conserved in dipterans and lepidopterans and organized around a strictly conserved Leu-Arg-Val motif. The structures of USP show that these residues play an architectural role in mediating interactions with helices 3, 11 and 12 and with the loop L11-12 (Fig. 3c). These conservations show a strong evolutionary pressure that suggests the functional relevance of the observed conformation for the lepidopteran and dipteran USPs. Furthermore, the observation that both USP structures exhibit the same features suggests that this loop conformation is relevant for the physiological state of the receptor, and not the result of crystal packing interactions.

2.4.3 A Fortuitous Ligand in the Ligand Binding Pocket

The ligand binding pocket of USP is filled by a fortuitous phospholipid co-purified and co-crystallized with the USP LBD that was further characterized by mass-spectroscopic and chemical analysis [57]. In a similar way, recent crystallographic investigations of the retinoid-acid related orphan receptor β (ROR β) [58] and of the heterodimeric complex RARα/RXRα [30] revealed an *E.coli* endogeneous fatty acid in the ROR β and in the RXRα subunit, respectively.
The ligand is stabilized inside the pocket by hydrophobic interactions with residues in helices 3, 5, 7 and 11 and in the loops L1-3, L6-7 and L11-12. The position of the ligand inside the USP pocket is different from that adopted by the 9-cis RA inside hRXRα [24]. The 9-*cis* RA is buried deep inside the pocket of RXRα LBD, where its carboxylate group establishes a salt bridge with the conserved arginine residue belonging to helix 5. In contrast, the phospholipid does not penetrate as deep inside the cavity. The USP ligand binding pocket is much bigger by about a factor 2.5 with respect to hRXRα. This is a consequence of the unusual receptor topology. In fact, helix 3 is straighter and longer

by one turn with respect to RXRα and it forms, together with helices 6 and 11 and with the loop L1-3, a bulky tube which extends to the solvent-exposed region. The USP ligand-binding pocket is a large hydrophobic cavity. Most of the residues in hvUSP and dmUSP have a similar type and nature. However, the major discrepancy between the residues that line the LBP of both receptors concerns the arginine in H5 participating in RXRα to the anchoring of the ligand carboxylate group, which is replaced by a cysteine residue in dmUSP, while it is conserved in hvUSP.
Due to the unusual type of ligand binding, the N-terminal part of helix H3 has to adapt suggesting a mechanism for the ligand entry in the cavity. For RXRα, the comparison of liganded versus unliganded conformations indicates a substantial movement of H3 upon ligand binding in an induced-fit mechanism to clamp the ligand and lock it inside the pocket [24,26]. In the case of USP, the displacement of the N-terminal of H3 which accompanies the binding of the bulky ligand results in a more widely open clamp. As it is apparent from the structure, the phospholipid most likely penetrates the ligand cavity from the channel formed by helices H3, H6, H11 and loops L1-3 and L6-7. This is in agreement with the path followed by the 9-*cis* RA when entering the receptor cavity as described for RXRα [24,26,59]. According to this dynamic model, the ligand would enter the cavity through a cleft formed by the breathing of H3 and H10/H11 and then be trapped in the observed antagonist conformation.

2.4.4 Discussion

As already suggested by the sequence comparison between USP and RXR LBDs and observed in the USP crystal structures, the loop connecting helices 1 and 3 has a peculiar structural role that was never observed before for other nuclear receptors. Key residues highly conserved in highly derived holometabolous insects that belong to this secondary structure element firmly lock helix 12 in an antagonist position irrespective of the bound ligand. Therefore, LXXLL-containing coactivators that have been shown to bind in the groove formed by the helix 12 in the agonist position as well as helices 3, 4, 5 [29,32,33,60] cannot bind to USP. In other words, the mechanism by which ligands bind to NRs causing the dissociation of bound corepressors and facilitating the binding of coactivators does not seem to apply in the case of USP [61]. The mode of action of USP seems to be different from other NRs, but at the same time the structures of hvUSP and dmUSP reveal the existence of a ligand-binding pocket that implies that USP almost certainly must have a natural ligand. However, the exact nature of the natural ligand as well as its mode of action cannot be elucidated from the structural studies.

2.5 Could USP Adopt an Agonist Conformation?

Both crystal structures of hvUSP and dmUSP show that the helix 12 bearing the ligand-dependent activation function is locked in the antagonist conformation. It is therefore of interest to understand whether the agonist conformation of H12 in USPs is structurally possible. For this analysis, an homology model of an agonist-like USP was constructed by taking as a template the agonist conformation of hsRXRα [47]. In the agonist conformation of hsRXRα, H12 leans over the ligand binding pocket, and is stabilized by a

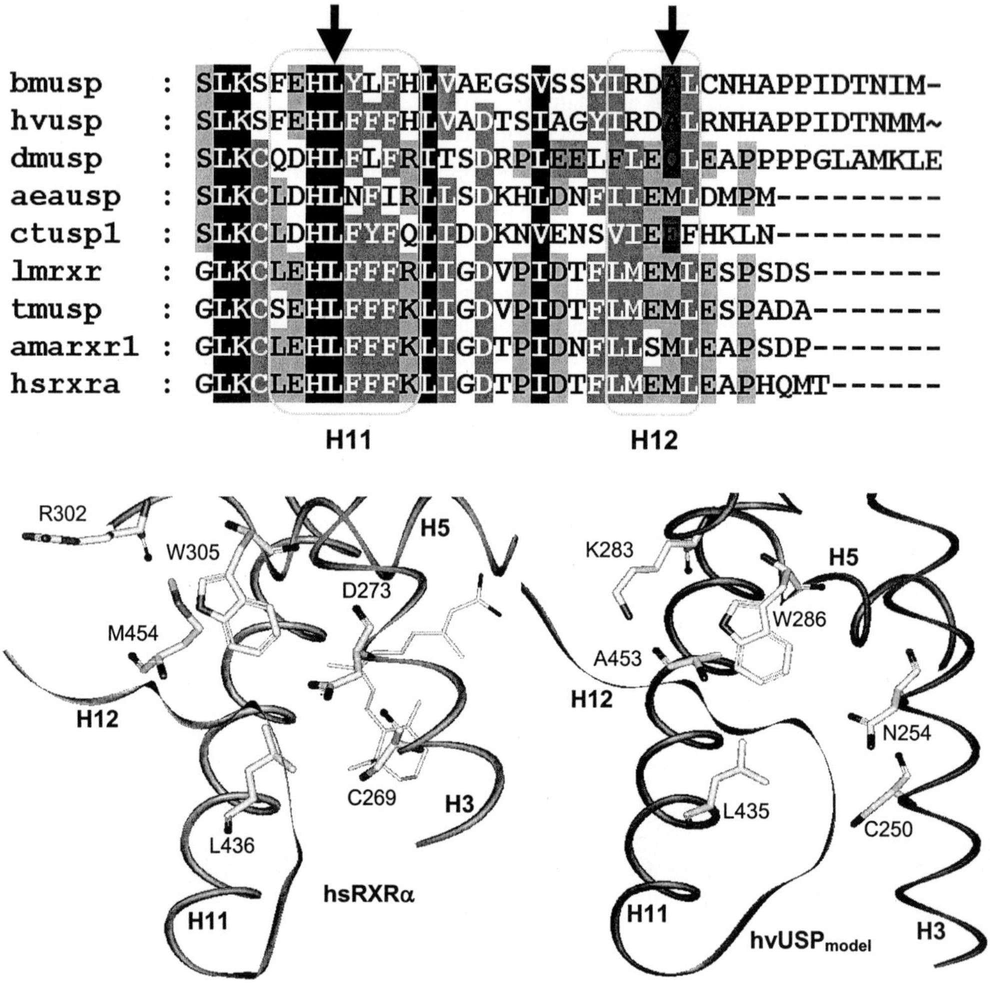

Figure 4. The agonist conformation of the transactivation helix H12 in hsRXRα and in the agonist-like hvUSP homology model.

a) Sequence alignment of residues in H11 and H12 for several USPs and RXRs. Lepidopterans are represented by *Bombyx mori* (bm) and *Heliothis virescens* (hv) and dipterans by *Drosophila melanogaster* (dm), *Aedes aegypti* (ae) and *Chironomus tentans* (ct). For the other arthropods, the RXR sequences of the *Locusta migratoria* (lm) *Tenebrio molitor* (tm) and *Amblyomma americanum* (ama) are given, while for vertebrate the *Homo sapiens* (hs) RXR is chosen as representative for all vertebrate RXRs.

b) Details of the interactions stabilizing helix H12 in the agonist position in the RXRα LBD. Several conserved residues in helices 3, 5 and 11 participate to the interaction with helix 12.

c) The same region as in b) for the agonist-like hvUSP LBD homology model.

In b) and c), the protein is depicted as a backbone C_{α}-trace. Protein atoms are colored in light grey for carbon, in grey for nitrogen and in black for oxygen. Residues are labeled accordingly.

series of hydrophobic contacts involving residues of H3, H4, H5 and H11 [24]. As shown in detail in [47], most of these residues are highly conserved or replaced by amino acids of similar nature.

Focusing at the amino acid sequence of H12, the alignment of the residues in H12 of USPs and RXRs shows a poor sequence homology (Fig. 4a). In particular, by adopting the hsRXRα residue numbering, the motif Glu 453 - Met 454 – Leu 455 – Glu 456 is very well conserved for the RXR sequences, but rather divergent for the USP sequences. These residues are involved in the interaction of H12 with the protein itself and with cofactors. The Met 454 (H12) is in hydrophobic contact with Trp 305 (H5) and its backbone carbonyl group is hydrogen-bonded through a water molecule to the indole NH group of Trp 305 (H5) (Fig.4b). The hydrophobic contact is maintained for *Aedes* USP and lepidopteran USPs (Fig. 4c). The methionine is replaced by glutamine in dmUSP (EQLE) and by glutamate in ctUSP(EEFH). Examining the three-dimensional structure of the homology model indicates that the glutamine and glutamate side chains point towards the surface close to the lysine at the C-terminus of H4 and could interact with its amine group. Clearly, for ctUSP, the carboxylate moiety of the glutamate residue would confer an additional negative charge to the coactivator interaction surface when compared to the other USPs. An additional observation concerns the residues Glu 453 (H12) and Lys 284 (H3) have been shown to clamp the LXXLL motif of the coactivator in the groove formed by helices 3, 4, 5 and 12 [29,32,33,60]. Both residues are conserved or replaced by very similar residues in the various USP sequences. However, the coactivator interaction surface generated by helices 3, 4, 5 and 12 bears residues which are of different type and nature from those seen in hsRXRα. Therefore, we might expect that the coactivators that would bind to USPs are different compared to those that bind to RXRs. This is not unexpected since different cofactors are found in insects and in vertebrates [62,63].

In summary, the structural analysis of the residues involved in the interaction of H12 with the receptor suggests that the pattern of interactions for the positioning and the stabilization of H12 in the agonist conformation is rather similar to that seen in RXRα. However, the positioning of helix 12 in the agonist conformation is only possible if the loop L1-3 adopts a different path as observed in the structures of hvUSP and dmUSP. Obviously, in the agonist-like USP homology model, this loop was modeled after RXRα. The question arises about the role of the conserved residues in the loop L1-3. As discussed above, the crystal structures of USP seem to indicate that the residues have acquired during evolution a structural and functional role that is to lock helix 12 in the antagonist position. However, no biological data are available yet to support this hypothesis and helps answering this fundamental issue.

3 Conclusion and Perspectives

The insect nuclear receptors EcR and USP are main targets for the agrochemical research. The structures of their ligand-binding domain is an essential step in the development of new types of environmentally safe insecticides. For EcR, a new pharmacophore model based on the spatial superposition of 20E and RH-5849 was proposed and discussed. This model gives new insight into the binding modes of ecdysteroids and non-steroidal agonists of the bisacylhydrazine class of compounds. The models based on the crystal structure of VDR are in rather good agreement with biological data for synthetic agonists, while only partial agreement is observed for natural agonists. This suggests that EcR might adopt a different structural arrangement than that observed for RAR and VDR LBDs. Obviously, the crystal structure of EcR is required for a complete understanding of the mode of ligand-receptor binding.

For USP, the structural studies suggest that USP is most probably a ligand-regulated nuclear receptor and not a constitutively active true orphan receptor. Furthermore, these studies have shown that despite the high sequence homology between USPs and RXRs, USP differs considerably from its mammalian counterpart. This divergence is found in the highly derived holometabolous insects, dipterans and lepidopterans, but not in other arthropods whose USP-RXR protein sequences are closer to those of vertebrate RXRs. The structures of both hvUSP and dmUSP indicate that residues in the loop connecting helices 1-3 modulate the structure, and hence the activity, of USP in a very significant fashion. These residues are conserved in the highly derived holometabolous insects, suggesting that these residues have gained a new structural and functional role. This structural studies also raise several fundamental questions. The structure-based analysis of USP homology models based on the agonist RXRα crystal structure indicates that the agonist position of USP is possible provided that the loop L1-3 adopts a position similar to that observed for RXRα. Since this hypothesis contrasts with structural observations and sequence analysis, its validity can be seriously questioned. In case the observed loop conformation was a consequence of the bound lipid, the USP ligand binding pocket would resemble more that of RXRα. In this case, recent docking studies indicate that juvenile hormones would fit rather well inside the USP cavity [47]. Clearly, the structure of EcR/USP will represent a major breakthrough towards a better understanding of the mechanism by which USP acts on its heterodimerization partner.

4 References

[1] H. F. Nijhout. Insect hormones. Princeton University Press, Princeton, 1994.
[2] L. M. Riddiford. Receptor 1999, 3, 203-209.
[3] L. M. Riddiford, K. Hiruma, Q. Lan & B. Zhou. Amer. Zool. 1999, 39, 736-746.
[4] B. L. Hall. Amer. Zool. 1999, 39, 714-721.
[5] L. M. Riddiford, P. Cherbas & J.W. Truman. Vitamins and Hormones 2001, 60, 1-73.
[6] C. S. Thummel. Dev. Cell 2001, 1, 453-465.
[7] V. Laudet. Cell 1999, 97, 161-163.
[8] H. Escriva, F. Delaunay & V. Laudet. Bioessays 2000, 22, 717-727.
[9] M. R. Koelle et al. Cell 1991, 67, 59-77.
[10] T.-P. Yao, W. A. Segraves, A. E. Oro, M. McKeown & R. M. Evans. Cell 1992, 71, 63-72.
[11] T.-P. Yao *et al.* Nature 1993, 366, 476-479.
[12] B.L. Hall & C.S. Thummel, Development (Camb.) 1998, 125, 4709-4717
[13] J. D. Sutherland, T. Kozlova, G. Tzertzinis & F. C. Kafatos. Proc. Natl. Acad. Sci. USA 1995, 92, 7966-7970.

[14] A. M. de Urquiza *et al.* Science 2000, 290, 2140-2144.
[15] M. Buszczak & W. A. Segraves. Curr. Biol. 1998, 8, R879-R882.
[16] G. Jones & P. A. Sharp. Proc. Natl. Acad. Sci. USA 1997, 94, 13499-13503.
[17] G. Jones, M. Wozniak, Y. Chu, S. Dhar & D. Jones. Insect Biochem. Mol. Biol. 2002, 32, 33-49.
[18] D. J. Mangelsdorf *et al.* Cell 1995, 83, 835-839.
[19] W. S. Talbot, E. A. Swyryd & D. S. Hogness. Cell 1993, 73, 1323-1337.
[20] W. Bourguet, P. Germain & H. Gronemeyer. TiPS 2000, 21, 381-388.
[21] J.-M. Wurtz *et al.* Nat. Struct. Biol. 1996, 3, 87-94.
[22] D. Moras & H. Gronemeyer. Curr. Opin. Cell Biol. 1998, 10, 384-391.
[23] W. Bourguet, M. Ruff, P. Chambon, H. Gronemeyer & D. Moras. Nature 1995, 375, 377-382.
[24] P. F. Egea *et al.* EMBO J. 2000, 19, 2592-2601.
[25] P. F. Egea, B. P. Klaholz & D. Moras. FEBS Lett. 2000, 476, 62-67.
[26] J.-P. Renaud *et al.* Nature 1995, 378, 681-689.
[27] J. Torchia, C. Glass & M. G. Rosenfeld. Curr. Opin. Cell Biol. 1998, 10, 373-383.
[28] A. M. Brzozowski *et al.* Nature 1997, 389, 753-758.
[29] A. K. Shiau *et al.* Cell 1998, 95, 927-937.
[30] W. Bourguet *et al.* Mol. Cell 2000, 5, 289-298.
[31] A. C. W. Pike *et al.* EMBO J. 1999, 18, 4608-4618.
[32] R. T. Nolte *et al.* Nature 1998, 395, 137-143.
[33] B. D. Darimont *et al.* Genes. Dev. 1998, 12, 3343-3356.
[34] T. S. Dhadialla, G. R. Carlson & D. P. Le. Annu. Rev. Entomol. 1998, 43, 545-569.
[35] K. D. Wing, R. A. Slawecki & G. R. Carlson. Science 1988, 241, 467-469.
[36] G. Smagghe, T. S. Dhadialla & M. Lezzi. Insect Biochem Mol. Biol. 2002, 32, 187-192.
[37] Y. Nakagawa, G. Smagghe, L. Tirry & T. Fujita. Pest Manag. Sci. 2001, 58, 131-138.
[38] Y. Nakagawa, G. Smagghe, M. Van Paemel, L. Tirry & T. Fujita. Pest Manag. Sci. 2001, 57, 858-865.
[39] J.-M. Wurtz *et al.* Protein Sci. 2000, 9, 1073-1084.
[40] M. A. Harmon, M. F. Boehm, R. A. Heyman & D. J. Mangelsdorf. Proc. Natl. Acad. Sci. U. S. A. 1995, 92, 6157-6160.
[41] J. W. Truman & L. M. Riddiford. Annu. Rev. Entomol. 2002, 47, 467-500.
[42] I. M. L. Billas, L. Moulinier, N. Rochel & D. Moras. J.Biol.Chem. 2001.276, 7465-7474.
[43] G. M. Clayton, S. Y. Peak-Chew, R. M. Evans & Schwabe. Proc. Natl. Acad. Sci. USA 2001, 98, 1549-1554.
[44] X. Guo *et al.* Mol. Cell. Endocrinology 1998, 139, 45-60.
[45] A. E. Oro, M. McKeown & R. M. Evans. Nature 1990, 347, 298-301.
[46] F. Bonneton, D. Zelus, T. Iwema, M. Robinson-Rechavi & V. Laudet. to be published 2002.
[47] S. Sasorith, I. M. L. Billas, T. Iwema, D. Moras & J.-M. Wurtz. to be published 2002.
[48] V. Laudet & Gronemeyer, H. The Nuclear Receptor Factsbook. Academic Press, 2002.
[49] D. Thormeyer, S. P. Tenbaum, R. Renkawitz & A. Baniahmad. J. Steroid Biochem Mol. Biol. 1999, 68, 163-169.
[50] G. A. Peters & S. A. Khan. Mol. Endocrinol. 1999, 13, 286-296.
[51] M. Nichols, J. M. Rientjes & A. F. Stewart. EMBO J. 1998, 17, 765-773.
[52] M. M. Montano, V. Muller, A. Trobaugh & B. S. Katzenellenbogen. Mol. Endocrinol. 1995, 9, 814-825.
[53] N. Rochel, J.-M. Wurtz, A. Mitschler, B. P. Klaholz & D. Moras. Mol. Cell 2000, 5, 173-179.
[54] G. Smagghe et al. Arch. Insect Biochem Physiol 1999, 41, 42-53.
[55] G. Smagghe, T. S. Dhadialla & M. Lezzi. Insect Biochem. Mol. Biol. 2002, 32, 187-192.
[56] R. L. Wagner et al. Nature 1995, 378, 690-697.
[57] N. Potier et al. to be published 2002.

[58] C. Stehlin et al. EMBO J. 2001, 20, 5822-5831
[59] A. Blondel, J.-P. Renaud, S. Fischer, D. Moras & M. Karplus. J. Mol. Biol. 1999, 291, 101-115
[60] W. Feng et al. Science 1998, 280, 1747-1749
[61] C. K. Glass & M. G. Rosenfeld. Genes. Dev. 2000, 14, 121-141
[62] J. Bai, Y. Uehara & D. J. Montell. Cell 2000, 103, 1047-1058
[63] R. Beckstead et al. Mol. Cell 2001, 7, 753-765

16 Molecular Design of Neonicotinoids: Past, Present and Future

Shinzo Kagabu

Department of Chemistry, Faculty of Education, Gifu University, Gifu 501-1193, Japan,
E-mail: kagabus@cc.gifu-u.ac.jp

1 Introduction

With the debut of imidacloprid the last decade opened the era of a new insecticide class. This new class, called neonicotinoid or chloronicotinyl, with its excellent insecticidal performance, is rapidly changing the traditional market situation. Until just 10 years ago, the insecticide market had been dominated by two modes of action. These are the organophosphates and carbamates, which act by inhibition of acetylcholinesterases, and the pyrethroids, which act on the voltage-gated sodium channel; the action on the nicotinic acetylcholine receptors (nAChR) had been of only minor economic significance. Various attempts had been made to exploit a new class over several decades and 230 insecticidal ingredients with eleven modes of action were registered [1], yet in 2000 the two major modes still accounted for more than 75% of all the insecticides [2,3]. Now imidacloprid and the subsequent neonicotinoid products are boosting the market share of the nAChR mode enormously. People are expecting it to achieve a 15 % share of the world insecticide market in 2005. In this review I would like to look at how the tremendous success of this new insecticide class came about, what its present state is and glimpse how it will develop in future from the viewpoint of molecular design.

2 nAChR Agents before Imidacloprid

Nicotine (**1**) is an old insecticide which acts on the postsynaptic nAChR. This botanic insecticide appears to have interested a few chemists as a lead for a new insecticide. More than 30 years ago Yamamoto recognized the 3-pyridylmethyl amine residue as the essential part for the insecticidal activity by referring to the model of nAChR agonists by Beers and Reich [4], and prepared quite a number of 3-pyridylmethylamines [5,6]. Nicotine, being an acetylcholine mimic, can bind to the receptor by two elements: (a) a

Coulombic interaction involving the positively charged pyrrolidinium nitrogen, and (b) a hydrogen bond through the pyridine nitrogen (Fig. 1). According to this model, the more basic the amine is, the stronger its Coulombic interaction with the nAChR should be, because the amine is fully ionised to the ammonium ion in the insect fluids. However, nicotine and nAChR agonists act on ganglia of the central nervous system in insects, and ionic molecules reach the ganglia only slowly through the lipophilic cuticles and the ion barriers. In other words, the more basic the molecule is, the less effective the penetration must be. This dilemma could never be solved by molecular designs based on nicotine.

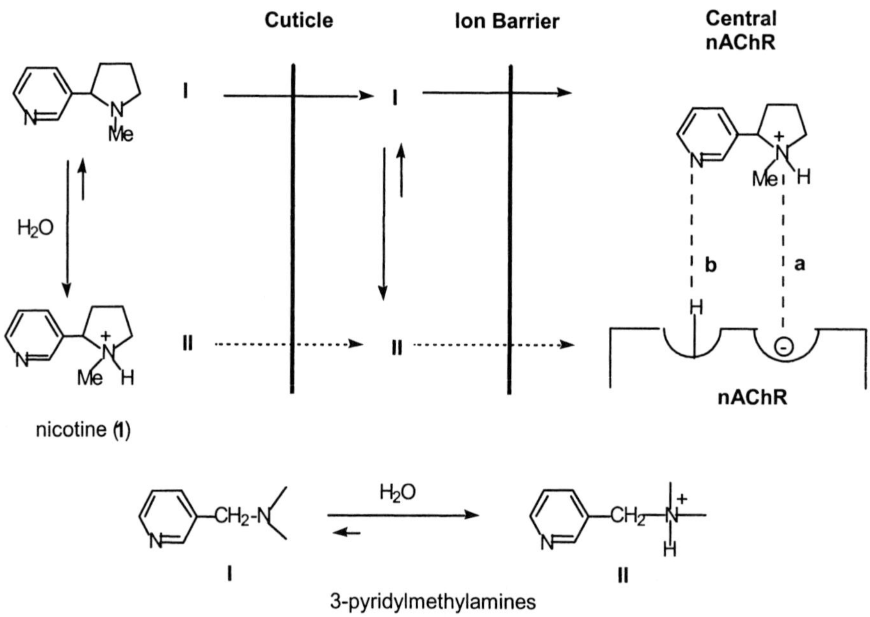

Figure 1. Action of Nicotine and 3-Pyridylmethylamines at Insect Central Nicotinic Acetylcholine

In 1978 Shell research group announced a new nAChR insecticide, nithiazine (**6**). Table 1 shows an abstract from their research, citing the two insect species, pea aphid and corn earworm [7]. Starting from a low-level lead (**2**) they worked their way to nithiazine, which has 1700 times higher potency against the target insect, corn earworm, in laboratory evaluation. The trail of their molecular design appears rational and straightforward. However, there were two misfortunes in their pursuit. Firstly, the chemistry had been largely concentrated on the nitromethylene amidine skeleton until they recognized the fatal photolability of this chromophore, which was first noticed in the field tests. Secondly, they seem to have neglected other species as the target than a lepidopteran, corn earworm, even though a divergence in the insecticidal spectrum was detectable in a few of the compounds prepared, such as **4**.

Table 1. Development of Nithiazine by Shell - 1970 ~ 1978 - [a]

No.	Compound	Toxicity index (TI)[b]	
		pea aphid	corn earworm
2	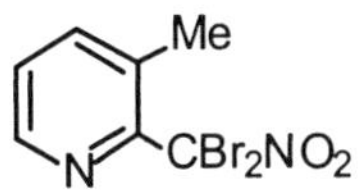	~5	0
3	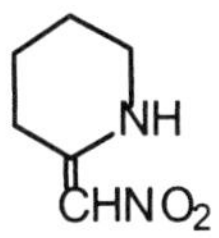	~3	155
4	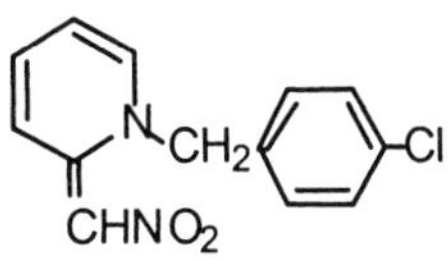	91	15
5	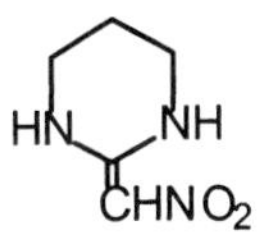	+	5.2
6	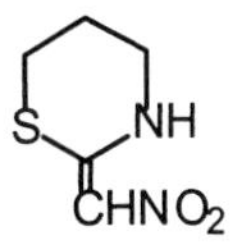	24	1662

[a] Extracted from ref 7. [b] TI = (LC_{50} parathion / LC_{50} test compound) X 100

3 Imidacloprid and the Subsequent Neonicotinoids

Our attention was caught by the remarkable structure and the unique insecticidal spectrum of nithiazine and we started our project in 1979 [8-13]. We chose the green rice leafhopper, a major rice pest, as our target insect for screening and optimising biological activity. Figure 2 shows our development flow of the milestone compounds along with the insecticidal potency LC_{90}, the concentration in ppm to kill 90% of the tested plant hoppers.

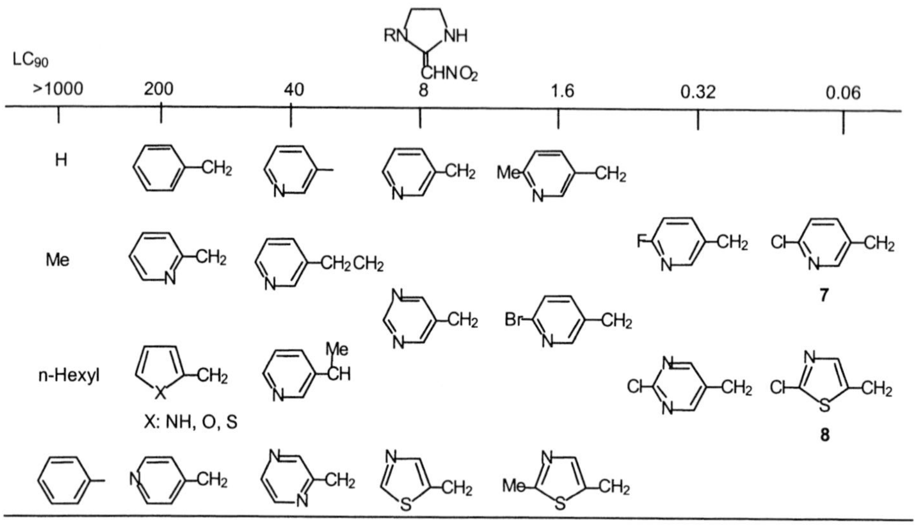

[a] LC_{90}: dose in ppm to kill 90% of the tested green rice leafhoppers

Figure 2. Development Flow towards Chloronicotinyl Insecticides

Starting with 1-methyl-2-nitromethyleneimidazolidine, application of bioisosterism as a tool of molecular design and optimisation through the introduction of new substituents led us to the discovery of 1-(6-chloro-3-pyridylmethyl)-2-nitromethyleneimidazolidine (7). With perfect hindsight, there were two key steps to success. The identification of the chloropyridylmethyl residue was the first. This substituent enormously enhanced the insecticidal potency. Look at the hierarchy according to the potency levels. How astounding! This group achieved a 3000-fold increase in the activity of low-level simply alkylated nitromethylenes. The finding of this activator also provided a further incentive to explore other potential bioisosteric heterocycles like chlorothiazolylmethyl (**8**), and ultimately shaped the core structure of this new insecticide class.

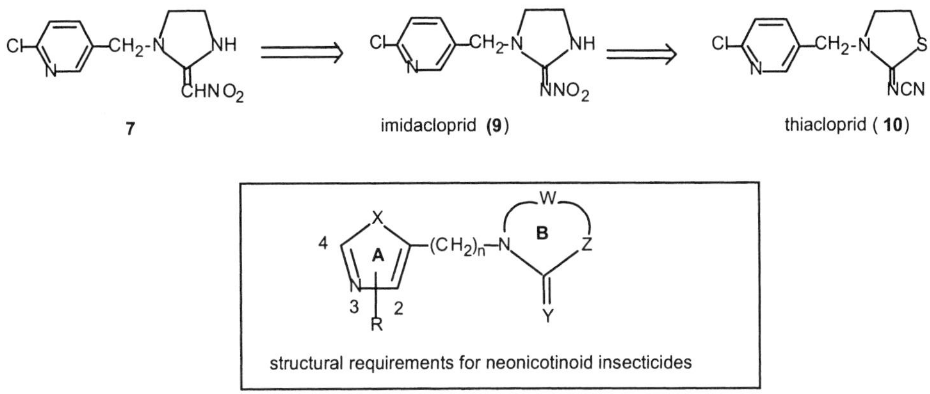

A ring

R: 4-Cl > 2-Cl
4-Cl > 4-Me > 4-OMe > H
N position: 3-N > 4-N >> 2-N
X: CH=CH, CH=N, O, S

Link

n: 1 >> 0 > 2

B ring

W: CH_2CH_2, $CH_2(X)CH_2$, CH=CH or open ring
Y: NNO_2, $CHNO_2$, NCN
Z: NH, NR, S, O, CH_2

Figure 3. Imidacloprid, Thiacloprid and the General Structure of Neonicotinoids

We continued to change the structure next for the functional group and found 1-(6-chloro-3-pyridylmethyl)-2-nitroimino-imidazolidine, that is, imidacloprid (**9**), with further improved activity. A parallel change to other electron-withdrawing substituents like 2-cyanoimino led to the discovery of thiacloprid (**10**). The second key step of our project was the observation that the fatal flaw of the original 2-nitromethylene chromophore, its photolability, could be overcome by replacing it with the 2-nitroimino or 2-cyanoimino chromophore.

As the research proceeded, the general structural requirements for active molecules became gradually clear. Thus, the basic molecule is made up of a chloro-substituted nitrogen heteroaromatic ring (A) joined through a methylene with a saturated nitrogen ring system (B). The conjugated nitromethylene, nitroimino, or cyanoimino group is essential for the activity (Fig. 3).

We should not forget that our success was also based on the screening approach to seek after compounds active against hemipteran rice pests, when we consider that the molecules we first prepared were mostly low in activity against lepidopterans, which are the target of nithiazine.

4 A few Notes on our Optimisation Process

In the optimisation process, we realized that the concept of bioisosterism and the available data for quantitative structure-physical property relationships were potentially a great help to pursue the optimisation effectively. But at the same time I should mention that even if comprehensive lists of bioisosteres and physicochemical parameters of substituents are available, it is not always easy to have an idea in advance about the actual properties of the new structure. I would like to present a few examples.

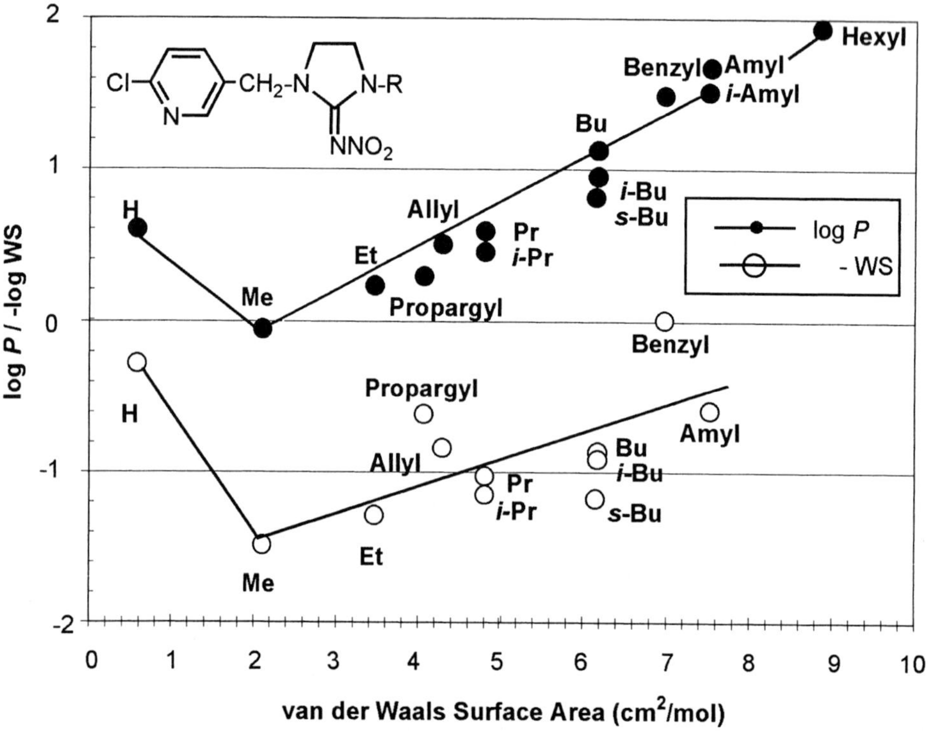

Figure 4. Relationship of log *P* and Water Solubility (WS, mmol/L) with van der Waals Surface Area of *N*-Alkylated Imidacloprid

We wanted to enhance the lipophilicity of imidacloprid and introduced alkyl groups to the imidazolidine nitrogen atom as a routine approach. Figure 4 relates the water solubility (WS) and log *Pow* values of 3*N*-alkylated imidacloprid derivatives. We can see clearly that molecules substituted with lower alkyl groups are more hydrophilic than imidacloprid itself. We can now understand such an anomaly by considering the disruption of the intramolecular hydrogen bonding [14]. Through this study we have learned that alkyl substitution can also enhance the hydrophilicity in certain cases.

Table 2. Partition Coefficients (log *P*) of Representative Chloronicotinyl Compounds [a)]

R: 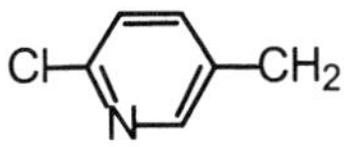	log *P*		log *P*
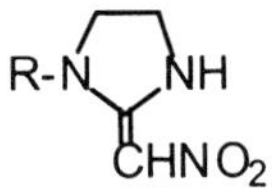 **7**	-0.19	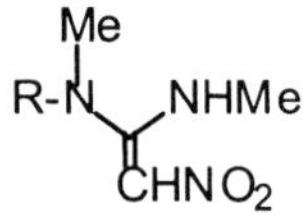 **12**	-0.71
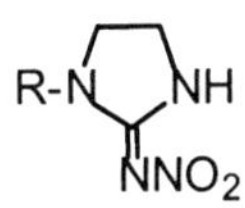 **9**	0.57	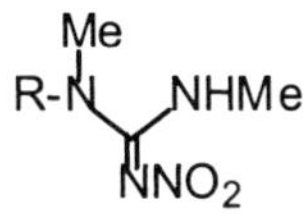 **13**	0.25
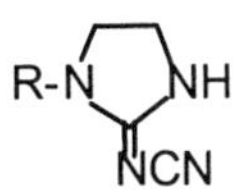 **11**	0.81	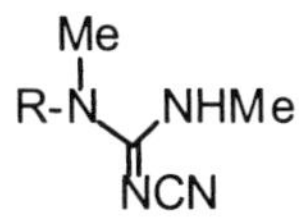 **14**	0.78

a) Flask shaking method at 25 °C.

The next case is the one where we could hardly predict the properties beforehand. Table 2 lists the log *Pow* values of representative neonicotinoid molecules [15]. We can see that nitroimine (**9**) and cyanoimine (**11**) are more lipophilic than the corresponding nitromethylene (**7**). I cannot explain why even now, anymore than we could foretell this physical property before measurements. From this table we see also that acyclic molecules are less lipophilic than the corresponding cyclic molecules.

Table 3. Electronic Absorption of Chloronicotinyl Compounds in Water and Half-life ($t_{1/2}$) Due to Photolyses [a)]

R: 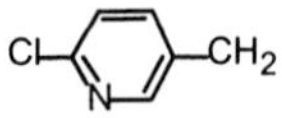	λ_{max} (nm) / log ε (log ε at 290 nm)	$t_{1/2}$ (hr)		λ_{max} (nm) / log ε (log ε at 290 nm)	$t_{1/2}$ (hr)
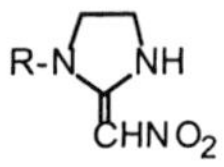 **7**	323 / 4.09	1	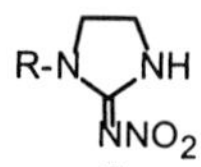 **9**	269 / 4.17 (3.73)	3
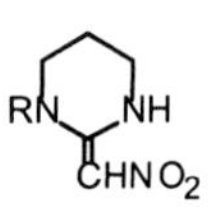 **15**	313 / 4.17	1	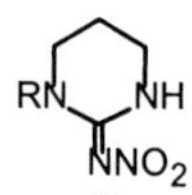 **17**	255 / 4.10 (3.21)	3.5
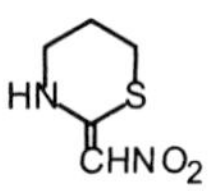 **16**	348 / 4.07	0.5	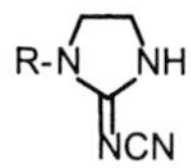 **11**	268 / 3.61 (<0.01)	>24
			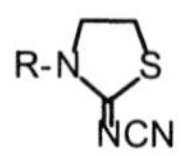 **10**	242 / 4.11 (<0.01)	>24

a) The $t_{1/2}$ data were obtained by irradiation of the compound in acetonitrile-water (8:2, v/v) solution by a 250 watt sun lamp at 30 °C.

Another example concerns the photostability. The half-lifes ($t_{1/2}$) using a simulated sunlamp for representative neonicotinoid compounds are given in Table 3 together with their electronic maximum absorptions, and the molar absorptions at 290 nm [16]. There are remarkable differences among these compounds. Nitromethylene compounds (**7**,**15**, **16**) decompose rapidly under the sunlamp, whereas the half-lifes of nitroimines (**9**,**17**) and cyanoimines (**10**,**11**) are significantly longer. We can understand now this photostability from their electronic absorptions. Thus, due to absorption in the upper atmosphere only sunlight of the wavelength between 290 nm and 400 nm is relevant at ground level. Accordingly, nitromethylenes with the maximum absorptions longer than 300 nm strongly absorb sunlight and decompose rapidly in the field, while nitroimines and cyanoimines, having essentially shorter maximum electronic absorptions than 290 nm, are expected to have greater photostability. This photostability difference between these chromophores can even be explained by calculation of the HOMO-LUMO energy gaps [17]. However, would someone have had an idea of such photolytic behaviour before the appearance of imidacloprid? I believe that we product seekers should learn the theoretical approaches as much as possible, but we should not swallow the outcome as such. For me, the first thing is actually to prepare by hand whatever comes into mind.

5 Consideration of Neonicotinoid Products from Molecular Design

Chain molecules can exert a similar biological action to the original ring compounds by forming a quasi-cyclic conformation when binding with the receptor. Thus, market products nitenpyram (**18**), [18] acetamiprid (**20**) [19] and clothianidin (**22**) [20] can be regarded as the open-ring structures of the original ring compounds (**7**,**19**,**21**), as pointed out in Figure 5. As mentioned above, acyclic molecules are generally more hydrophilic than the corresponding cyclic molecules. The excellent systemic property due to the enhanced hydrophilicity is one of the selling points for these products.

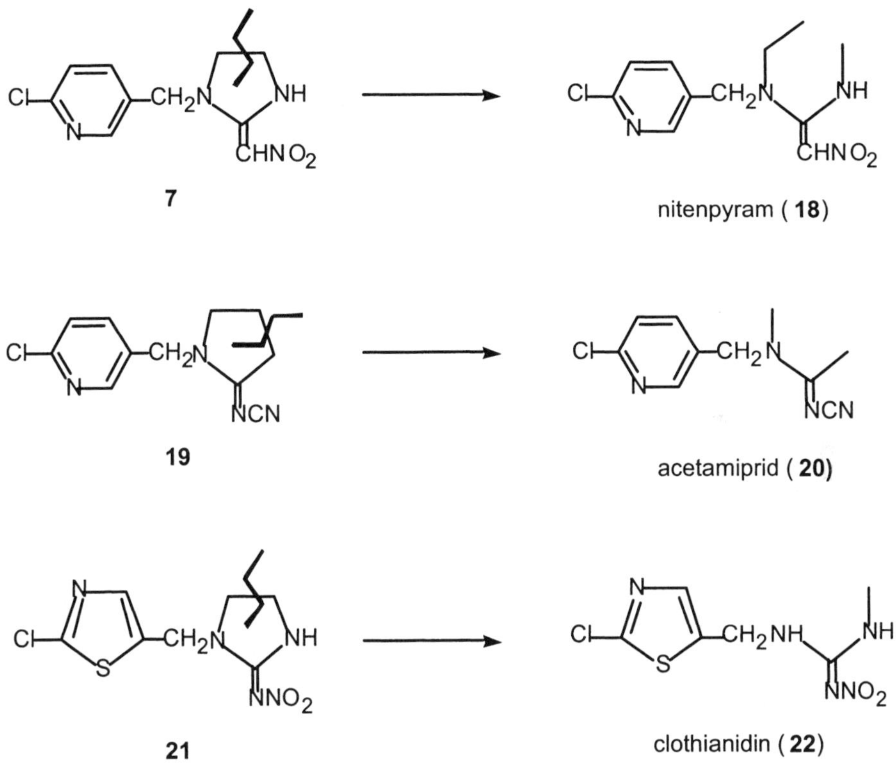

Figure 5. Molecular Design of Market Products by Acyclic ⟷ Cyclic Bioisosterism

Three-dimensional analyses of a set of ligands using the steric and electronic descriptors will furnish reliable information to drug designers. Figure 6, prepared by referring to a comparative molecular field analysis (CoMFA) for imidacloprid, nitenpyram and acetamiprid [21], shows the preferable region for negative electrostatic potentials around the nitro-oxygen atoms of imidacloprid and the corresponding areas of the other two molecules, and the sterically forbidden region beyond the imidazolidine 3-nitrogen atom or the like specified by Z. Also a permissible region is noted around the 6-chlorine atom of the pyridine ring. Such maps facilitate our insight into the binding mode from the ligand side. In general the maps based on the electrostatic and steric superimposition of molecules will be more exquisite with the increasing number of input molecules and disclose to us the form of active ligands in more detail, and the assumed picture advances stepwise to a persuasive model. For product seekers, however, it means that there may remain less chance to be able to contrive new molecules in a sense of market products. One way may be, paradoxical though it may seem, to concentrate on the ambiguous points of these maps, or to design a structure deviating from the model. I would like to name dinotefuran in this context.

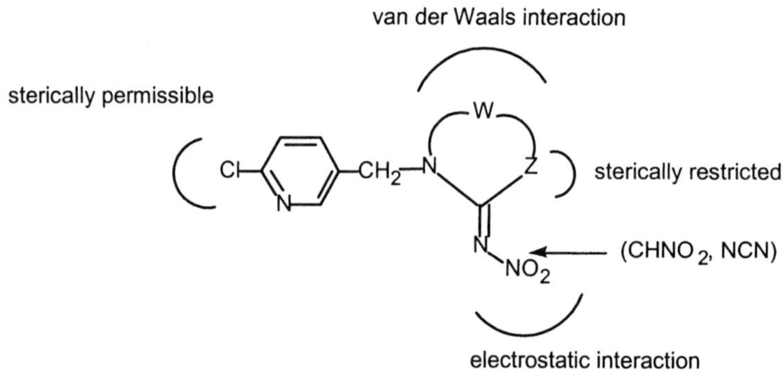

Figure 6. Predicted Properties of the Binding Site for Chloroniconinyl Insecticides Reproduced by Referring to the CoMFA of Okazawa *et al.*[21] (Specification of letters W, X, Z; see Fig. 3)

Dinotefuran (**23**) can be regarded as a molecule deliberately designed to deviate from the established general model for neonicotinoids (Fig. 7). It had been assumed thus far that the hydrogen bonding due to pyridyl or thiazolyl nitrogen plays an important role for the activity. Mitsui scientists replaced the prevailing nitrogen heteroaromatic rings with a simple saturated oxygen-containing heteroring at the cost of undermining the assumed hydrogen bonding and found a new product. The designers deserve to take pride in its unique insecticidal spectrum, distinguishing it from existing neonicotinoid products [22].

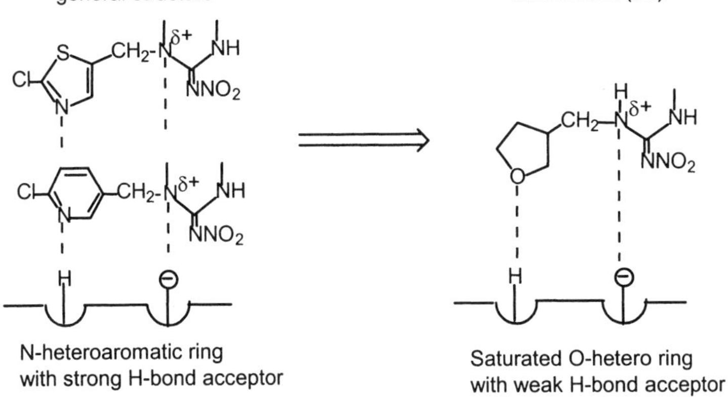

Figure 7. Dinetofuran: A View of Molecular Design to Deviate from the Assumed General Structure

Successful designs of a molecule which has to be transformed in order to show or to enhance its biological activity, a so called prodrug, are found time and again in developed pesticides [23]. We can see some examples also in the neonicotinoid field. If we can link both the amine ends of a potent insecticide like clothianidin through a bond which can be cleaved to release clothianidin before it reaches the site of action, a new, potentially insecticidal, molecule will be formed (Fig. 8). From this viewpoint we studied the hydrolysis of thiadiazine, triazine and oxadiazine derivatives (**25**,**26**) [24-27] in a physiological salt solution [15]. Table 4 compares their half-lifes. Oxadiazines, **25-O** and **26-O**, have notably longer half-lifes than the other aminals. Widmer *et al.* also reported about the hydrolysis experiment of **26-O** at various pH regions over a long period of time showing no degradation to the conceptual prodrug [28]. Judging from these experiments insecticide thiamethoxam **26-O** seems to be stable under hydrolytic conditions [29]. It is curious how the oxidative cleavage occurs in these aminals.

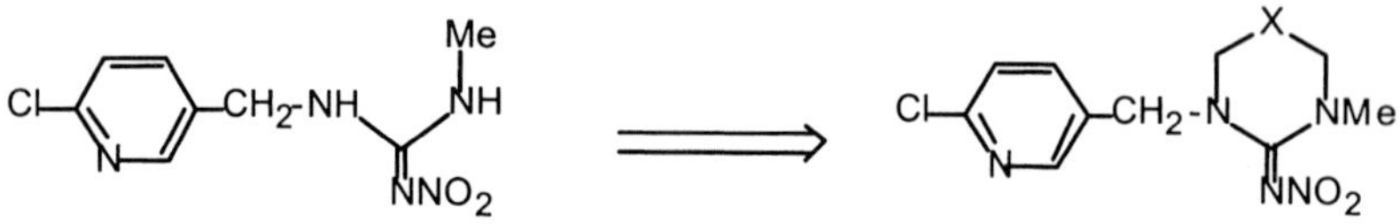

24 **25** X= O, S, NR

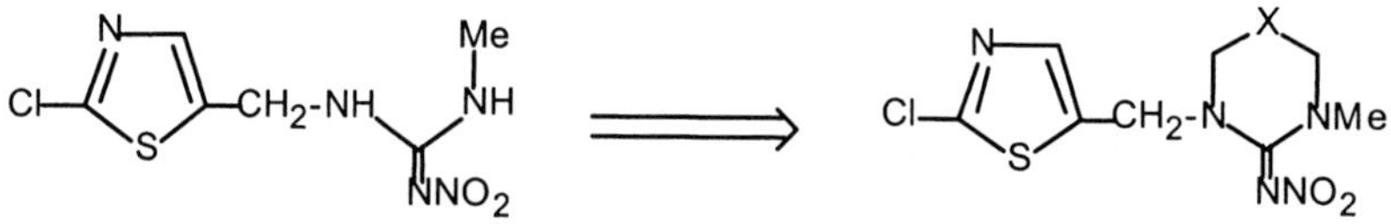

22 (clothianidin) **26** X= O (thiamethoxam), S, NR

Figure 8. Molecular Design by Prodrug Concept

Table 4 . Half-life $t_{1/2}$ (days) of compounds **25/26** in salt solution at 25 °C (pH=7.34)

Compound	**25-O**	**25-S**	**25-NMe**	**26-O**	**26-S**	**26-NMe**
$t_{1/2}$	132	12	5.3	72	6	8

Another example of a prodrug-based molecular design is the Mannich adduct (**27**) of nitromethylene compound (**7**) with amine and formaldehyde, which were prepared independently by Kishida *et al.* [30] and Krüger *et al.* (Fig. 9) [31]. Nauen *et al.* reported that these adducts are hydrolyzed to the original material in acidic milieu [2].

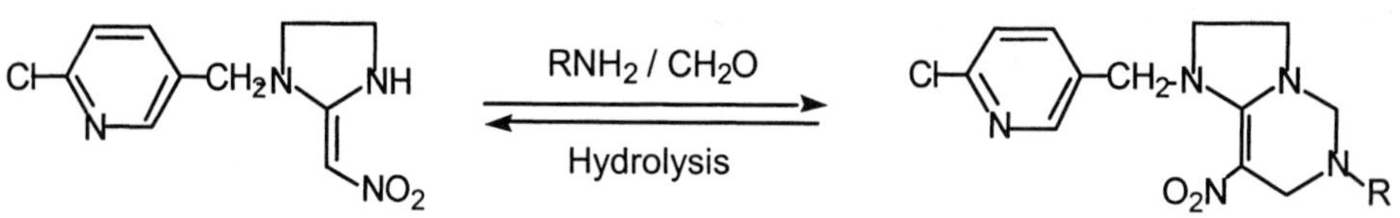

7 **27**

Figure 9. Molecular Design of Prodrug: Mannich Adducts

Metabolites of pesticidal products in the environment often give useful hints for molecular designs. Studying the metabolic pathway in plants we saw that imidacloprid (**9**) decomposed to the olefinic metabolite (**29**) through the hydroxylated intermediate (**28**). Interestingly, olefin (**29**) showed higher pI_{50} value for housefly nAChR than imidacloprid [32]. Bayer scientists presented a design of potent insecticides like methoxides (**30,31**) from knowledge of the metabolites (Fig. 10) [33].

metabolites of imidacloprid

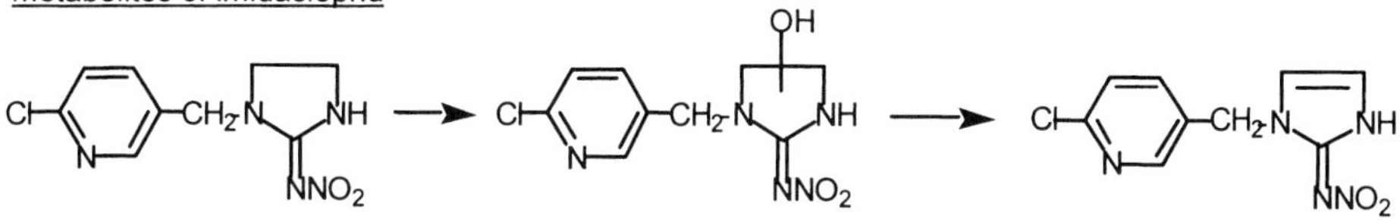

imidacloprid (9) 28 29

Affinity to nAChR from housefly head membranes (pI_{50})

pI_{50}	9.3	8.1	9.6
Compound	**9**	**28**	**29**

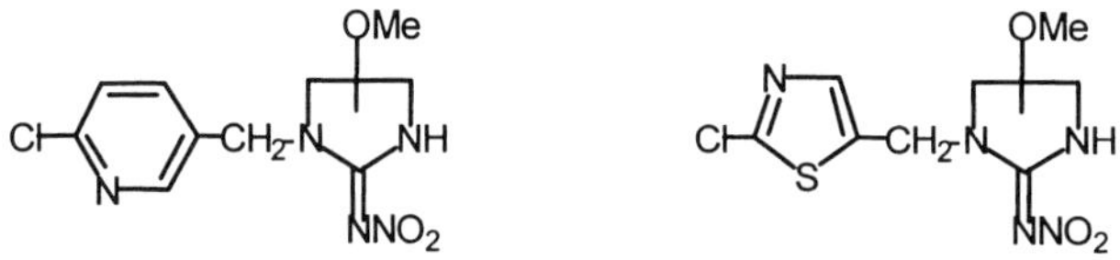

30 **31**

Figure 10. Molecular Design Based on Active Metabolites

Molecular design includes topological construction of an active molecule so as to facilitate its binding to the receptor site. A molecule of divalent or polyvalent ligands linked through tethers is such a case. Polyvalent molecules can bind to the receptor readily with a thermodynamic profit. Thus, the hydrophobic tethers will play a measurable role to reduce the entropy loss that would occur upon binding the independent monomeric units [34]. With this aim we prepared a set of alkylene tethered bivalent imidacloprid derivatives (**32**). These molecules can be expected to bind simultaneously to the dual sites in the receptor. An interesting result was found; the hexamethylene tether molecule has high insecticidal activity, even higher than imidacloprid, for the American cockroach (Table 5) [35].

Table 5. Insecticidal Activity of Divalent Imidacloprid Derivatives[a)]

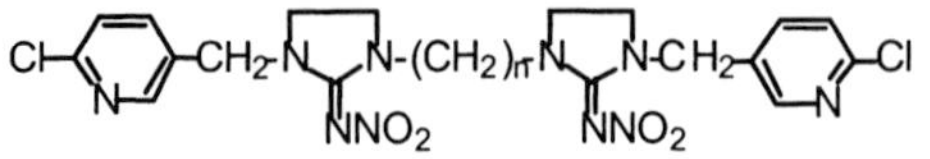

32

n	* c)	2	3	4	5
MLD[b)]	2.5	8.3	4.9	11	3.9
n	6	7	8	9	10
MLD[b)]	2.2	2.8	4.5	2.8	3.7

[a)] Injection treatment to *P. americana.* [b)] Minimum lethal dose (nmol). [c)] Imidacloprid.

6 Neonicotinoid Molecules for the Elucidation of nAChR Structure

The selective toxicity of neonicotinoids is mainly conferred by their outstanding specificity for insect nAChRs. Molecular design based on the structural differences in binding subsites among pests and/or between mammals may lead to a new era of more effective and safer insecticides. The functional architecture and diversity of insect nAChRs are poorly understood relative to those of mammals. The identification and characterization of insect nAChR subtypes is a current interest. Neonicotinoids are probes for exploring structures and diversity among the insect nAChRs not only because of their high affinity to the insect nAChR but also because of their suitable physicochemical properties such as non-ionizability and moderate water solubility. Casida's group developed a neonicotinoid-agarose matrix (**33**) and successfully applied it to isolate three putative subunits of the native *Drosophila* and *Musca* nAChR from head membrane preparations (Fig. 11) [36,37].

Photoaffinity labelling is another approach to access and characterize the ligand-binding site environment in the receptor. The same group constructed two promising azido-substituted photoaffinity probes, [^{125}I]benzoylneonicotinoid (**34**) [36] and [^{3}H]neonicotinoid (**35**) [38,39]. Along with the azido a diazirinyl group is another prevailing photogenerator [40]. Matsuda *et al.* designed ^{3}*N*-[*m*-(trifluoromethyl)diazirinyl benzyl]imidacloprid as a photoaffinity probe [41].

Neonicotinoid Agarose Matrix

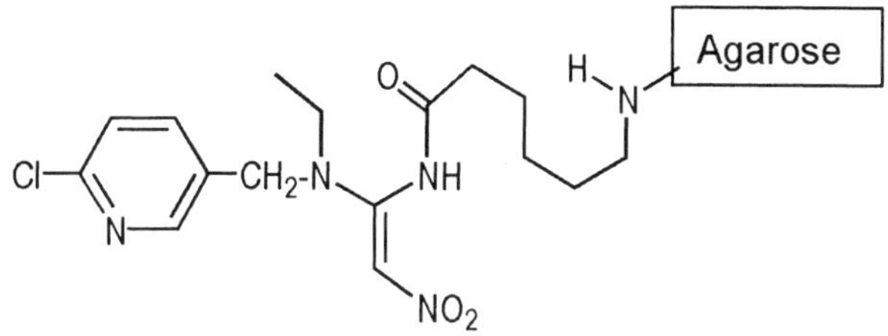

33

Neonicotinoid Photoaffinity Radioligands

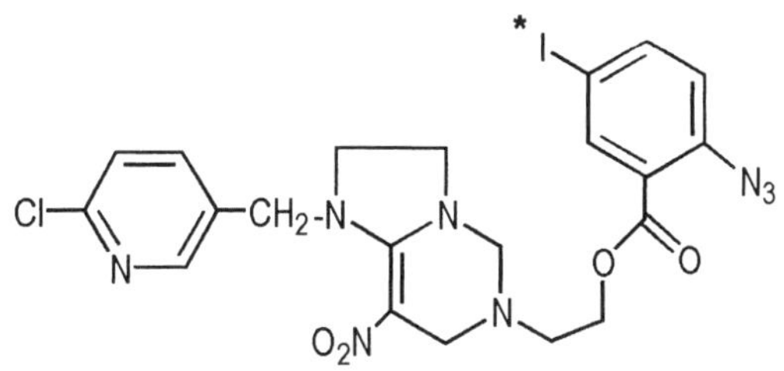

34

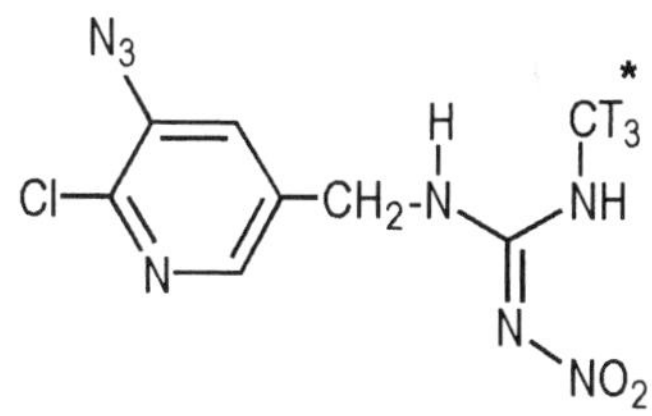

35

Figure 11. Molecular Design for Elucidation of the nAChR Structure

7 Neonicotinoids as Mammalian nAChR Agents

An exciting topic in the area of neonicotinoids is the discovery of the natural analgesic epibatidine (**36**). Epibatidine is one of the most potent nAChR ligands known to date [42]. Epibatidine possesses some structural features resembling those of neonicotinoids, that is a 6-chloro-3-pyridyl moiety and an amino nitrogen atom at almost the same internitrogen distance. On the other hand, it is assumed that the nitro group of imidacloprid and the corresponding electron-withdrawing groups bearing a hydrogen acceptor tip are playing a vital role for the high selectivity of neonicotinoid insecticides to the insect nAChR receptor [43-46]. So we can expect that the displacement of these anchors to construct a similar structure to epibatidine and nicotine will impart to the insecticidal neonicotinoid molecules an affinity to mammalian nAChRs (Fig. 12).

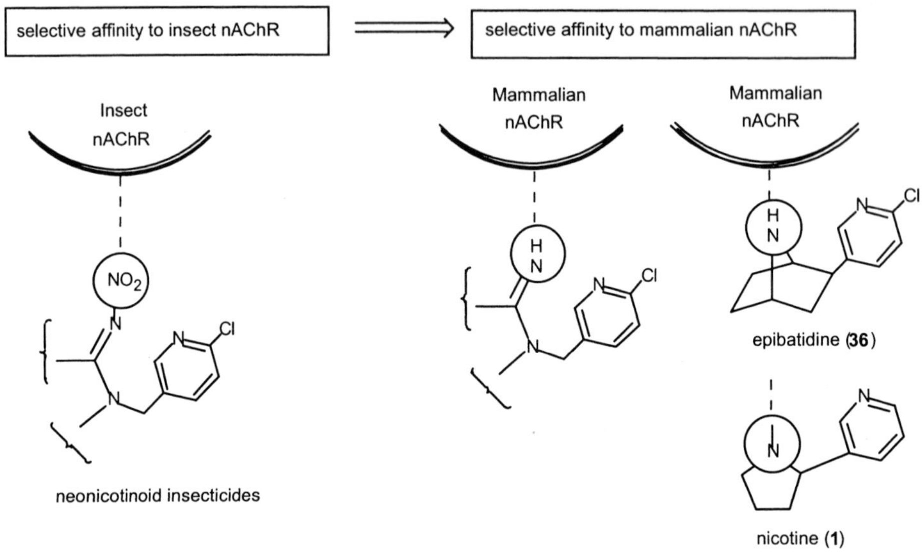

Figure 12. Design of Possible Ligands for Mammalian nAChRs

Table 6 compares the toxicity LD_{50} values of nitro and desnitro compounds to housefly and mouse. Actually desnitro imidacloprid (**37**) and the analogous imine (**38**) show strong toxicity to mouse, whereas the original insecticides (**9**,**10**) show the high selectivity to the housefly [45]. At present neonicotinoids and related compounds including *N*-unsubstituted imines will be a focus in the intensive search for potential agents for neuropathic disorders, cognitive enhancement and analgesia.

Table 6. Toxicity of Neonicotinoids and Analogues to Housefly and Mouse[a)]

Compound No.		housefly LD_{50} (μg/g)	mouse LD_{50} (mg/kg)
9	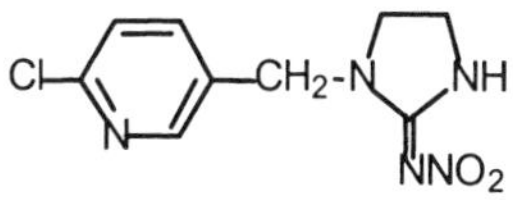	0.02	168[b)]
10	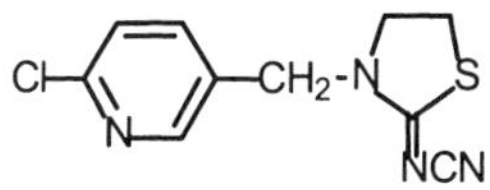	0.03	147[b)]
37	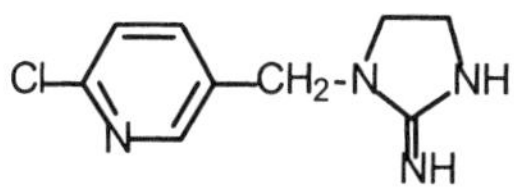	>5	8
38	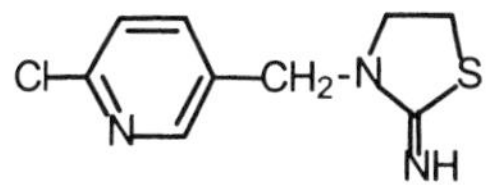	>5	1.1
36	epibatidine	>25	0.08
1	nicotine	>50	7

[a)] Extracted from ref 45. [b)] Females (fasted), oral application; data taken from Bayer CropScience.

8 Future Topics of Neonicotinoids

Finally I would like to consider briefly the future topics of neonicotinoids that would interest product seekers. First is to find new neonicotinoid members. Imidacloprid has extremely high intrinsic insecticidal potency. Nevertheless there still could be many possibilities to develop new insecticides by modifying its functional and prosthetic groups, leading to different physicochemical properties. A newly designed molecule could have different aspects from imidacloprid involving penetration, translocation to the target site or unique binding features with the receptor, and consequently a new insecticidal profile.

Additional interest for product seekers will be a search after systemic lepidopterocides from this class. The present neonicotinoid products are primarily targeting the sucking and chewing insects, and are rather weak against biting insects like lepidopterans, while the lead compound nithiazine has an opposite tendency. Both compounds, however, have the common structural features and exert their activity by similar mechanisms of action. It can be expected that some lepidoptericidal molecules exist somewhere on the line between the both structures.

Second is the resistance problem. To our knowledge there are no examples of synthetic insecticides that have been totally immune from insect resistance development. Should neonicotinoid be an exception? The resistance is emerging in most cases at laboratory levels but only in some cases in field practices too for the time being [47,48]. It is high time for molecular designers to think of resilient compounds against resistant strains.

As a third topic I would like to add the exploration of the nAChR as a target for pharmaceuticals for human health. Such target research on nAChR has received enormous input not only from genomic and physiological studies but also from investigations with compounds acting at this target site, compounds coming out of research directed towards insecticides as well as their metabolites, and compounds originating from chemical modification of the structure of the original naturally-occurring insecticide acting at the nAChR, nicotine.

I did not mention above the genome-based and computer-assisted molecular designs of insecticides. In future with the remarkable progress in computer science and gene technology along with the introduction of combinatorial synthesis and high through-put screening, the highly integrated network system will undoubtedly be of a great help for product seekers. However, no matter how the systematized methodology may advance, I believe, the decisive step to the discovery will be made only by the serendipity of the individual scientist.

In closing this review I would like to thank former colleagues in the Bayer group. Over seven years I was able to share with them such an exciting challenge as digging out a jewel in a sand hill. Indeed imidacloprid and thiacloprid found are glittering now in cropland.

9 References

[1] C. D. S. Tomlin (Ed.), The Pesticide Manual, 11th ed. BCPC, Farnham, 1997.

[2] R. Nauen, U. Ebbinghaus-Kintscher, A. Elbert, P. Jeschke and K. Tietjen in Biochemical Sites of Insecticide Action and Resistance (Ed.: I. Ishaaya), Springer, Tokyo, 2000, pp. 77-105.

[3] D. Wollweber and K. Tietjen in Nicotinoid Insecticides and the Nicotinic Acetylcholine Receptor (Eds.: I. Yamamoto and J. E. Casida), Springer, Tokyo, 1999, pp. 109-125.

[4] W. H. Beers and E. Reich, Nature 1970, **228**, 917-922.

[5] I. Yamamoto in Advances in Pest Control Research (Ed.: R. L. Metcalf), Intersci. Pub. New York, 1965, pp. 231-260.

[6] I. Yamamoto and M. Tomizawa in Pesticide/Development: Molecular Biological Approaches (Eds.: T. Mitsui, F. Matsumura and I. Yamaguchi), Pesticide Sci. Soc. Japan, 1993, pp. 67-83.

[7] W. D. Kollmeyer, R. F. Flattum, J. P. Foster, J. E. Powell, M. E. Schroeder and S. B. Soloway in Nicotinoid Insecticides and the Nicotinic Acetylcholine Receptor (Eds.: I. Yamamoto and J. E. Casida), Springer, Tokyo, 1999, pp. 71-89.

[8] K. Shiokawa, S. Tsuboi, S. Kagabu and K. Moriya (Nihon Tokushu Noyaku Seizo Co., Ltd), EP 0 192 060 A1, 1986 [Chem. Abstr. 1987, **106**, 28848].

[9] K. Shiokawa, S. Tsuboi, S. Kagabu, S. Sasaki, K. Moriya and Y. Hattori (Nihon Tokushu Noyaku Seizo Co. Ltd), Jpn. Kokai Tokkyo Koho JP 62-207266, 1987 [Chem. Abstr. 1987, **108**, 21897].

[10] A. Elbert, B. Becker, J. Hartwig, and C. Erdelen, Pflanzen. Nachr. Bayer 1991, **44**, 113-136.

[11] S. Kagabu, K. Shibuya, Y. Hattori, S. Tsuboi K. Shiokawa and K. Moriya, Biosci. Biotechnol. Biochem. 1992, **56**, 362-363.

[12] K. Moriya, K. Shibuya, Y. Hattori, S. Tsuboi, K. Shiokawa and S. Kagabu, Biosci. Biotechnol. Biochem. 1992, **56**, 364-365.
[13] S. Kagabu, Review. Toxicol. 1997, **1**, 75-129.
[14] S. Kagabu, K. Yokoyama, K. Iwaya and M. Tanaka, Biosci. Biotechnol. Biochem. 1998, **62**, 1216-1224.
[15] S. Kagabu and R. Hibino, unpublished data.
[16] S. Kagabu and S. Medej, Biosci. Biotechnol. Biochem. 1995, **59**, 980-985.
[17] S. Kagabu and T. Akagi, J. Pesticide Sci. 1997, **22**, 84-89.
[18] A. Akayama and I. Minamida in Nicotinoid Insecticides and the Nicotinic Acetylcholine Receptor (Eds.: I. Yamamoto and J. E. Casida), Springer, Tokyo, 1999, pp. 127-148.
[19] T. Yamada, H. Takahashi and R. Hatano in Nicotinoid Insecticides and the Nicotinic Acetylcholine Receptor (Eds.: I. Yamamoto and J. E. Casida), Springer, Tokyo, 1999, pp. 149-176.
[20] H. Uneme, K. Iwanaga, N. Higuchi, Y. Kando, T. Okauchi, A. Akayama and I. Minamida, 9th IUPAC Con. on Pestic. Chem. 1998, Abstract 1: 1D-009.
[21] A. Okazawa, M. Akamatsu, H. Nishiwaki, Y. Nakagawa, H. Miyagawa, K. Nishimura and T. Ueno, Pest. Manag. Sci. 2000, **56**, 509-515.
[22] K. Kodaka, K. Kinoshita, T. Wakita, E. Yamada, N. Kawahara and N. Yasui, Proc. 1998 Brighton Crop Prot. Conf. Pests and Disease, BCPC, Farnham, 1998, pp. 21-26.
[23] J. Drabek and R. Neumann in Progress in Pesticide Biochemistry and Toxicology, Vol. 5 (Eds.: D. H. Hutson and T. R. Roberts), Wiley, New York, 1985, pp. 35-86.
[24] K. Shiokawa, S. Tsuboi, K. Moriya, Y. Hattori, I. Honda and K. Shibuya (Nihon Tokushu Noyaku Seizo Co. Ltd.) Jpn Kokai Tokkyo Koho JP 02,235,881 A, 1990; EP 386,565 A1, 1990 [Chem Abstr. 1991, **114**, 185521].
[25] K. Ishimitsu, J. Suzuki, T. Kishimoto, H. Ohishi, T. Yamada, R. Hatano and N. Takakusa (Nippon Soda Co., Ltd.), WO 91-1978 A1, 1990 [Chem. Abstr. 1991, **115**: 8833].
[26] W. Frank, A. Kariya, N. Katsuyama, A. Tsuji, S. Takasuka and S. Segami (AgroKanesho Co. Ltd.) Jpn Kokai Tokkyo Koho JP 03,218,370 A2, 1991 [Chem. Abstr. 1992, **116**: 123318].
[27] P. Maienfisch, O. Kristiansen and L. Gsell (Ciba-Geigy AG) Jpn Kokai Tokkyo Koho JP 04,273,863 A, 1992; P. Maienfisch and L. Gsell, EP 580,553 A2, 1994 [Chem. Abstr. 1994, **120**: 21774].
[28] H. Widmer, A. Steinemann and P. Maienfisch, 218th ACS National Meeting (New Orleans), 1999, AGRO-134.
[29] P. Maienfisch, F. Brandl, W. Kobel, A. Rindlisbacher and R. Senn, in Nicotinoid Insecticides and the Nicotinic Acetylcholine Receptor (Eds.: I. Yamamoto and J. E. Casida), Springer, Tokyo, 1999, pp. 177-209.
[30] H. Kishida, N. Sakamoto, K. Umeda and H. Fujimoto (Sumitomo Chemical Co., Ltd.) Jpn. Kokai Tokkyo Koho JP 04,173,788, 1992 [Chem. Abstr. 1992, **118**, 22251].
[31] B.-W. Krüger, H. Uhr, J. Kanellakopulos, E. R. F. Wolf, A. Turberg, N. Mencke, C. Erdelen, U. Wachendorff-Neumann and J. Harwig (Bayer AG), Ger. Offen. DE 4401635 A1, 1995 [Chem. Abstr. 1995, **124**, 8836].
[32] R. Nauen, U. Ebbinghaus-Kintscher and R. Schmuck, Pest. Manag. Sci. 2001, **57**, 577-586.
[33] K. Wagner, C. Elderlen, J. Hartwig, W. Leicht and W. Stendel (Bayer AG), Ger. Offen. DE 4236204 A1, 1994 [Chem. Abstr. 1994, **121**, 35611].
[34] M. Mammen, S.-K. Choi and G. M. Whitesides, Angew. Chem., Int. Ed. Engl. 1998, **37**, 2754-2794.
[35] S. Kagabu, K. Iwaya, H. Konishi, A Sakai, Y. Itazu, K. Kiriyama and K. Nishimura, J. Pesticide Sci. 2002, **27**, 249-256.
[36] M. Tomizawa, B. Latli and J. E. Casida, J. Neurochem. 1996, **67**, 1669-1676.
[37] B. Latli, M. Tomizawa and J. E. Casida, Bioconjugate Chem. 1997, **8**, 7-14.

[38] S. Kagabu, P. Maienfisch, A. Zhang, J. Granda-Minones, J. Hattenschwiler, H. Kayser, T. Maetzke and J. E. Casida, J. Med. Chem. 2000, **43**, 5003-5009.
[39] M. Tomizawa, Z. Wen, H.-L. Chin, H. Morimoto, H. Kayser and J. E. Casida, J. Neurochem. 2001, **78**, 1359-1366.
[40] H. Bayley, Photogenerated Reagents in Biochemistry and Molecular Biology, Elsevier, Amsterdam, 1983.
[41] K. Matsuda, M. Ihara, K. Nishimura, D. B. Sattelle and K. Komai, Biosci. Biotechnol. Biochem. 2001, **65**, 1534-1541.
[42] M. W. Holladay, M. J. Dart and J. K. Lynch, J. Med. Chem. 1997, **40**, 4169-4194.
[43] S. Kagabu and H. Matsuno, J. Agric. Food Chem. 1997, **45**, 276-281.
[44] A. Nakayama, M. Sukekawa and Y. Eguchi, Pestc. Sci. 1997, **51**, 157-164.
[45] M. Tomizawa, D. L. Lee and J. E. Casida, J. Agric. Food Chem. 2000, **48**, 6016-6024.
[46] K. Matsuda, S. D. Buckingham, D. Kleier, J. J. Rauh, M. Grauso and D. B. Sattelle, Trends Pharmacol. Sci. 2001, **22**, 573-579.
[47] A. Elbert, R. Nauen, M. Cahill, A. L. Devonshire, A. W. Scarr, S. Sone and R. Steffens, Pflanzen. Nachr. Bayer 1996, **49**, 5-54.
[48] M. Cahill and I. Denholm in Nicotinoid Insecticides and the Nicotinic Acetylcholine Receptor (Eds.: I. Yamamoto and J. E. Casida), Springer, Tokyo, 1999, pp. 253-270.

17 Mode of Action of *Bacillus thuringiensis* Toxins

Meibao Zhuang and Sarjeet S. Gill

Department of Cell Biology and Neuroscience, and Environmental Toxicology Graduate Program, University of California, Riverside, CA 92521, E-mail: sarjeet.gill@ucr.edu

1 Introduction

Bacillus thuringiensis, a gram-positive bacterium was first isolated from the silkworm, *Bombyx mori* in 1902 by Ishiwata [1]. *B. thuringiensis* strains have specificity toward insects from different orders, but most strains are active against lepidopteran, dipteran and coleopteran insects [2] (Table 1). The insecticidal activity of *B. thuringiensis* is attributed primarily to parasporal cystalline inclusions, which are produced during its sporulation stage of growth. These inclusions generally contain proteins, which have been called crystalline (Cry) and cytolytic (Cyt) proteins [3]. Due to their highly toxic specificity for target insects, Cry toxins are widely used as biological agents for the control of important insect pests in agriculture and of disease vectors [4-7]. More recently, vegetative insecticidal proteins (VIPs) produced during the vegetative phase, have been isolated from *B. thuringiensis* [8]. These VIPs are structurally distinct from the Cry proteins. In addition many *B. thuringiensis* also produce binary toxins that are structurally unrelated to Cry toxins [9-11]. Moreover, the toxicity of *B. thuringiensis* strains to insects is also dependent on virulence factors that play crucial roles in insect pathogenicity [12-14]. In this review we will focus primarily on the Cry and Cyt toxins.

2 *B. thuringiensis* Toxins and Genes

Since the first *B. thuringiensis* Cry protein was identified and cloned in 1981 by Schnepf and Whiteley [15], many toxin genes have been identified, and the number of known genes is continually increasing. In order to better organize these toxin genes, Hofte and Whiteley [3] created the first toxin structural similarity- and specificity-based nomenclature in 1989. This nomenclature classified the toxins into four groups CryI-IV, and a fifth group, Cyt. However, while this system provided a valuable framework, the ever-expanding sets of genes required a new nomenclature, based entirely on the similarity of toxin amino acid sequences [2] (Tables 1 and 2). To date, more than 243 Cry and Cyt genes have been identified, and a current list of these genes is available online (http://www.biols.susx.ac.uk/Home/Neil_Crickmore/Bt/).

Genes encoding these Cry and Cyt proteins are generally located in large stable extra-chromosomal plasmids, which range from 2.1-kb to 273-kb [16, 7]. Loss of these large plasmids usually results in loss of insect toxicity [18,19]. More than one toxin gene may be present in one *B. thuringiensis* strain, on the same or different plasmids. In *B. thuringiensis* subsp. *israelensis*, the *cry*4A, *cry*4B, *cry*10A, *cry*11A, and *cyt*1Aa genes are all located on a 110-kb plasmid, while the *cry*1Aa and *cry*1Ab genes in *B. thuringiensis* subsp. *kurstaki* HD1 are harbored by 225-kb and a 66-kb plasmids, respectively [20-22]. Interestingly, *cry* gene probes frequently hybridize to *B. thuringiensis* chromosomal DNA; although it is unclear whether and how these chromosomal homologs contribute to production of toxins [23].

Early studies showed *B. thuringiensis* species contained a large number of transposable elements including insertion sequences. Structural analysis revealed that *cry* genes are frequently flanked by insertion sequences. For example, the *cry*1A gene is flanked by two sets of inverted repeated sequences [24, 25]. These insertion sequences were designated IS231 and IS232, which belong to the IS4 and IS21 family of insertion sequences respectively [26]. The IS240 sequence, belonging to the IS6 family, is usually found associated with *cry* genes from dipteran-active *B. thuringiensis* subsp. *israelensis* [27], *jegathesan* [28], and *fukuokaensis* [29]. It is hypothesized that these transposable elements allow for horizontal transfer of *cry* genes between different *B. thuringiensis* strains.

Table 1. *Bacillus thuringiensis* crystal toxins and their specificity.

Specificity (Order)	Toxins
Lepidopteran	Cry1Aa1-Cry1Ka1, Cry1-like, Cry9Aa1-Cry9Eb1, Cry9-like, Cry15Aa1
Dipteran	Cry4Aa1-Cry4Ba4, Cry10Aa1-Cry10Aa2, Cry11Aa1-Cry11Bb1, Cry16Aa1, Cry19Aa1-Cry19Aa2, Cry20Aa1, Cry24Aa1, Cry25Aa1, Cry39Aa1, Cry40Aa1
Lepidopteran + Dipteran	Cry1C, Cry2Aa1-Cry2Ad1
Coelopteran	Cry3Aa1-Cry3Ca1, Cry7Ab1-Cry7Ab2, Cry8Aa1-Cry8Ca1, Cry14Aa1, Cry22Ab1, Cry23Aa1, Cry34Aa1-Cry34Ba1, Cry35Aa1-Cry35Ba1, Cry36Aa1, Cry37Aa1, Cry38Aa1
Rhabditida (Nemato)	Cry5Aa1-Cry5Ba1, Cry6Aa1-Cry6Ba1Cry12Aa1, Cry113A1
Hymenopteran	Cry22Aa1
Dipteran + Cytolytic	Cyt1Aa1-Cyt2Bb1
Coleopteran + Cytolytic	Cyt2Ca1

3 Toxin Structure

Structures of five representative *B. thuringiensis* toxin proteins, Cry3A [30], Cry1Aa [31], Cry2Aa [32], Cry3Bb1 [33], and Cyt2A [34] are known. The Cry3Aa and Cry1Aa have 36% amino acid sequence identity, while the Cry2Aa and Cry1Aa proteins have 24% amino acid sequence identity. Nevertheless all Cry proteins have similar three-domain structures as demonstrated by X-ray crystallography.

Table 2. *B. thuringiensis* toxin domains and their proposed functions.

	Toxin/Region	Function	Ref.
Domain 1	Cry1	Toxin formation and stability	[174]
	Cry1Aa and Cry3Aα7	Toxin stability	[30,31]
	Cry1Ab helices α1-α2	Toxin stability	[60]
	Cry1Ab α3-α4	Irreversible binding, toxicity	[116]
	Cry1Ab5/Cry1Ac1α1	Toxin oligomerization	[175]
	Cry1Ab/Cry1Ac	Insertion, irreversible binding	[80,176]
	Cry1Ab/Cry1Ac, α5	Pore formation, toxicity, stability	[100-103]
	Cry1Ac α2-α3, α4-α5	Insertion, orientation	[104]
	Cry1Ac α7	Ion transport, toxicity, stability	[177-179]
	Cry1Ac α4-loop-α5	Toxicity, membrane insertion, pore formation, toxin oligomerzation	[102,180]
	Cry2Aa	Pore formation, ion efflux	[181]
	Cry3A	Crystallization	[182]
Domain II	Cry1Aa	Receptor binding, toxicity, specificity	[74,183,184]
	Cry1A loop2	Cadherin receptor binding, toxin activation, oligomeric and pre-pore formation, toxin insertion	[61,81]
	Cry1Aa loop3	Cadherin receptor binding	[61,81]
	Cry1Aa/Cry1Ab loop3	Initial binding, toxicity	[77]
	Cry1Aa/Cry3A β3	Toxin stability	[30,31]
	Cry1Ab loop2	Receptor binding, toxicity	[78,185]
	Cry1Ab/Cry1Ac, loop2	Ion transport, toxicity, APN binding	[79,186]
	Cry1Ac	Aminopeptidase binding, high affinity	[89]
	Cry1Ac apical loop	Orientation, irreversible binding	[104]
	Cry1Ac loop8	Aminiopeptidase binding, toxicity	[187]
	Cry1C loops 2 and 3	Receptor binding, toxicity	[188]
Domain III	CryAaβ17	Ion channel activity, toxicity	[31,105]
	Cry1Ac	GalNAc-mediated aminopeptidase binding, toxicity	[85,87,89,90]
	Cry1Ac	Structural integrity, toxicity	[189]
	Cry1C	Specificity	[86]

Domain I consists of an antiparallel α-helical bundle in which helix 5 is located in the center. Domain II consists of three-antiparallel β-sheets, while domain III is a β-sandwich with a "jelly roll" topology formed by two twisted, antiparallel β-sheets. The core of the conserved Cry structure, encompassing central helix α5 and all domain interfaces, are built up from five sequence blocks that are highly conserved throughout the *cry* gene family [3,5,35]. Because helices α3-α7 are long enough to span a membrane bilayer, and they have an amphipathic surface along their length, domain I was recognized as being responsible for pore formation, while domain II and domain III were thought to be important for receptor binding and specificity [30] (Table 2). Phylogenetic analyses showed the three domains of these Cry proteins have evolved independently, and domain recombination has taken place between pairs of different toxins [36]. These events have resulted in similar mode of action of Cry toxins but with distinct toxic specificities. Domain I seem to have a common origin, while domains II and III are common only among small subgroups of Cry proteins. This speculation was consistent with the roles of these domains. Furthermore, phylogenetic division among domain I sequences showed a correlation with insect susceptibility, suggesting domain I has been selected for functioning in distinct membrane environments in different insects [36].

Unlike the Cry proteins, Cyt toxins have a much broader toxicity spectrum. They are able to lyse a wide range of cell types *in vitro*, corresponding to the fact that Cyt proteins are structurally and functionally distinct from Cry toxins. To date, 18 *cyt* genes have been identified, and most of them were isolated from dipteran-active *B. thuringiensis*. X-ray crystallography of Cyt2A revealed that this toxin consists of a single domain, in which two outer layers of α-helices wrap around mixed β-sheets [34].

4 Mode of Action

B. thuringiensis crystals are first solubilized in the midgut of susceptible insects, followed by activation of the protoxins to active toxins by midgut proteases. The activated toxins then bind to midgut membrane receptors, insert into the apical membrane and form pores. Formation of the pores causes loss of osmotic regulation, and eventually leads to cell lyses, which is thought to be responsible for insect death [4,5].

4.1 Toxin Solubilization

Crystal solubilization is facilitated by an alkaline pH of susceptible insects. The typical midgut pH is between pH 9-11 in lepidopteran larvae [37-39]. In mosquito larvae, the pH inside the posterior midgut/gastric caeca is between 7-8, while the pH inside the anterior midgut is close to 11 [40]. Thus alkaline buffers are usually used for *in vitro* solubilization of lepidopteran and dipteran active *B. thuringiensis* crystals. Differential crystal solubility can be useful in partial separation of toxins. For example, Cry1A toxins are fully soluble at pH 9.5, while the Cry2 proteins require a pH of 12 for complete solubilization [41]. Moreover, pH has different effects on Cry toxin pore-formation activities [42], and differences in the level of solubilization can contribute to toxicity differences among Cry proteins [43]. Thus a reduction in solubility may account for low level insect resistance [44]. Coleopteran insects have a gut pH (5-6) and it was first considered that processing of coleopteran-specific protoxins was unnecessary for toxicity because these protoxin sizes are naturally truncated (70-75 kDa). However, Koller et al [45] demonstrated that Cry3 protoxins dissolved only at pH above 10 or below 4, and the soluble toxin was more toxic than crystals to the cottonwood leaf beetle, *Chrysomela scripta*, suggesting that protoxin solubility and/or activation may correlate with toxin susceptibility in different coleopteran insects [46].

4.2 Activation

After solubilization, many protoxins are processed by midgut proteases to become active toxins (Figure 1). In general, Cry1 and Cry4 protoxins are 130- to 140-kDa in size, and are activated by proteases to toxic cores of 65- to 70-kDa. Toxins such as Cry2, Cry3, Cry10, Cry11, and Cry20 appear to be naturally truncated, and they present in crystals as 65- to 86-kDa proteins [47]. However, even in these cases proteolytic processing is critical for toxicity as discussed above for Cry3.

The major midgut proteases involved in toxin processing are trypsin-like [39], chymotrypsin-like [48,49], or thermolysin-like [48,50]. Activation of the 130- to 140-kDa Cry1A protoxins occurs through specific cleavages starting at the C-terminus and proceeding toward the N-terminus, leaving a 60- to 65-kDa protease-resistant core [41,51]. Although naturally truncated, the 65- to 86-kDa Cry2, Cry3 and Cry11 toxins also require proteolytic activation, and result in 60- to 65-kDa active cores [52]. The stability of this core varies with different toxins, with lepidopteran-active toxins having more stable cores than those from dipteran-active toxins [53]. Interestingly, DNA, which has been found to be tightly associated with the crystal, could also play a role in proteolytic activation [54]. In general, the C-terminal portion of protoxins is not essential for toxicity/specificity. This cysteine-rich region likely contributes to crystal stability by forming intermolecular disulfide bonds [55]. The N-terminal portion of the protoxin, however, is essential to cell viability, toxin stability, and/or expression [56].

The exact proteolytic cleavage point varies with different toxins. The Cry1A toxins are cleaved at R28 in the N-terminus, and at least in Cry1Ac, it is cleaved at K623 at the C-terminus [57]. The site of N-terminus processing varies with specific insects as well. For example, cleavage of Cry1Ac protoxin by midgut juices from *Bombyx mori* and *Manduca sexta* resulted in an active toxin beginning with arginine-28, while incubation with *Plutella xylostella* and *Spodoptera litura* midgut juices gave N-termini with leucine-46 and glycine-66, respectively [58].

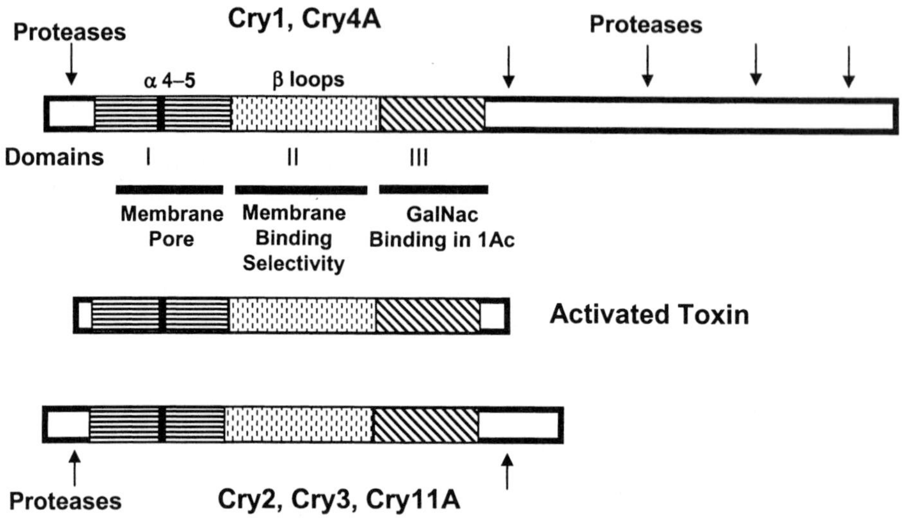

Figure 1. Processing of Cry toxins in the midgut of susceptible insects. Both N-and C-terminal processing by midgut proteases are essential for proper activation of the toxin. Activation by trypsin is often insufficient for the full range of biological activity of the toxins.

Proteolytic activation of Cry toxins is critical not only for protoxin activation, but has also implications for toxin specificity [59,60], receptor binding [61], and insect resistance [62,63]. The absence of a major gut protease in *Plodia interpunctella* correlated with its resistance to Cry1Ac [62]. Moreover, rapid degradation of Cry toxins was associated with the loss of sensitivity of 5th instar *S. litoralis* larvae to Cry1C [64], and serine protease inhibitors enhanced the toxicity of some Cry proteins up to 20-fold [65].

Recent studies show that intramolecular processing might occur *in vivo* and can be important for toxin activation [60,66]. In *Pieris brassica*, midgut juice treatment of Cry1Ac removed a small part of domain I and resulted in an active form, while treatment of Cry1Ac with midgut juice from a non-susceptible insect, *Mamestra brassicae*, cleaved the toxin at Arg423 in domain II, and the resultant toxin fragments were not toxic to *M. brassicae* [66]. Similarly, trypsin-activated Cry1A protoxins exhibited lower pore formation activity than the *M. sexta* midgut juice activated Cry1A, despite the fact that they retained toxicity. Cry1Ab [67] and Cry1Ac [68] mutants, which had the N-terminal processing sites removed, showed reduced toxicity. Treatment of Cry1Ab with *M. sexta* midgut juice removed helices α1 and α2 of domain I, resulting in an activated core of 58-kDa with higher pore formation activity and toxicity than trypsin activated toxin [60]. Moreover, Gomez et al [61] showed that Cry1Ab toxin binding to the cadherin-like receptor in *M. sexta* promoted this proteolytic activation and formation of a pre-pore complex, facilitating toxin insertion, which led to the formation of functional ionic pores *in vitro*.

4.3 Binding

Early studies showed a correlation between the biological toxicities of Cry proteins and their abilities to bind to midgut brush border membrane vesicles (BBMV) of susceptible larvae [69-71]. Toxin binding to membrane is a two-stage process involving reversible and irreversible steps, with irreversible binding generally associated with membrane insertion [69,70,72,73].

Usually, the binding affinity (K_d) and/or the binding site concentration (B_{max}) is correlated with Cry toxicity. Cry1Aa binds to *B. mori* BBMV with high affinity in a saturable form, while Cry1Ac, which is 200 times less toxic to *B. mori*, exhibits a much lower affinity to BBMV isolated from this insect [74]. The K_d values of Cry toxin binding to membrane receptors vary from relatively low to high, with a range between 10^{-7} to 10^{-10} M [4]. It is clear that insect cells may have multiple toxin receptors present on BBMV, thus in some cases, the toxicity of the Cry toxins appears to be correlated with receptor number rather than receptor affinity [70,71]. For example, the toxicity of Cry1Aa, 1Ab and 1Ac towards *H. virescens* is highly dependent on the binding site concentration in BBMV [70].

However, high affinity or high binding site concentration does not always result in high toxicity. In the gypsy moth, *Lymantria dispar*, there is a negative correlation between receptor affinity and toxicity [75]. Later studies showed that irreversible binding is better correlated with toxicity [73]. Therefore, a straight correlation between toxicity and toxin binding, according to either receptor affinity or receptor concentration, may not always be a generally applicable concept [4]. This also suggests that steps of irreversible binding, insertion and pore formation may be more important than affinity binding in its correlation with toxicity [76], although clearly initial receptor binding is important for these latter steps.

X-ray crystallography structures of Cry proteins predicted that the receptor binding sites of toxins are predominantly locate in domain II [30,31]. Indeed, a *Bombyx mori* receptor binds to Cry1Aa on amino acids 332-450, a hypervariable region of domain II [74]. Subsequent studies by Dean and colleagues showed that loop regions of domain II of Cry1A toxins are critical for receptor binding [77-79]. Moreover, truncated Cry1Ab protein containing only domains II and III can still bind to membrane receptors, but only reversibly [80]. This data suggests that irreversible binding requires the insertion of domain I. More recent data reveals in Cry1A toxins, loop 2 of domain II recognized the conserved epitope of the *M. sexta* cadherin receptor Bt-R1 ^{869}HITDTNNK^{876}, and loop 3 of Cry1Aa was also apparently important in the interaction with the Bt-R1 receptor [81].

Domain swapping between Cry toxins resulted in changes in insect specificity [82,83]. Binding experiments with these hybrid toxins showed that domain III is involved in binding to putative insect membrane receptors, suggesting that this domain may exert its role in specificity determination through receptor recognition [83-89]. Binding of domain III is particularly critical for GalNAc-mediated receptor binding [87], but may not directly correlate with insect toxicity [88]. Cry1Ac mutants of amino acids ^{509}QNR^{511} showed decreased aminopeptidase binding; however, they retained significant insect toxicity [88]. Structural simulation studies showed these amino acids are part of a pocket that binds GalNAc [87,88]. Thus binding of Cry1Ac to aminopeptidases may be a two-step reaction in which domain II and domain III are involved. The first, a reversible step, involves interaction between domain III and the GalNAc moiety on the receptor, and a second step, which involves an irreversible binding step, is initiated by binding of domain II to a separate receptor [88.90].

4.4 Pore Formation

Following initial receptor-mediated binding, toxin molecules insert into the apical membrane of columnar epithelial cells, and become resistant to proteases and monoclonal antibodies [91]. Toxin insertion subsequently induces formation of a nonspecific pore in the target membrane. Voltage-clamping studies of lipid bilayers [92 and the midgut sections [93,94 support the functional role of toxin in pore formation. The size and selectivity of the formed pore varies with toxins and insect species, but the nature of these pores is still controversial. Alternatively, it is described as a non-specific pore that has no ion selectivity or as an ion-specific channel that disrupts the membrane potential [5].

Like some other membrane active proteins and peptides [95-97], Cry toxins bind to the cell surface in a water soluble form, followed by an irreversible conformational change converting into a form capable of inserting into the membrane [35]. The Cry toxin structures revealed that putative membrane-spanning amphipathic helices located in domain I might be involved in pore formation [30,31]. Since the amphipathic helices predicted to span the membrane are buried in the helical bundle in domain I, a conformational change was predicted to expose a relatively non-polar/hydrophobic hairpin composed of helices α4 and α5 to initiate membrane insertion [30,31].

Subsequent studies have borne these early predictions. For example introduction of charged amino acid residues into the conserved hydrophobic domain (amino acid residues 162-171) of Cry1Ac results in lower toxicity [98]. Charged amino acids on Cry1Aa helix α4 were also shown to be essential for toxicity and ion transport, which suggested Cry1Aa α4 containing asparagine 136 faces the lumen of the ion channel as predicted by the crystal structure [99]. Mutations introduced into Cry1Ab and Cry1Ac conserved amino acid residues abolished toxin pore-formation activity and toxicity, supporting the speculation that α helix 5 is located in the center, and considered as the region that inserts into membrane [99-103]. Moreover, mutagenesis of α2b-α3 and α4-α5 loops of domain I and in the apical loops of domain II were found to inhibit irreversible binding, which supports a toxin orientation during membrane insertion with the α4-α5 loop towards the cytoplasm [104].

Several models have been proposed for toxin insertion into the membrane that leads to pore formation. An original "penknife" model [108] was replaced by an "umbrella" model after the crystal structure was determined. The latter model proposes that helices α4 and α5 drop down into the plasma membrane as a helical hairpin, and the remaining helices flatten out on the membrane surface with significant conformational changes to form an umbrella [30].

Domain III has also been shown to participate in ion channel activity. Cry1Aa and Cry3A show highly structural similarity in domain III. On the inner (facing domain I) sheet of the β-sandwich, four arginines on the highly conserved strand β17 participate in a network of charge interactions [31]. In Cry1Aa, substitution of the central arginines in this β-sheet resulted in decreased ion channel activity, accompanied by reduction of toxicity [105]. Single amino acid (R521) mutations on this arginine-rich region of Cry1Aa reduced channel conductance on planar lipid bilayers, suggesting this region of domain III interacts with domain I, and confirming the role of domain III in membrane permeabilization [106]. Mutation of the outer arginines to lysines was shown to inhibit permeability increases across the midgut membrane without affecting binding [105,107]. It was suggested that while the membrane pore is formed by elements from domain I, the arginines on β17 regulate the ion conductance through the pore by a gating mechanism [105].

It is accepted that both Cry and Cyt the toxin pores are formed by aggregation of several monomers [4,91,110]. Aggregated forms of the Cry1A toxins were extracted from *H. virescens* and *M. sexta* BBMV by mildly solubilization of membranes [91. Recent data supports these early predictions. By using atomic force microscopy, Vie et al [111] showed that the Cry1Aa pore consists of four subunits forming a pore of 1.5nm diameter. It is speculated that Cry toxin aggregation and pore formation occurs in a two-step process as observed with *S. aureus* α-hemolysin [112]. An initial aggregation likely occurs due to interactions among the hydrophobic helix α5, followed by formation of functional pores lined by the helix α4 [99]. Experiments using fluorescent-labeled synthetic peptides to measure the kinetics of fluorescent emission in BBMV implied that toxin aggregation might occur after integration of toxin monomers within the membranes, rather than occur at the surface of the membrane [113,114].

Cry toxins do not form spontaneous oligomers in solution *in vitro* [115], suggesting that oligomer formation *in vivo* may occur either after the toxin binds or by lateral movement of monomers in the membrane. Gomez et al. [61] recently showed that toxin binding to receptor promotes proteolytic activation of Cry1Ab by removing helices α1 and α2 of domain I, a prerequisite for toxin oligomerization.

Moreover, site-directed mutagenesis showed that loops between β sheets 2 and 3 of Cry1Ac domain II are important for reversible and irreversible binding [73,104]. The irreversible step is usually ignored when binding data are analyzed by Scatchard [69,70], thus data analysis accounting for both reversible and irreversible binding seems to be more accurate and more useful in understanding changes in toxin binding [73,116]. Measuring of protein-protein or lipid-protein interactions by surface plasmon resonance thus may be more accurate and can measure both association and dissociation rates [117-119]. More qualitative assessments of initial binding, especially of the variety of proteins involved, however, can be obtained by immunoblotting [114], although this data has to be interpreted with caution since the binding observed may not be correlated with affinity for the toxin.

Cyt toxins, however, are significantly different from Cry proteins. The Cyt1Aa protoxin is proteolytically activated by both N- and C-terminal cleavage to produce a protease-resistant core of 24-kDa, which lyses erythrocytes and is toxic to mosquito larvae [120,121]. The active Cyt protein has a single domain with an α/β architecture. Unlike Cry toxins, the Cyt toxins are thought to form pores through the β-sheet enriched C-terminus, with the α-helix enriched N-terminus remaining on the outside of the lipid membrane [34,109]. These α-helices of Cyt toxin are not involved in pore formation because they are too short to span a membrane lipid bilayer, and mutations reducing their amphiphilicity had no effect on toxicity [34]. In contrast, mutations that affect toxicity and phospholipid binding are located on loops linking the β-sheets. Thus the Cyt toxin pore formation is thought to involve a hinge movement of helices around these loops, followed by insertion of the underlying long amphiphilic β-sheets (β5-β7) into the bilayer, leaving the α helices exposed on the membrane surface [34].

5 Toxin Binding Receptors

5.1 Aminopeptidases

Aminopeptidases have been characterized as putative receptors for Cry1 toxins [122]. After the first aminopeptidases were cloned and identified from *H. virescens* and *M. sexta* [123,124], numerous aminopeptidases have been identified from various insects (Figure 2). In *H. virescens*, four aminopeptidase have been cloned and identified: APN120 [123], APN170 [125], APN110 [126] and APN180 (Oltean et al., unpublished data). These aminopeptidases are named after their protein migration sizes on SDS-PAGE, and they bind the Cry1Aa, Cry1Ab and Cry1Ac toxins differentially. APN120 [123] binds only Cry1Ac toxin, while the APN170 [125] binds Cry1Ac, Cry1Ab and Cry1Aa differentially. APN170 was designated as receptor A according to the three-site model proposed by Van Rie et al. [70,127], and it was suggested as a putative altered site in the resistant strain [128]. Reconstitution of APN170 into artificial vesicles mediated Cry1Ac pore formation [127]. However, APN170 co-purifies with its less glycosylated form, APN130. Although these two forms are products of the same gene, they have differential affinities for Cry1Ac with APN130 having a higher affinity [125]. These results explain the presence of two Cry1Ac binding sites that was observed with the 170-kDa protein complex [127]. The higher affinity of APN130 versus that of APN170 is not surprising since the Cry1Ac toxin recognizes N-acetylgalatosamine residues [87,89], and excessive glycosylation would be more difficult to displace in assays used to measure affinity parameters. Additionally a recent study identified a 110-kDa *H. virescens* aminopeptidase whose interaction with Cry1Ac was not mediated by GalNAc [126]. Oltean et al. (unpublished) identified another *H. virescens* aminopeptidase, APN180, whose binding to Cry1A toxins is dependent on GalNAc as well.

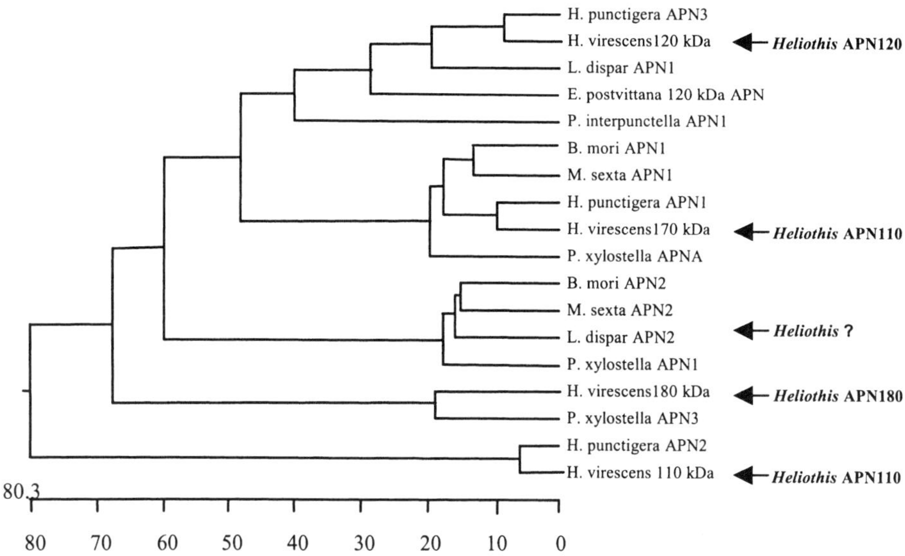

Figure 2. Phylogenetic relationship of lepidopteran insect aminopeptidases.

Aminopeptidases have also been identified from other insect species as Cry1 binding protein. Aminopeptidase cDNAs have been cloned from *M. sexta* [124], *Plutella xylostella* [129], *Plodia interpunctella* [130,131], *Helicoverpa punctigera* [132], *B. mori* [133-135], *Epiphyas postvittana* [136], *Trichoplusia ni* [137], and *L. dispar* [138,139]. Based on homology analysis of aminopeptidases clones from these insects it appears that lepidopteran insects each have at least four aminopeptidases with different affinities for Cry1A toxins (Figure 2).

All aminopeptidases cloned to date have several common sequence motifs (Figure 3). These include a signal peptide, zinc binding sites, and N- or O-linked glycosylation sites, and a GPI anchor attachment sequence motif. The presence of glycosylated moieties on aminopeptidases is required for binding to Cry1Ac toxin [87,89], with the exception of APN110 from *H. virescens* [126]. GalNAc was uniquely recognized by Cry1Ac but not Cry1Aa or Cry1Ab [140,141]. This data suggested that the latter toxins bind the protein directly rather than through glycoslyated moieties. Indeed the Cry1Aa toxin-binding domain for a *B. mori* aminopeptidase has been mapped to amino acids 135-198 [142].

Surface plasmon resonance analysis of Cry1Ac binding to *M. sexta* APN120 in a supported lipid monolayer showed that only the reversible binding was GalNAc sensitive [141]. Irreversible binding, which has a higher binding affinity, was not sensitive to GalNAc or reagents that disrupt protein-protein interactions, suggesting that GalNAc-mediated binding to the aminopeptidases is followed immediately by a rate-limiting step in the mode of action, enabling the toxin to insert into the lipid [35,89].

All aminopeptidases identified so far contain a glycosylphosphatidyl inositol (GPI) moiety through which they anchor to membranes [123,124,143,144]. These GPI-anchored proteins may play an essential role in targeting Cry toxins to lipid raft membrane microdomains and leading to toxin aggregation and pore formation [145]. Removal of the GPI-anchor with bacterial or endogenous phospholipase C converts the membrane-bound protein to become soluble [144]. Carbohydrate modification of these proteins may be critical in insect resistance to Cry1 toxins, since impaired glycosylation was a factor in Cry5B resistant *Caenorhabditis elegans* [146].

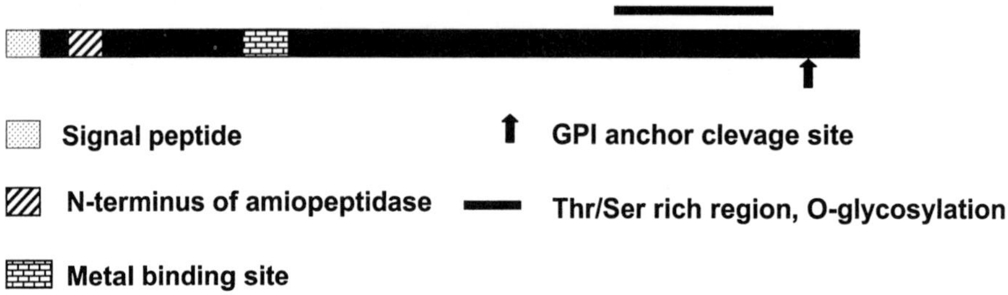

Figure 3. Common protein structural features of lepidopteran insect aminopeptidases.

5.2 Cadherins

Vadlamudi et al. [133] purified from *M. sexta* a 210-kDa protein (Bt-R[1]), which was termed a cadherin-like protein (Figure 4). This protein contained a putative signal peptide of 20-residues and a transmembrane domain of 22-residues at amino acids 1406-1428. It also had 11 cadherin-like repeats in the putative extracellular domain and a 110-residue C-terminus, which is shorter than the corresponding region in vertebrate cadherins (160 residues). These differences in insect and vertebrate cadherins may account for its unique function in Cry toxin binding [133]. The *M. sexta* cadherin-like receptor bound Cry1Ab with high binding affinity, and binds Cry1Aa and Cry1Ac with lower affinities. Cry1A binding cadherin-like proteins have also been isolated from *B. mori* [147,148]. Although similar in molecular size to *M. sexta* cadherin-like proteins, the *B. mori* cadherin protein migrates at about 175-kDa in SDS-PAGE.

In *H. virescens*, loss of expression of a cadherin-like protein was found to be associated with Cry1A toxin resistance and consequently this protein plays a crucial role in *B. thuringiensis* toxicity *in vivo* [149]. In the laboratory-selected resistant *H. virescens* strain YHD2, a retrotransposon insertion in the cadherin gene was linked to high levels of Cry1Ac resistance. More recently, disruption of a cadherin gene may also be linked to the development of field resistance to Cry1A toxins in the pink bollworm, *Pectinophora gossypiella* [150].

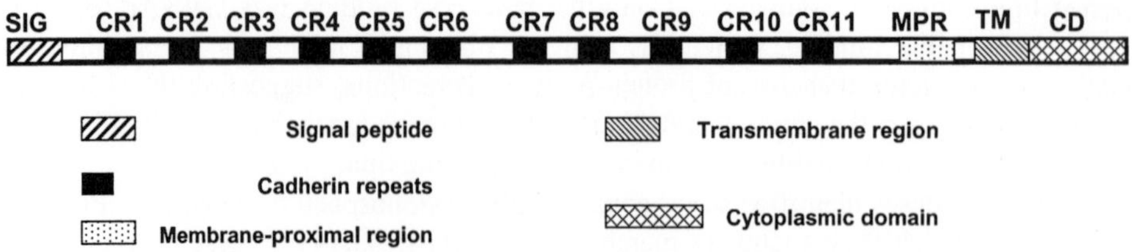

Figure 4. *M. sexta* cadherin-like protein showing predicted sequence motifs

The cadherin-like proteins have been successfully expressed in heterologous systems and the recombinant proteins have similar toxin-binding capabilities as the native proteins [133,151-153]. Insect cells expressing the cadherin-like protein also exhibited Cry1Aa toxin-induced cell swelling and whole cell patch clamp measurements indicating possible pore formation [147].

The toxin binding region of the cadherin-like proteins from *M. sexta* and *B. mori* have been mapped. In *B. mori* BtR175 a 146 amino acid region was shown to be involved in Cry1Aa binding [154]. In contrast by using phage display strategy, Gomez et al [155] identified a smaller region involved in toxin binding. They showed that an eight amino acid epitope in *M. sexta* Bt-R1 (^{869}HITDYNNK^{876}) and two amino acid epitope in *B. mori* BtR175 (^{873}IIDTNNK^{880} and ^{1296}LDETTN^{1301}) were involved in binding of Cry1A toxins, likely through hydropathic complementarity [81].

5.3 Other Binding Molecules

With a combination of chromatography and SDS-PAGE, more toxin-binding proteins that could not be separated by one-dimensional electrophoresis have been detected. Indeed, toxin overlay assays of chromatography fraction samples from *H. virescens* BBMV demonstrated that the Cry1Ac toxin binds *H. virescens* BBMV proteins of various sizes between 50- and 180-kDa [156]. Except for the aminopeptidases and cadherin proteins that have been identified, there are several Cry1A binding proteins with sizes of 55-, 90-, 100-, and 190-kDa. It would be interesting to detect what these proteins are, and whether they are involved in Cry1A toxin function. Moreover, in *L. dispar* a large highly gylcosylated protein that binds Cry1A toxins with high affinity has recently been identified [157]. The significance of this macromolecule in mediating Cry1A toxicity needs further investigation.

Neutral lipids have been co-purified with APN115 from *M. sexta* [122,158], and these lipids consist mostly of diacylglycerol and long, unsaturated free fatty acids. *In vitro* reconstitution of these lipid aggregates into phosphatidylcholine vesicles containing APN115 significantly facilitated the binding and pore formation ability of Cry1Ac to the vesicles [158]. More recently, Zhuang et al [145] showed that some toxin binding proteins and Cry1A toxins are associated with lipid raft microdomains, and lipid rafts integrity was essential for Cry1A toxin pore-formation. It was speculated that lipid rafts are important for toxin aggregation; however, it does not exclude the fact that some lipid molecules located in lipid rafts could bind Cry toxins and play a role in Cry toxicity.

Recently a novel p85 protein that binds Cry1Ac was identified (Lee and Gill, unpublished). This protein has a K_D of 15.7 nM with Cry1Ac, and interestingly is an intracellular protein. The significance of these findings is not clear, and further studies are needed to determine this protein's role in Cry toxicity action.

6 Resistance to *Bacillus thuringiensis*

When *B. thuringiensis* was first used for pest control, it was hoped that insects would not develop resistance to these toxins because *B. thuringiensis* and insects have apparently coevolved. However, starting in the mid-1980s, a number of insect populations from several different species were shown to have the capacity to develop resistance to Cry toxins under laboratory and field conditions [159,160]. Development of resistance to Cry toxins in the field poses the most serious threat to the continued use of this novel agent for control of insect pests.

McGaughey [161] reported the first case of insect resistance to *B. thuringiensis* in 1985. They showed that laboratory selection of the Indian meal moth, *Plodia interpunctella* with a *B. thuringiensis* subsp. *kurstaki* based commercial formulation, Dipel, led to a 100-fold decrease in insect susceptibility after 15 generations, and after 36 generations the resistance levels reached 250-fold [44,161,162]. Since then, many instances of insect resistance to Cry toxins have been reported. So far significant resistance against Cry1Aa, Cry1Ab, Cry1Ac, Cry1Fa, and Cry1Ja has been observed in a number of insect species, while resistance to Cry1Ba, Cry1Bb, Cry1Ca, Cry1Da, Cry1Ia and Cry2Aa was not considered significant [159,163]. While most resistant insects have been obtained from laboratory selections, there is increasing evidence that resistance development also occur under field conditions. For example, a diamondback moth, *Plutella xylostella*, population collected from Hawaiian fields heavily sprayed with Dipel showed 30-fold resistance to Cry1A [164]. Since the genetic diversity is lesser in laboratory populations, it is unlikely to predict the resistance mechanisms may develop in the field. However, such laboratory studies provide primary information on elucidation of resistance mechanism and development of resistance management strategies [165].

According to the multiple stage mode of action of *B. thuringiensis* Cry toxins, it is not surprising that insects selected for resistance to one toxin are still highly susceptible to other Cry proteins. For example, a field developed *P. xylostella* colony from Philippines showed over 200-fold resistance to Cry1Ab [166], while it was still fully susceptible to Cry1Ba and Cry1Ca. Following additional selection with Cry1Ab and a hybrid Cry1A protoxin under laboratory condition, another Cry1Ab resistant *P. xylostella* colony collected from the field gained resistance to Cry1Aa and Cry1Ac as well, but not to Cry1Ca, Cry1Fa, and Cry1Ja [167]. Another *P. xylostella* colony collected from Florida showed more than 200-fold resistance to Cry1Aa, Cry1Ab, and Cry1Ac toxins, but they were still susceptible to Cry1Ba, Cry1Ca, Cry1Da and Cry9Ca [168,169]. Similarly, a *H. virescens* colony (YHD2) selected with Cry1Ac protoxin displayed more than 10,000-fold resistance to Cry1Ac, but had only moderate cross-resistance to Cry2Aa and was almost nonresistant to Cry1Ca and Cry1Ba, even though this colony was highly cross-resistant to Cry1Ab and Cry1Fa. With continuous selection on Cry1Ac, the colony became over 230,000-fold resistant to this toxin, but the cross-resistance against Cry2Aa remained low [170,171]. The lack of cross-resistance encourages the continuous use of commercial products containing multiple toxins, especially if the toxins have different modes of action. Commercial formulations with Cyt protein containing *B. thuringiensis* subsp. *israelensis* have been used for mosquito and black fly control throughout the world for over twenty years, but little resistance has been found. This lack of resistance is due to the presence of multiple toxins with different modes of action in this strain [4]. Indeed, Wirth and colleagues [172] showed that the Cyt toxin in this strain plays a critical role in this case. The mechanism of how the Cyt toxins delay the development of resistance is not known, but probably Cyt1Aa facilitates the binding and/or membrane insertion of Cry4/Cry11 proteins in mosquito midgut.

Presently two major mechanisms of resistance have been demonstrated: altered proteolysis [62], and altered expression of the receptor [149]. However, since *B. thuringiensis* toxin act through a multiple-stage mechanism, resistance could develop during each of the steps in the mechanism of action. Table 3 shows potential sites and sites that have been found to be associated with resistance development for *B. thuringiensis*. At each of these sites resistance development could occur via mutations in multiple genes that modulate these specific sites. This example is best illustrated by elegant studies performed by Raffi Aroian and colleagues [146,173] with selection of *C. elegans* for Cry5B. Although five mutated genes were identified in this screen, it appears that modification in carbohydrate pathways may be a common mechanism of resistance in more than one of these genes. If these mutations lead to changes in Cry5B receptor modification, these data would suggest that even if resistance occurs at one site, e.g. receptor modification, multiple resistance genes could be involved.

Table 3. Sites associated with potential for development of insect resistance to *B. thuringiensis*.

Steps in the Mode of Action	Sites of Resistance Developed	Ref.
Solubilization	Reduced crystal solubilization	[44]
Activation	Lack of gut protease	[46]
	Changes in protease activity	[190]
	Reduced activation	[191]
Membrane receptor binding	Cadherin receptor mutation	[149]
	Altered APN170	[128]
	Reduced receptor binding	[192]
	Absence of membrane binding	[166]
	Lowered BBMV binding affinity	[193]
	Altered interaction between insect and toxin domain II	[163]
	Loss of carbohydrate modification enzyme	[146]
Conformational change	No published data	
Membrane insertion	No published data	
Toxin oligomerization	No published data	
Pre-pore formation	No published data	
Pore formation	No published data	
After pore formation	No published data	

7 General Conclusions

Much has been learned in the last decade on how *B. thuringiensis* toxins act. We have an excellent idea of the structure of these toxins, and the critical domains involved in receptor binding and pore formation in cell membranes. Information from resistance studies has also identified critical protein molecules involved in toxin binding. However, our understanding of how *B. thuringiensis* and its toxins actually kill insects is less illuminating. Characterization of virulence factors involved in insect toxication would go a long way to more thorough understanding of *B. thuringiensis* pathogenesis. Although resistance development has been observed in the field with lepidopteran pests, our experience with *B. thuringiensis* subsp. *israelensis* use in the field should provide clear direction to governmental agencies and companies on the utility of using toxin mixtures with multiple modes of action for the sustained use of *B. thuringiensis* in agriculture and in public health control programs.

8. References

[1] Luthy, P., and Ebersold, H. R. (1981) In: Pathogenesis of Invertebrate Microbial Disease. Davison, E. W. (Ed). pp 235-267. Allenheld, Osmun & Co. Publishers, Inc., Totowa. N.J.

[2] Crickmore, N., Zeigler, D. R., Feitelson, J., Schnepf, E., Van Rie, J., Lereclus, D., Baum, J., and Dean, D. H. (1998) Microbiol Mol Biol Rev 62, 807-813.

[3] Hofte, H., and Whiteley, H. R. (1989) Microbiol Rev 53, 242-255.

[4] Gill, S. S., Cowles, E. A., and Pietrantonio, P. V. (1992) Annual Rev Entomol 37, 615-636

[5] Schnepf, E., Crickmore, N., Van Rie, J., Lereclus, D., Baum, J., Feitelson, J., Zeigler, D. R., and Dean, D. H. (1998) Microbiol Mol Biol Rev 62, 775-806.

[6] Van Rie, J. (2000) Int J Med Microbiol 290, 463-469.

[7] de Maagd, R. A., Bravo, A., and Crickmore, N. (2001) Trends Genet 17, 193-199.

[8] Estruch, J. J., Warren, G. W., Mullins, M. A., Nye, G. J., Craig, J. A., and Koziel, M. G. (1996) Proc Natl Acad Sci USA 93, 5389-5394.

[9] Moellenbeck, D. J., Peters, M. L., Bing, J. W., Rouse, J. R., Higgins, L. S., Sims, L., Nevshemal, T., Marshall, L., Ellis, R. T., Bystrak, P. G., Lang, B. A., Stewart, J. L., Kouba, K., Sondag, V., Gustafson, V., Nour, K., Xu, D., Swenson, J., Zhang, J., Czapla, T., Schwab, G., Jayne, S., Stockhoff, B. A., Narva, K., Schnepf, H. E., Stelman, S. J., Poutre, C., Koziel, M., and Duck, N. (2001) Nat Biotechnol 9, 668-672.

[10] Ellis, R. T., Stockhoff, B. A., Stamp, L., Schnepf, H. E., Schwab, G. E., Knuth, M., Russell, J., Cardineau, G. A., and Narva, K. E. (2002) Appl Environ Microbiol 68, 1137-1145.

[11] Herman, R. A., Scherer, P. N., Young, D. L., Mihaliak, C. A., Meade, T., Woodsworth, A. T., Stockhoff, B. A., and Narva, K. E. (2002) J Econ Entomol 95, 635-639.

[12] Lereclus, D., Agaisse, H., Grandvalet, C., Salamitou, S., and Gominet, M. (2000) Int J Med Microbiol 290, 295-299.

[13] Salamitou, S., Ramisse, F., Brehelin, M., Bourguet, D., Gilois, N., Gominet, M., Hernandez, E., and Lereclus, D. (2000) Microbiology 146, 2825-2832.

[14] Gohar, M., OA, O. K., Gilois, N., Sanchis, V., Kolst Solidus In Circle, A. B., and Lereclus, D. (2002) Proteomics 2, 784-791.

[15] Schnepf, E., and Whiteley, H. R. (1981) Proc Natl Acad Sci USA 78, 2893-2897.

[16] Gonzalez, J. M., Jr., and Carlton, B. C. (1980) Plasmid 3, 92-98.
[17] Kronstad, J. W., Schnepf, H. E., and Whiteley, H. R. (1983) J Bacteriol 54, 419-428.
[18] Gonzalez, J. M., Jr., Dulmage, H. T., and Carlton, B. C. (1981) Plasmid 5, 351-365.
[19] Gonzalez, J. M., Jr., and Carlton, B. C. (1984) Plasmid 11, 28-38.
[20] Ward, E. S., Ellar, D. J., and Todd, J. A. (1984) FEBS Lett 175, 377-382.
[21] Thorne, L., Garduno, F., Thompson, T. E., Decker, M., Zounes, M., Wild, M., Walfield, A. M., and Pollock, T. (1986) J Bacteriol 166, 801-811.
[22] Kronstad, J. W., and Whiteley, H. R. (1986) Gene 43, 29-40.
[23] Carlson, C. R., and Kolsto, A. B. (1993) J. Bacteriol 175, 1053-1060.
[24] Kronstad, J. W., and Whiteley, H. R. (1984) J Bacteriol 160, 95-102.
[25] Lereclus, D., Ribier, J., Klier, A., Menou, G., and Lecadet, M. M. (1984) Embo J 3, 2561-2567.
[26] Menou, G., Mabillon, J., Lecadet, M., and Lereclus, D. (1990) J Bacteriol 172, 6689-6696.
[27] Bougouin, C., Delecluse, A., Ribier, J., Klier, A., and Rapoport, G. (1988) J Bacteriol 170, 3575-3583.
[28] Delecluse, A., Rosso, M. L., and Ragni, A. (1995) Appl Environ Microbiol 61, 4230-4235.
[29] Dunn, M. G., and Ellar, D. J. (1997) Plasmid 37, 205-215.
[30] Li, J. D., Carroll, J., and Ellar, D. J. (1991) Nature 353, 815-821.
[31] Grochulski, P., Masson, L., Borisova, S., Pusztai-Carey, M., Schwartz, J. L., Brousseau, R., and Cygler, M. (1995) J Mol Biol 254, 447-464.
[32] Morse, R. J., Yamamoto, T., and Stroud, R. M. (2001) Structure 9, 409-417.
[33] Galitsky, N., Cody, V., Wojtczak, A., Ghosh, D., Luft, J. R., Pangborn, W., and English, L. (2001) Acta Crystallogr D Biol Crystallogr 57, 1101-1109.
[34] Li, J., Koni, P. A., and Ellar, D. J. (1996) J Mol Biol 257, 129-152.
[35] Li, J., Derbyshire, D. J., Promdonkoy, B., and Ellar, D. J. (2001) Biochem Soc Trans 194, 571-577.
[36] Bravo, A. (1997) J Bacteriol 179, 2793-2801.
[37] Lecadet, M., and Dedonder, R. (1967) J Invertebr Pathol 9, 310-321.
[38] Tojo, A., Samasanti, W., Yoshida, T., and Aizawa, K. (1986) Agric Biol Chem 50, 575-580.
[39] Milne, R., and Kaplan, H. (1993) Insect Biochem Mol Biol 23, 663-673.
[40] Dadd, R. H. (1975) J Insect Physiol 21, 1847-1853.
[41] Bietlot, H., Carey, P. R., Choma, C., Kaplan, H., Lessard, T., and Pozsgay, M. (1989) Biochem J 260, 87-91.
[42] Tran, L. B., Vachon, V., Schwartz, J. L., and Laprade, R. (2001) Appl Environ Microbiol 67, 4488-4494.
[43] Aronson, A. I., S., H. E., McGaughey, W. H., and Johnson, D. (1991) Appl Environ Microbiol 57, 981-986.
[44] McGaughey, W. H., and Whalon, M. E. (1992) Science 258, 1451-1455.
[45] Koller, C. N., Bauer, L. S., and Hollingworth, R. M. (1992) Biochem Biophys Res Commun 184, 692-699.
[46] Oppert, B. (1999) Arch Insect Biochem Physiol 42, 1-12.
[47] Lee, H. K., and Gill, S. S. (1997) Appl Environ Microbiol 63, 4664-4670.
[48] Dai, S. M., and Gill, S. S. (1993) Insect Biochem Mol Biol 23, 273-283.
[49] Novillo, C., Castanera, P., and Ortego, F. (1997) Arch Insect Biochem Physiol 36, 181-201.
[50] Convents, D., Cherlet, M., Van Damme, J., Lasters, I., and Lauwereys, M. (1991) Eur J Biochem 195, 631-635.
[51] Choma, C., Surewicz, T., Carey, W. K., Pozsgay, M., and Raynor, T. (1990) Eur J Biochem 189, 523-527.
[52] Aronson, A. (1993) Mol Microbiol 7, 489-496.
[53] Chungjatupornchai, W., Hofte, H., and Seurinck, J. (1988) Eur J Biochem 173, 9-16.

[54] Bietlot, H. P., Schernthaner, J. P., Milne, R. E., Clairmont, F. R., Bhella, R. S., and Kaplan, H. (1993) J Biol Chem 268, 8240-8245.
[55] Bietlot, H. P. L., Vishmulhatla, J., Carey, P. R., Pozsgay, M., and Kaplan, H. (1990) Biochem J 267, 309-315.
[56] Strizhov, N., Keller, M., Konez-Kalman, Z., Regev, A., Sneh, B., Schell, J., Koncz, C., and Zilberstein, A. (1996) Mol Gen Genet 253, 11-19.
[57] Bietlot, H. P., Carey, P. R., Pozsgay, M., and Kaplan, H. (1989) Anal Biochem 181, 212-215.
[58] Ogiwara, K., Indrasith, L. S., Asano, S., and Hori, H. (1992) J Invertebr Pathol 60, 121-126.
[59] Haider, M. Z., Smith, G. P., and Ellar, D. J. (1989) FEMS Microbiol Lett 49, 157-163.
[60] Miranda, R., Zamudio, F. Z., and Bravo, A. (2001) Insect Biochem Mol Biol 31, 1155-1163.
[61] Gomez, I., Sanchez, J., Miranda, R., Bravo, A., and Soberon, M. (2002) FEBS Lett 513, 242-246.
[62] Oppert, B., Kramer, K. J., Beeman, R. W., Johnson, D., and McGaughey, W. H. (1997) J Biol Chem 272, 23473-23476.
[63] Shao, Z., Cui, Y., Liu, X., Yi, H., Ji, J., and Yu, Z. (1998) J Invertebr Pathol 72, 73-81.
[64] Keller, M., Sneh, B., Strizhov, N., Prudovsky, E., Regev, A., Koncz, C., Schell, J., and Zilberstein, A. (1996) Insect Biochem Mol Biol 26, 365-373.
[65] MacIntosh, S. C., Kishore, G. M., Perlak, F. J., Marrone, P. G., Stone, T. B., Sims, M. A., and Fuchs, R. L. (1990) J Agric Food Chem 38.
[66] Lightwood, D. J., Ellar, D. J., and Jarrett, P. (2000) Appl Environ Microbiol 66, 5174-5181.
[67] Martens, J. W., Visser, B., Vlak, J. M., and Bosch, D. (1995) Mol Gen Genet 247, 482-487.
[68] Bravo, A., Sanchez, J., Kouskoura, T., and Crickmore, N. (2002) J Biol Chem 277, 23985-23987.
[69] Hofmann, C., Luthy, P., Hutter, R., and Pliska, V. (1988) Eur J Biochem 173, 85-91.
[70] Van Rie, J., Jansens, S., Hofte, H., Degheele, D., and Van Mellaert, H. (1989) Eur J Biochem 186, 239-247.
[71] Van Rie, J., Jansens, S., Hofte, H., Degheele, D., and Van Mellaert, H. (1990) Appl Environ Microbiol 56, 1378-1385.
[72] Ihara, H., Kuroda, E., Wadano, A., and Himeno, M. (1993) Biosci Biotechnol Biochem 57, 200-204.
[73] Liang, Y., Patel, S. S., and Dean, D. (1995) J Biol Chem 270, 24719-24724.
[74] Lee, M. K., Milne, R. E., Ge, A. Z., and Dean, D. H. (1992) J Biol Chem 267, 3115-3121.
[75] Wolfersberger, M. G. (1990) Experientia 46, 475-477.
[76] Pietrantonio, P. V., and Gill, S. S. (1996) in Biology of the Insect Midgut (Lehane, M. J., and Billingsley, P. F., Eds.), Chapman and Hall, London.
[77] Rajamohan, F., Hussain, S. R., Cotrill, J. A., Gould, F., and Dean, D. H. (1996) J Biol Chem 271, 25220-25226.
[78] Rajamohan, F., Cotrill, J. A., Gould, F., and Dean, D. H. (1996) J Biol Chem 271, 2390-2396.
[79] Lee, M. K., Rajamohan, F., Jenkins, J. L., Curtiss, A. S., and Dean, D. H. (2000) Mol Microbiol 38, 289-298.
[80] Flores, H., Soberon, X., Sanchez, J., and Bravo, A. (1997) FEBS Lett 414, 313-318.
[81] Gomez, I., Miranda-Rios, J., Rudino-Pinera, E., Oltean, D. I., Gill, S., Bravo, A., and Soberon, M. (2002) J Biol Chem.
[82] Ge, A. Z., Rivers, D., Milne, R., and Dean, D. (1991) J Biol Chem 266, 17954-17958.
[83] de Maagd, R. A., Kwa, M. S., van der Klei, H., Yamamoto, T., Schipper, B., Vlak, J. M., Stiekema, W. J., and Bosch, D. (1996) Appl Environ Microbiol 62, 1537-1543.
[84] Lee, M. K., Young, B. A., and Dean, D. H. (1995) Biochem Biophys Res Commun 216, 306-312.
[85] de Maagd, R. A., Bakker, P. L., Masson, L., Adang, M. J., Sangadala, S., Stiekema, W., and Bosch, D. (1999) Mol Microbiol 31, 463-471.

[86] de Maagd, R. A., Weemen-Hendriks, M., Stiekema, W., and Bosch, D. (2000) Appl Environ Microbiol 66, 1559-1563.
[87] Burton, S. L., Ellar, D. J., Li, J., and Derbyshire, D. J. (1999) J Mol Biol 287, 1011-1022.
[88] Jenkins, J. L., Lee, M. K., Sangadala, S., Adang, M. J., and Dean, D. H. (1999) FEBS Lett 462, 373-376.
[89] Jenkins, J. L., Lee, M. K., Valaitis, A. P., Curtiss, A., and Dean, D. H. (2000) J Biol Chem 275, 14423-14431.
[90] Lee, M. K., You, T. H., Gould, F. L., and Dean, D. H. (1999) Appl Environ Microbiol 65, 4513-4520.
[91] Aronson, A. I., Geng, C., and Wu, L. (1999) Appl Environ Microbiol 65, 2503-2507.
[92] Slatin, S. L., Abrams, C. K., and English, L. (1990) Biochem Biophys Res Commun 169, 765-772
[93] Harvey, W. R., and Wolfersberger, M. (1979) J Exp Biol 83, 293-304
[94] Liebig, B., Stetson, D. L., and Dean, D. (1995) J Insect Phsiol 41, 17-22
[95] Parker, M. W., Pattus, F., Tucker, A. D., and Tsernoglou, D. (1989) Nature 337, 93-96.
[96] Song, L., Hobaugh, M. R., Shustak, C., Cheley, S., Bayley, H., and Gouaux, J. E. (1996) Science 274, 1859-1866.
[97] Wiener, M., Freymann, D., Ghosh, P., and Stroud, R. M. (1997) Nature 385, 461-464.
[98] Wu, D., and Aronson, A. (1990) in Proc. 5th Int. Colloquim on Invertebrate Pathology and Microbial Control, pp. 273-277, Soc. Invertebrate Pathology, Adelaide, Australia
[99] Masson, L., Tabashnik, B. E., Liu, Y. B., Brousseau, R., and Schwartz, J. L. (1999) J Biol Chem 274, 31996-32000.
[100] Wu, D., and Aronson, A. (1992) J Biol Chem 267
[101] Gazit, E., Bach, D., Kerr, I. D., Sansom, M. S., Chejanovsky, N., and Shai, Y. (1994) Biochem J 304, 895-902.
[102] Kumar, A. S., and Aronson, A. I. (1999) J Bacteriol 181, 6103-6107.
[103] Nunez-Valdez, M., Sanchez, J., Lina, L., Guereca, L., and Bravo, A. (2001) Biochim Biophys Acta 1546, 122-131.
[104] Dean, D. H., Rajamohan, F., Lee, M. K., Wu, S. J., Chen, X. J., Alcantara, E., and Hussain, S. R. (1996) Gene 179, 111-117.
[105] Chen, X. J., Lee, M. K., and Dean, D. (1993) Proc Natl Acad Sci USA 90, 9041-9045
[106] Schwartz, J. L., Potvin, L., Chen, X. J., Brousseau, R., Laprade, R., and Dean, D. H. (1997) Appl Environ Microbiol 63, 3978-3984.
[107] Carroll, J., and Ellar, D. J. (1997) Eur J Biochem 245, 797-804.
[108] Hodgman, T. C., and Ellar, D. J. (1990) DNA Seq Mapp 1, 97-106
[109] Du, J., Knowles, B. H., Li, J., and Ellar, D. J. (1999) Biochem J 338, 185-193
[110] Soberon, M., Perez, R. V., Nunez-Valdez, M. E., Lorence, A., Gomez, I., Sanchez, J., and Bravo, A. (2000) FEMS Microbiol Lett 191, 221-225.
[111] Vie, V., Van Mau, N., Pomarede, P., Dance, C., Schwartz, J. L., Laprade, R., Frutos, R., Rang, C., Masson, L., Heitz, F., and Le Grimellec, C. (2001) J Membr Biol 180, 195-203.
[12] Walker, B., and Bayley, H. (195) J Biol Chem 270, 23-65-23071
[113] Gazit, E., La Rocca, P., Sansom, M. S. P., and Shai, Y. (1998) Proc Natl Acad Sci USA 95, 12289-12294
[114] Aronson, A. I., and Shai, Y. (2001) FEMS Microbiol Lett 195, 1-8.
[115] Guereca, L., and Bravo, A. (1999) Biochim Biophys Acta 1429, 342-350.
[116] Chen, X. J., Curtiss, A., Alcantara, E., and Dean, D. H. (1995) J Biol Chem 270, 6412-6419.
[117] Masson, L., Mazza, A., Brousseau, R., and Tabashnik, B. (1995) J Biol Chem 270, 11887-11896.
[118] Masson, L., Mazza, A., and De Crescenzo, G. (2000) Methods Mol Biol 145, 189-201

[119] Masson, L., Mazza, A., Sangadala, S., Adang, M. J., and Brousseau, R. (2002) Biochim Biophys Acta 1594, 266-275.
[120] Armstrong, J. L., Rohrmann, G. F., and Beaudreau, G. S. (1985) J Bacteriol 161, 39-46
[121] Gill, S., and Hornung, J. M. (1987) J Invertebr Pathol 50, 16-25
[122] Sangadala, S., Walters, F. S., English, L., and Adang, M. J. (1994) J Biol Chem 269, 10088-10092
[123] Gill, S. S., Cowles, E. A., and Francis, V. (1995) J Biol Chem 270, 27277-27282.
[124] Knight, P. J., Knowles, B. H., and Ellar, D. J. (1995) J Biol Chem 270, 17765-17770.
[125] Oltean, D. I., Pullikuth, A. K., Lee, H. K., and Gill, S. S. (1999) Appl Environ Microbiol 65, 4760-4766.
[126] Banks, D. J., Jurat-Fuentes, J. L., Dean, D. H., and Adang, M. J. (2001) Insect Biochem Mol Biol 31, 909-918.
[127] Luo, K., Sangadala, S., Masson, L., Mazza, A., Brousseau, R., and Adang, M. J. (1997) Insect Biochem Mol Biol 27, 735-743.
[128] Lee, M. K., Rajamohan, F., Gould, F., and Dean, D. H. (1995) Appl Environ Microbiol 61, 3836-3842.
[129] Nakanishi, K., Yaoi, K., Shimada, N., Kadotani, T., and Sato, R. (1999) Biochim Biophys Acta 1432, 57-63.
[130] Mohammed, S. I., Johnson, D. E., and Aronson, A. I. (1996) Appl Environ Microbiol 62, 4168-4173.
[131] Zhu, Y. C., Kramer, K. J., Oppert, B., and Dowdy, A. K. (2000) Insect Biochem Mol Biol 30, 215-224.
[132] Emmerling, M., Chandler, D., and Sandeman, M. (2001) Insect Biochem Mol Biol 31, 899-907.
[133] Vadlamudi, R. K., Weber, E., Ji, I., Ji, T. H., and Bulla, L. A., Jr. (1995) J Biol Chem 270, 5490-5494.
[134] Yaoi, K., Kadotani, T., Kuwana, H., Shinkawa, A., Takahashi, T., Iwahana, H., and Sato, R. (1997) Eur J Biochem 246, 652-657.
[135] Hua, G., Tsukamoto, K., and Ikezawa, H. (1998) Comp Biochem Physiol B Biochem Mol Biol 121, 213-222.
[136] Simpson, R. M., and Newcomb, R. D. (2000) Insect Biochem Mol Biol 30, 1069-1078.
[137] Lorence, A., Darszon, A., and Bravo, A. (1997) FEBS Lett 414, 303-307
[138] Lee, M. K., You, T. H., Young, B. A., Cotrill, J. A., Valaitis, A. P., and Dean, D. H. (1996) Appl Environ Microbiol 62, 2845-2849.
[139] Garner, K. J., Hiremath, S., Lehtoma, K., and Valaitis, A. P. (1999) Insect Biochem Mol Biol 29, 527-535.
[140] Knowles, B. H., Knight, P. J., and Ellar, D. J. (1991) Proc R Soc Lond B Biol Sci 245, 31-35.
[141] Masson, L., Lu, Y. J., Mazza, A., Brousseau, R., and Adang, M. J. (1995) J Biol Chem 270, 20309-20315.
[142] Yaoi, K., Nakanishi, K., Kadotani, T., Imamura, M., Koizumi, N., Iwahana, H., and Sato, R. (1999) FEBS Lett 463, 221-224.
[143] Lorence, A., Darszon, A., and Bravo, A. (1997) FEBS Lett 414, 303-307.
[144] Luo, K., McLachlin, J. R., Brown, M. R., and Adang, M. J. (1999) Protein Expr Purif 17, 113-122.
[145] Zhuang, M., Oltean, D. I., Gomez, I., Pullikuth, A. K., Soberon, M., Bravo, A., and Gill, S. S. (2002) J Biol Chem 277, 13863-13872.
[146] Griffitts, J. S., Whitacre, J. L., Stevens, D. E., and Aroian, R. V. (2001) Science 293, 860-864

[147] Nagamatsu, Y., Toda, S., Koike, T., Miyoshi, Y., Shigematsu, S., and Kogure, M. (1998) Biosci Biotechnol Biochem 62, 727-734.
[148] Nagamatsu, Y., Toda, S., Yamaguchi, F., Ogo, M., Kogure, M., Nakamura, M., Shibata, Y., and Katsumoto, T. (1998) Biosci Biotechnol Biochem 62, 718-726.
[149] Gahan, L. J., Gould, F., and Heckle, D. G. (2001) Science 293, 857-860
[150] Morin, S., Sisterson, M., Biggs, R., Brown, J. K., Dennehy, T. J., Carriere, Y., Heckel, D. G., and Tabashnik, B. (2002) in Fourth International Synposium on Molecular Insect Science, pp. 74, Tucson, Arizona, USA
[151] Francis, B. R., and Bulla, L. A., Jr. (1997) Insect Biochem Mol Biol 27, 541-550.
[152] Keeton, T. P., and Bulla, L. A., Jr. (1997) Appl Environ Microbiol 63, 3419-3425.
[153] Meng, J., Candas, M., Keeton, T. P., and Bulla, L. A., Jr. (2001) Protein Expr Purif 22, 141-147.
[154] Nagamatsu, Y., Koike, T., Sasaki, K., Yoshimoto, A., and Furukawa, Y. (1999) FEBS Lett 460, 385-390.
[155] Gomez, I., Oltean, D. I., Gill, S., Bravo, A., and Soberon, M. (2001) J Biol Chem 30, 30
[156] Cowles, E. A., Yunovitz, H., Charles, J. F., and Gill, S. S. (1995) Appl Environ Microbiol 61, 2738-2744.
[157] Valaitis, A. P., Jenkins, J. L., Lee, M. K., Dean, D. H., and Garner, K. J. (2001) Arch Insect Biochem Physiol 46, 186-200.
[158] Sangadala, S., Azadi, P., Carlson, R., and Adang, M. J. (2001) Insect Biochem Mol Biol 32, 97-107
[159] Tabashnik, B. E., Finson, N., Groeters, F. R., Moar, W. J., Johnson, M. W., Luo, K., and Adang, M. J. (1994) Proc Natl Acad Sci USA 91, 4120-4124.
[160] Ferre, J., Escriche, B., Bei, Y., and Van Rie, J. (1995) FEMS Microbiol Lett 132, 1-7
[161] McGaughey, W. H. (1985) Science 229, 193-195
[162] McGaughey, W. H., Gould, F., and Gelernter, W. (1998) Nat Biotechnol 16, 144-146.
[163] Tabashnik, B. E., Malvar, T., Liu, Y. B., Finson, N., Borthakur, D., Shin, B. S., Park, S. H., Masson, L., de Maagd, R. A., and Bosch, D. (1996) Appl Environ Microbiol 62, 2839-2844.
[164] Tabashnik, B., Cushing, N. L., Finson, N., and Johnson, M. W. (1990) J Econ Entomol 83, 1671-1676
[165] Ferre, J., and Van Rie, J. (2002) Annu Rev Entomol 47, 501-533
[166] Ferre, J., Real, M. D., Van Rie, J., Jansens, S., and Peferoen, M. (1991) Proc Natl Acad Sci USA 88, 5119-5123.
[167] Tabashnik, B. E., Liu, Y. B., Malvar, T., Heckel, D. G., Masson, L., Ballester, V., Granero, F., Mensua, J. L., and Ferre, J. (1997) Proc Natl Acad Sci USA 94, 12780-12785.
[168] Tang, Z., Scherer, P. E., Okamoto, T., Song, K., Chu, C., Kohtz, D. S., Nishimoto, I., Lodish, H. F., and Lisanti, M. P. (1996) J Biol Chem 271, 2255-2261.
[169] Lambert, B., Buysse, L., Decock, C., Jansens, S., Piens, C., Saey, B., Seurinck, J., Van Audenhove, K., Van Rie, J., Van Vliet, A., and Peferoen, M. (1996) Appl Environ Microbiol 62, 80-86.
[170] Gould, F., Anderson, T. G., Reynolds, A., Bumgarner, L., and Moar, W. (1995) J Econ Entomol 88, 1545-1559
[171] Kota, M., Daniell, H., Varma, S., Garczynski, S. F., Gould, F., and Moar, W. J. (1999) Proc Natl Acad Sci USA 96, 1840-1845.
[172] Wirth, M. C., Georghiou, G. P., and Federici, B. A. (1997) Proc Natl Acad Sci USA 94, 10536-10540.
[173] Marroquin, L. D., Elyassnia, D., Griffitts, J. S., Feitelson, J. S., and Aroian, R. V. (2000) Genetics 155, 1693-1699.
[174] Rang, C., Vachon, V., Coux, F., Carret, C., Moar, W. J., Brousseau, R., Schwartz, J. L., Laprade, R., and Frutos, R. (2001) Curr Microbiol 43, 1-6.

[175] Tigue, N. J., Jacoby, J., and Ellar, D. J. (2001) Appl Environ Microbiol 67, 5715-5720.
[176] Hussain, S. R., Aronson, A. I., and Dean, D. H. (1996) Biochem Biophys Res Commun 226, 8-14.
[177] Gazit, E., and Shai, Y. (1995) J Biol Chem 270, 2571-2578.
[178] Chandra, A., Ghosh, P., Mandaokar, A. D., Bera, A. K., Sharma, R. P., Das, S., and Kumar, P. A. (1999) FEBS Lett 4 58, 175-179.
[179] Alcantara, E. P., Alzate, O., Lee, M. K., Curtiss, A., and Dean, D. H. (2001) Biochemistry 40, 2540-2547.
[180] Gerber, D., and Shai, Y. (2000) J Biol Chem 275, 23602-23607.
[181] Von Tersch, M. A., Slatin, S. L., Kulesza, C. A., and English, L. H. (1994) Appl Environ Microbiol 60, 3711-3717.
[182] Park, H. W., and Federici, B. A. (2001) Mol Biotechnol 16, 97-107.
[183] Jenkins, J. L., and Dean, D. H. (2001) BMC Biochem 2, 12
[184] Jurat-Fuentes, J. L., and Adang, M. J. (2001) Appl Environ Microbiol 67, 323-329.
[185] Lee, M. K., You, T. H., Curtiss, A., and Dean, D. H. (1996) Biochem Biophys Res Commun 229, 139-146.
[186] Karim, S., and Dean, D. H. (2000) Curr Microbiol 41, 430-440.
[187] Lee, M. K., Jenkins, J. L., You, T. H., Curtiss, A., Son, J. J., Adang, M. J., and Dean, D. H. (2001) FEBS Lett 497, 108-112.
[188] Abdul-Rauf, M., and Ellar, D. J. (1999) Curr Microbiol 39, 94-98.
[189] Masson, L., Tabashnik, B. E., Mazza, A., Prefontaine, G., Potvin, L., Brousseau, R., and Schwartz, J. L. (2002) Appl Environ Microbiol 68, 194-200.
[190] Loseva, O., Ibrahim, M., Candas, M., Koller, C. N., Bauer, L. S., and Bulla, L. A., Jr. (2002) Insect Biochem Mol Biol 32, 567-577.
[191] Sayyed, A. H., Gatsi, R., Kouskoura, T., Wright, D. J., and Crickmore, N. (2001) Appl Environ Microbiol 67, 4372-4373.
[192] Van Rie, J., McGaughey, W. H., Johnson, D. E., Barnett, B. D., and Van Mellaert, H. (1990) Science 247, 72-74.
[193] MacIntosh, S. C., Stone, T. B., Jokerst, R. S., and Fuchs, R. L. (1991) Proc Natl Acad Sci USA 88, 8930-8939

Formulation and Application

Methodology, Performance, Safety

18 Foliar Interactions and Uptake of Agrichemical Formulations– Present Limits and Future Potential

Jerzy A. Zabkiewicz

Plant Protection Chemistry $_{NZ}$, Forest Research, PB 3020, Rotorua, New Zealand

1 Introduction

The question has been asked "will primary food producers be allowed to use new technology to feed the future world population, despite having the technology to do so". Clearly the question of safety and sustainability in the primary food and material producing sectors is critical today and into the future. Risk reduction and eco-friendly products are the current drivers, with economics coming a poor third. The introduction of genetically modified plants is a further area of philosophical and regulatory conflict.

Agrichemical users are caught in the middle of these developments, as pragmatically they need to balance their need to remain profitable, as well as comply with ever increasing environmental regulations that demand reduced pesticide inputs or residues. The apparently simple solution is to reduce the application rates or frequency of pesticide or agrichemical treatments, without reducing product quality. This in turn requires increases in treatment efficacy or integration with non-chemical management systems. Integrated pest management systems have been introduced successfully into several sectors, but they require intensive monitoring and a good understanding of the factors involved for successful implementation. Conventional chemical control of weeds, pests and diseases is still much simpler, faster, and in many cases the only option for catastrophic outbreaks. Perhaps the most complex and publicly visible operation is in the control of pests and diseases during the growing cycle, involving sprays. These operations immediately raise concerns over off-site drift, pollution of waterways and potential longer term bioaccumulation and chronic ecological side effects. And perhaps there is still a good reason for such concerns to be raised. Even with good application technology and reasonable retention by the target species, the final amount of a systemic active reaching the target site may be very small, as little as a few percent of the initial amount applied. In the case of a contact pesticide, overall retention is the main factor, but then the distribution and form of the deposit is critical, and generally smaller droplets are used for greater efficacy. Obviously the amount of product delivered to the target will vary with plant growth stage and crop type [1] as well as formulation [2].

The introduction of genetically modified plants will bring a changed pattern of pesticide use, either reducing the need for chemical control, or alternately restricting choice to a smaller number of pesticides. The number of weeds, pests and diseases will not fall, so the available pesticides will need to be more effective over a wider control spectrum. If the introduction of new chemistry is put to one side, then the requirement to increase biological efficacy and reduce detrimental environmental effects, can only be met by improved spray application and formulation technologies. Accordingly the focus of this overview will be on the factors controlling the efficacy of liquid formulations of agrichemicals.

Spray formulation efficacy can be expressed as a function of

[deposition : retention : uptake : translocation : active ingredient activity]

where deposition is the mass of pesticide applied to an area; retention is the mass captured and retained on the target plant; uptake is the amount absorbed into the plant; translocation is the movement (for systemic compounds) to the site of action; and active ingredient activity is the biochemical activity of the pesticide [3]. It is important to remember that the carrier may play a significant part in the overall efficacy and it must be considered as part of the formulation i.e. we are considering the efficacy of the spray formulation, not the product formulation. Adjuvants do not directly affect the inherent biological activity of an active, but they can significantly influence each of the other terms. Neither spray application technologies nor formulation effects can be considered in isolation. The behaviour of spray formulations is predominantly influenced by the formulants (adjuvants) which can alter spray solution flow rate, droplet size, dynamic surface tension, spreading over and wetting of plant leaf surfaces and micro-organisms, and uptake by plant foliage, pests and pathogens.

The study of the efficacy and behaviour of different products and their formulations has evolved through purely empirical field trials, into various controlled environment or laboratory screening methods, to more detailed fundamental and mechanistic studies, leading to the development of models for the processes of deposition, retention, uptake and translocation [3]. This overview will consider these processes, their limitations, and their future potential. Due to the fact that most studies have been with herbicides into plants, with little or no information on uptake of pesticides by insects or pathogens, the main focus will be on the uptake of herbicides into plant foliage.

2 Deposition

The nominal spray application rate, as kg active ingredient (a.i.) per hectare, does not define in any way the dose received by a target plant or organism, nor does it represent the actual distribution within the canopy or individual plant. This consideration does not apply to soil applications, but becomes important for weed control and even more so for pest or pathogen control [4,5]. Hence physical application and formulation factors need to be tailored to specific applications [6].

Spray drift is always an issue. In general, spray drift increases in the sequence ground spray application <airblast < aerial [7]. Typical air-assisted deposition in fruit crops can be 65% on foliage, 25% on the ground and 10% as drift when applied at full foliage. To date, improvements in application technology have not been able to increase foliar deposition significantly (up to 70%), but they can reduce the ground deposition (down to 8%) and drift to 2% with 20% material being recycled [1].
The complexity of interactions during spray formation, deposition and retention were recognised long ago [8] and are strongly influenced by the formulants [9]. Specifically, spray parameters such as swathe width, spray flow rates through the nozzle and as a consequence, efficacy will also change [10,11]. The change in droplet size (droplet spectrum) is one of the most important effects caused by the addition of adjuvants. However, it is difficult to forecast the influence of adjuvants on droplet size though some generalisations have been made [12] and are represented in Figure 1.

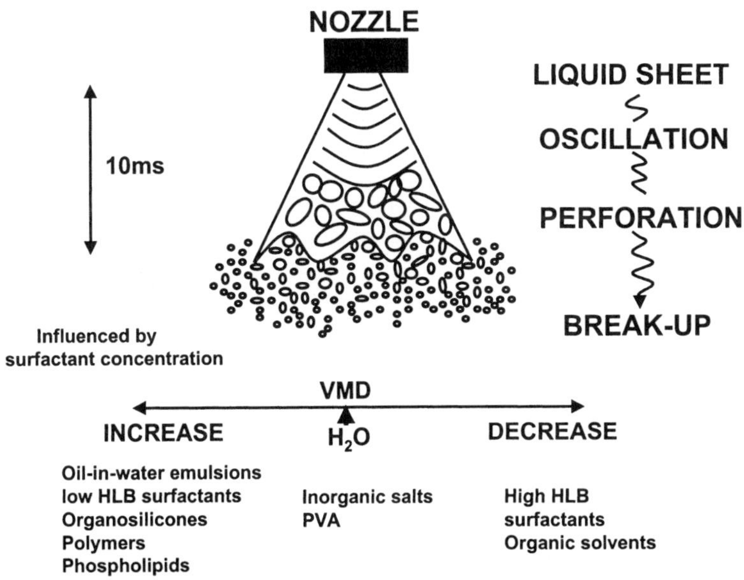

Figure 1. Spray break up and effect of formulant. Adapted from [12]

Although the droplet size distribution of agrichemical sprays cannot currently be forecast or modelled, the droplet spectra can be easily determined, and that information, together with other operational parameters, put into models of spray deposition. Models have been developed for ground and aerial spray application [13,14,15]. The advantage of such models is that they can be used pre-application to optimise deposition, and if biological efficacy or ecotoxicity information is available, to also estimate the efficacy of the deposited spray, and any required buffer distances, aquatic deposition and dose, or off-target bio-accumulation or persistence of the driftable fraction [16]. Such models are fully integrated within geographic information systems, using real geographic position inputs to plot travel lines, and provide deposition profiles within and outside the designated spray area. In future these will be real-time analyses of spray operations. Ground based systems will be able to plot the application of pesticide, not only by location, but by the amount as determined by sensors that identify the presence of plants and their size/area [1].

The presence of other plants, such as tree or vine rows, or shelter belts, has a considerable influence on the distance that small spray droplets travel, or in the interception of spray in general. Modelling distribution of spray deposits in such environments is much more complex and is not yet at a stage that it can be applied to operational situations. The capability to do so will lead to improvements in the deposition of spray within a spray area, as well as within the canopy, and it will be used to define spray drift interception. This could become an important part of spray mitigation management plans, especially relevant to the use in horticulture of relatively hazardous insecticides.

3 Adhesion and Retention

The physical properties and principles involved at droplet impaction have been studied extensively [17] and it is recognised that small droplets (20-50 μm diameter) will adhere to all plant surfaces but larger droplets will rebound or shatter on impact [18]. Hence there is a need to alter formulation solution properties, most conveniently with adjuvants, so that larger droplets (above 100 μm) will be retained [19]. The predominant factors controlling adhesion have been identified as plant leaf angle and surface character, solution dynamic surface tension, droplet velocity and droplet size. Since these are all physical parameters, they can be included in a model of adhesion [20] which has an explained deviance of 74%. The influence of different parameters such as droplet size and dynamic surface tension (Figure 2a) and the magnitude of leaf factors such as surface roughness and leaf angle (Figure 2b) can be easily estimated. However, it is recognised that pubescent leaves, wet surfaces and environmental conditions are still outside the scope of such a model.

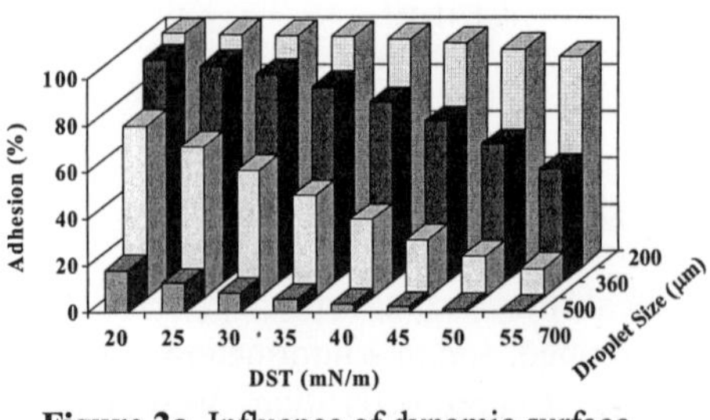

Figure 2a. Influence of dynamic surface tension and droplet size on adhesion

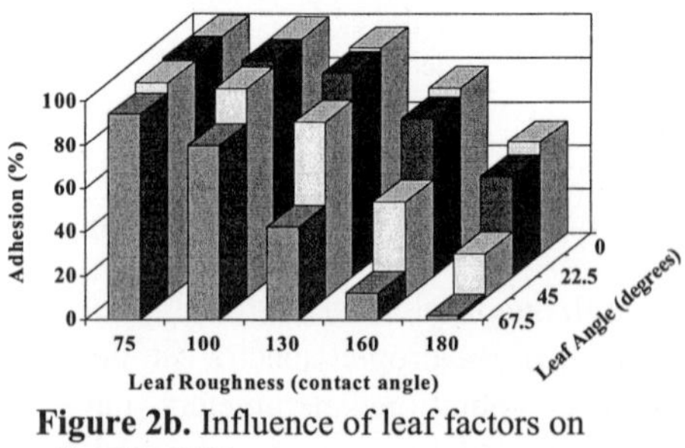

Figure 2b. Influence of leaf factors on adhesion [21]

Total spray retention also depends very much on crop or plant canopy architecture and is not as easy to model. If the canopy is even and "homogenous" then retention relationships can be derived and used in predictions, but in general it is difficult to estimate canopy retention with existing models [22,23,24]. The best model for retention [25] can explain variances of 98% under controlled conditions, but it is not a universal model, and cannot be used as it stands to predict retention for other crops and spray formulations. The existing models do have the potential to be useful tools for identifying selective adhesion or retention, but they cannot be used to predict re-distribution after retention, such as is being achieved though formulation modification. The addition of low surface tension surfactants, such as the organosilicones, can have a significant influence on the retention and final distribution of a.i. on a plant. If application volumes are tailored to surfactant concentration, increased retention is obtained on hard to wet plant surfaces. Not only that, but retained spray solution can move down vertical surfaces, and even creep over leaf surfaces to the leaf underside, where uptake may be better, or to reach cryptic pests or fungal spores. In the case of insecticides and fungicides, placement and distribution of the spray is as important as total retention [26].

4 Uptake

The uptake of agrichemical formulations into plants, insects and fungi is critical if there is to be any effect from the applied pesticides. Though uptake is predominantly required by systemic bioactives, non-systemic actives can also benefit by absorption into the surface or epicuticular wax, or cuticle of the organism. Uptake into plant leaves can occur by two pathways, either stomatal infiltration or by diffusion through the leaf cuticle [27,28]. Insects or their larvae may ingest insecticides (intestinal absorption), absorb them directly through their cuticle, or their spiracles can be infiltrated by solutions. Uptake of fungicides into fungal bodies is more complex, with little published information [29]. Since foliar uptake has been studied the most, and the mechanisms involved are still under debate, this will be discussed in some detail to illustrate the complexity and interactions involved.

4.1 Cuticle Structure

The common denominator in all organisms is that there is some type of membrane, skin or cuticle around the organised cellular tissues. The cuticular layer covering plant leaves is complex, and its characteristics vary with leaf age and environment. The structure and ontogeny of leaf cuticles has been reviewed extensively [30,31,32]. In general, there is a superficial wax layer (epicuticular wax), then a bilayer cuticular membrane (cuticle layer), the outermost layer (cuticle proper) composed of soluble (cuticular waxes) and polymerised lipids (cutin), and the innermost of polysaccharides which may contain high proportions of pectin. Other features which are important in relation to uptake, are the epidermal cell wall and stomata. The relative thickness of the cuticle layer *vs* the epidermal cell wall also varies enormously among species. In addition to detailed analyses of the epicuticular wax lipids [33], the cuticle has been shown to have weak acid ion exchange capability and a high affinity for calcium ions [34,35], as well as containing phenolic constituents [36] and reactive epoxy groups [37]. This variability illustrates the inhomogeneous nature of the cuticle and is a major factor in the failure to date to describe comprehensively the uptake of xenobiotics through the cuticle.

4.2 Stomatal Uptake

The requirements of good foliar wetting by sprays of low surface tension, and stomata of specific morphology, were identified as essential preconditions for stomatal infiltration in 1972 [38], but surfactants that could reduce the bulk surface tension of spray formulations to about 20 dynes/cm were not introduced until the mid-1980s [39]. Stomatal infiltration is a mass flow mechanism and if spray solutions have a low surface tension and plant surfaces have open stomata, high uptake (up to 80%) occurs very quickly, provided a threshold surfactant concentration is exceeded, or at high carrier volumes. An illustration of stomatal infiltration after a few minutes into bean and wheat leaves is given in Figure 3 (using confocal laser scanning microscopy (CSLM), and an aqueous solution of 0.5% Silwet L-77 and a fluorescent dye). At the left, the area inside the leaf under the stomata is fluorescing brightly; at the right only the stomata fluoresce, indicating a much more localised solution presence, and the significant influence of leaf structure and morphology on the internal redistribution of absorbed solution [40]. There are no models for this uptake mechanism.

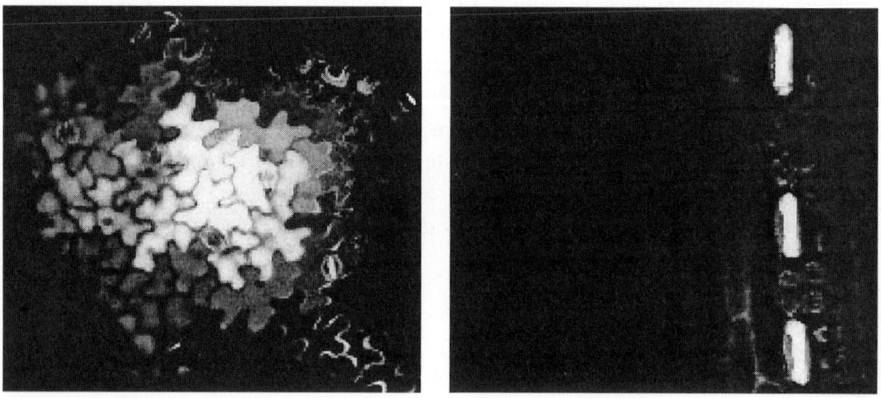

Figure 3. Confocal images of fluorescent dye in tissues of bean (left) and wheat (right) leaves approximately 25 microns below the leaf surface [40]

4.3 Cuticular Uptake

The process of xenobiotic uptake through leaf cuticles is an area of active study and debate as to the detailed mechanisms involved. An illustration of the wide differences that can occur in the movement and pathways is shown in Figures 4a, 4b and 4c. Using CLSM and fluorescent probes of different lipophilicity, it can be shown that the uptake and movement (in the presence of a surfactant) through the cuticle proper, and into the epidermal cells varies significantly. The most lipophilic compound (Figure 4, left) is held up completely in the cuticle; one that is less lipophilic diffuses evenly into the epidermal cells (Figure 4, center); while the most hydrophilic compound migrates through the epidermal cell walls (Figure 4, right).

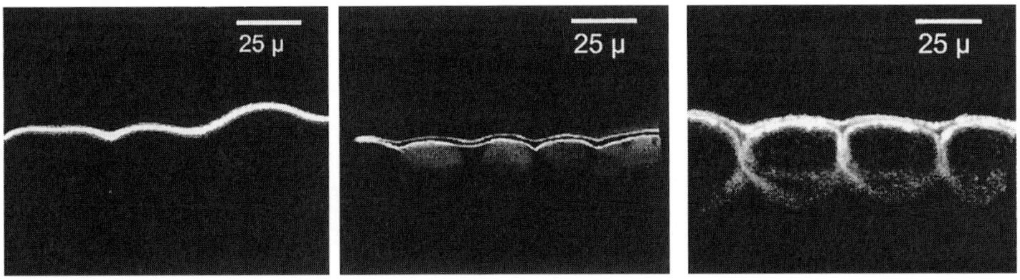

Figure 4. All treatments 0.05% dye concentration in presence of 0.2% surfactant after 24h into bean leaf (left: Nile red; center: Rhodamine 6G; right: Fluorescein; Liu, unpublished)

Quantification of uptake, either of the active or the adjuvant, has usually been as a simple percentage of amount applied, but if the mechanism of uptake is to be considered, this may be inappropriate. The most common experimental approach uses applications of formulation droplets directly onto living plant surfaces. This technique has been useful in illustrating the changes in uptake with either different types of actives, or plant species, and the influence of surfactants and surfactant concentrations, on the total percentage uptake of active or surfactant into plant leaves. This is very useful as a screening method for overall uptake comparisons, but mechanistically it can only give at best a generic set of guidelines that will be species or surfactant specific [41].

If uptake is considered to be a diffusion process through the cuticle, an alternative measurement may be mass or moles of compound taken up per unit area, per unit time. The diffusion of substances through a membrane is described by Fick's first law. Although the cuticle is a non- homogeneous membrane, as an approximation the flux J (per unit area in moles m^2 s^{-1}) can be derived from:

$$J = D.\left(\frac{\delta c}{\delta x}\right) \qquad (1)$$

and is the amount of solute that diffuses through a unit area per unit of time (i.e. M/At where M is moles or mass, A is area and t is time). It is proportional to the concentration gradient ($\delta c/\delta x$ in moles m^{-3} m^{-1}) and the diffusion coefficient (D in m^2s^{-1}). In reality the equation has to include further interactions [31]. The solute has to partition between the external solution and the membrane (essentially between an aqueous and a lipophilic phase) and this is denoted by K_p. Where the membrane (or cuticle) has thickness *x*, available area A, concentration difference $C_o - C_i$, partition coefficient K_p, and a "cuticle tortuosity factor" *l*, then the uptake of a solute after a given time t*x*, is given by:

$$J_{tx}A = \frac{D.K_p.A}{l.x}(C_o - C_i) \qquad (2)$$

More recent approaches [42], based on studies with isolated leaf cuticles *in vitro* with a wide range of formulations, considered transport through the cuticle to be a three stage diffusion mechanism, namely absorption into the cuticle, diffusion through the cuticle and finally desorption from the cuticle into the internal leaf cells. This experimental method used a constant cuticular area, constantly covered by a volume of spray solution, so variable spreading was not an issue. An alternate version of this approach uses droplets applied to cuticular membranes which is more realistic [43]. The modified Fick's equation used in these *in vitro* studies [41,42] has the following form:

$$J = k^{*}.l_{ls}.(K_oC_o - K_iC_i) \quad (3)$$

Where J is the flux, k* is the solute mobility in the cuticle (s^{-1}), l_{ls} is the "limiting skin" tortuosity factor (m), K_o is the partition coefficient between the cuticular membrane and the formulation residue, C_o (moles m^{-3}) is the concentration of a.i. in the formulation residue, K_i is the partition coefficient between the polymer matrix and cell wall water, and C_i is the concentration of a.i. in the epidermal cell wall. These elements and their respective interactions are represented in Figure 5.

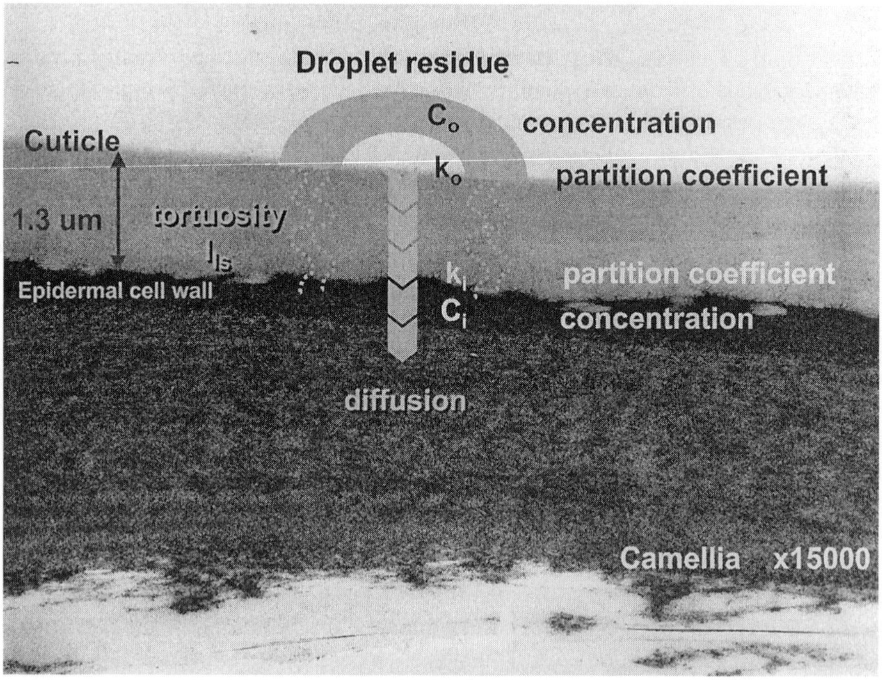

Figure 5. Relevant factors from the diffusion equation (3) represented within droplet residue and leaf cuticle system

According to this approach, the principal factors affecting uptake rates are:

Solute mobility (k*) which is affected by temperature, solute molar volumes, and cuticular wax composition.

Limiting skin or the limiting skin tortuosity factor (l_{ls}) which is the length of the diffusion path through the "limiting skin". This is only a section, not the entire cuticle, and its influence is considered to depend on the size and orientation of the cuticular wax crystals.

Driving force ($K_oC_o - K_iC_i$) which is affected by the starting and continuing concentration of a.i. in the solution on the cuticle surface, in the cuticular layers, and in the epidermal cell wall.

Unfortunately, it is very difficult, if not impossible, to measure these parameters for *in vivo* systems. The validation of the accuracy or applicability of these parameters to *in vivo* systems still requires testing. Adjuvants have been classified into "accelerator" and "passive" categories, and they can influence both the solute mobility (accelerator adjuvants only) and the driving force (both categories)[42].

The uptake of the surfactant components in a formulation has not been quantified extensively. They have been observed to have different uptake rates and amounts from the actives, and both of these can be at times further influenced by the active. Hence the simple categorisation into accelerator and passive categories with respect to their effect on the active, may not be an adequate description of potential interactions.

Overall, *in vivo* approaches using percentage uptake, at arbitrary intervals, and largely ignoring the initial physical processes occurring on the leaf surface after impaction, have not been helpful towards explaining the controlling factors involved. The *in vitro* studies with isolated cuticles, which have dilute instead of concentrated spray residue deposits, are not realistic either, though they have helped to identify the controlling factors in the diffusion process. Neither approach has resulted in a practical mechanistic model for xenobiotic uptake into living plant leaves.

4.3.1 Leaf Surfaces and Droplet Behaviour

The spreading of droplets is dictated by the nature and amount of surfactant in the spray formulation, as well as the microscopic nature of the leaf surface. Hence the same formulation can have quite different droplet spread areas, on leaves of different plant species, which in turn will affect the initial dose per unit area of the a.i. Increased droplet spread usually results in faster drying of the spray solution, which will further influence uptake, and will ultimately affect the solution concentration that is actually left on the leaf surface. As the droplet spreads, so the dose per unit area decreases; then as drying starts, the concentration increases to a much higher level (assuming most of the water evaporates or is rapidly absorbed into the leaf). This is important, as one of the drivers in the diffusion processes is the concentration of a.i. in the "spray droplet residual solution" from which the a.i. migrates into the cuticle. The dried droplet spread area can be easily measured and the mean dose per unit area calculated, but the spreading process may not leave an area evenly covered in formulation residue. Leaf micro-structures, such as veins, trichomes, glands and wax deposits will further influence the homogeneity of the deposit. The main questions are: how much water is left in the spray residue; what is the nature of this "solution"; and how is it that the uptake process effectively starts only after there is visual dry-down of droplets. It is known that surfactants can alter the permeability of the cuticle to water [44] but there has been no quantification to date of the water content (and hence concentration in solution) of spray droplet residues over time. In addition, the a.i. may undergo further partitioning between the aqueous and more lipophilic surfactant components, making it more, or less, capable of diffusion into the cuticle. It is well documented that the wax constituents of the cuticle form a major barrier to the diffusion of hydrophilic compounds, but conversely can also form a "sink" for highly lipophilic ones, so the physico-chemical properties of the spray droplet residues will be important, and quite different from the starting bulk spray solution properties.

A recognition of the importance of spray droplet residues is developing [45,46] and indeed the relationship of initial dose to uptake (expressed in molar amounts over a specific time) has identified some interesting correlations [41].

This manner of representing uptake as Jt *vs* M/A (at different time intervals) can provide a different, and mechanistically more useful, perspective of spray residue and leaf cuticle interactions. If uptake (Jt_x) is presented in moles mm^{-2} *vs* initial dose (M/A) also in moles mm^{-2}, then as illustrated in Figure 6 (left) there is a linear relationship over a wide range of initial dosages, with this a.i. and these formulations. At lower initial dosages, this relationship is curvilinear and is seen to vary with different formulations (Figure 6, right). The extent of the deviation from linearity will also depend upon the time interval after which uptake is measured, as rates of uptake vary more with shorter post-application intervals.

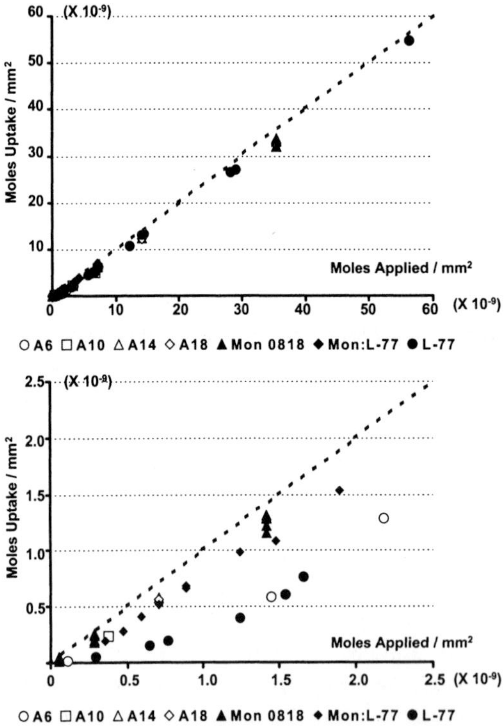

Figure 6. Uptake of glyphosate into wheat after 24h on a moles mm^{-2} initial dose basis. Formulations are with different surfactants at different concentrations of both glyphosate and surfactant. Surfactants are: A6, A10, A14, A18 which refer to the series of C12 alcohol ethoxylates with different EO chain lengths; Mon 0818 is an aliphatic amine with mean EO of 15 ; L-77 is Silwet L-77; and Mon:L-77 refers to mixtures of Mon 0818 and Silwet L-77. The dashed line represents 100% uptake.

It strongly suggests that the "driving force" (which includes the concentration values) component of equation (3) is dominant at higher dosages in these formulations. The fact that there is non-linearity at lower dosages or over shorter intervals (unpublished observations) and among different formulations, indicates that other factors (e.g. solute mobility) is more dominant or different at lower dosages. These results conform with the principle factors in uptake equation (3) and suggest it may be possible to represent them in a much simpler manner than has been done to date. However, it is still a composite representation of Jt *vs* initial M/A and does not reflect any one element individually. As both the diffusion coefficient and concentration gradient will vary over time, (affected further by the influence of absorbed adjuvant), digressions from linear relationships, or rates, may be expected, due to species and formulation differences. The actual molar quantities and ratios of surfactant to active can also be readily determined over the range of interest. These can provide much more useful and relevant comparisons of formulation influences than can be obtained by percentage uptake studies.

Other effects identified during cuticular uptake include the inhibition of surfactant uptake by an active, and the influence of cations in the cuticle on anionic compound diffusion. Different pathways have been postulated for lipophilic and hydrophilic compound movement. All these interacting factors require careful evaluation and identification of their effect on the generic concept of uptake through cuticular diffusion processes. Completely different analytical techniques may be required to elucidate these processes [41].

5 Translocation

The most important property controlling the movement of chemicals in plants is their lipophilicity, which determines their movement across membranes and determines the potential for long-distance transport via xylem or phloem [47]. A considerable body of information exists on the uptake and transport of xenobiotics from soil into plant roots, which is related to compound water solubility in the first instance, and then the ratio of compound in the root *vs* the external solution. Movement is then in the transpiration stream, via the xylem. It has been found that there is an optimum K_{ow} between 1 and 3 for translocation, which may be due to the fact that polar compounds may have difficulty crossing hydrophobic membranes, and that lipophilic compounds may have difficulty crossing the endodermis in plants. Xylem transport of xenobiotics taken up through leaves follows the same general relationships as uptake through the roots, though the movement of lipophilic compounds may be retarded by absorption in the cuticular lipid layer. Movement will be to the dominant sinks. The majority of phloem-mobile xenobiotics are weak acids and this is thought to be due to ion-trapping in the more basic phloem (pH ~8) relative to the apoplast (pH ~5.5), including the xylem.

Information to date suggests that surfactants do not have a direct effect on herbicide translocation [28] though the addition of adjuvants will increase the uptake, and hence the percentage translocation of applied herbicide. However, in cases where the uptake of the active and adjuvant is so rapid or so large, localised tissue death can occur (contact phytotoxicity), and consequently movement out of this zone can be reduced. Contact phytotoxicity has been related to surfactant structure, more so with ionic than non-ionic, and with lipophilic *vs* hydrophilic surfactants.

6 Generic Efficacy Modelling

Modelling of xenobiotic uptake and transport in whole plant systems has been focussed largely on root:water or leaf:air interfaces due to interest in uptake from the soil or the atmosphere [48,49], though in the latter case, leaf:liquid interfaces may be involved as well. The modelling of pesticide uptake into leaves , and the influence of adjuvants, has only recently been attempted. Early models were purely theoretical [50,51], but recently more applied systems were attempted, though interestingly the basis for the cuticular uptake part of the model is still based on the parameters in Equation (3) [52,53]. This whole plant approach is complex, and may not be strictly accurate in all its assumptions or inputs, but has been applied to a range of herbicides, plant species, times, and environmental conditions. Uptake percentages as well as distribution in, above and below the treated leaf, and in the root, were estimated and compared with known data. Although there were some large differences, and not all herbicide types could be accommodated within this model, there was a remarkably good fit of actual to estimated. A more specific modelling approach for uptake into foliage has combined the principles of compartmental models and diffusion processes [54]. It clearly recognises the importance of contact area between the droplet solution and the leaf surface and involves the diffusion process parameters developed by previous workers [42]. Comparisons of calculated percentage uptake *vs* known values, as presented over a narrow range of herbicides and species, are extremely good. Although this approach is well focussed on uptake, it still depends on inputs that are in turn largely obtained through studies with isolated cuticles.

7 Conclusions

There is still a continuing need to improve methods for getting pesticides to the target site, and reduce off-site or off-target deposition. Deposition into a target area can be modelled, but the ability to place spray deposits within specific plant parts is still limited. Adhesion and retention processes are well understood, and can be modelled, though these models need further development to include less common plant features and characteristics. The factors controlling plant foliar uptake are said to be primarily solute mobility, cuticle tortuosity and solute driving force, but their measurement still awaits simple methods or approximations for operational applicability. Translocation models exist and can be adapted readily for simulations after pesticide uptake through foliage.

More complex generic models of uptake or total pesticide efficacy are being developed and show great promise. However their universal application and acceptance depends on a substantial simplification of required inputs. At the present time, perhaps their greatest value lies in identifying the gaps in our knowledge and making us focus on the essential factors and their quantification, and the recognition that simple quantitative methods based on mass or molar quantities of formulants (actives and adjuvants as well as water), instead of percentage uptake, are essential for further progress.

8 Acknowledgement

This work was supported by the New Zealand Foundation for Research, Science and Technology. Thanks are due to W. A. Forster and R. E. Gaskin for constructive advice, Z. Q. Liu for confocal microscopy images, and M. Haslett for assistance with the preparation of this manuscript.

9 References

[1] H. Ganzelmeier in Pesticide Chemistry and Bioscience, Proceedingss 9th IUPAC (Eds.: G. T. Brooks, T. R. Roberts), Royal Society of Chemistry, Cambridge, 1999, pp. 100-119.
[2] F. R. Hall, A. C. Chapple, R. A. Downer, L. M. Kirchner, R. M. Thacker, Pestic Sci. 1993, 38, 123-140.
[3] J. A. Zabkiewicz, Weed Research, 2000, 40, 139-149.
[4] F. R. Hall,. HortScience, 1991, 26, 1012-1017.
[5] T. A. Ebert, R. A. J. Taylor, R. A. Downer, F. R. Hall, Pestic. Sci. 1999, 55, 793-798.
[6] P. C. H. Miller, A., Lane, H. Wheeler, Pesticide Outlook, 2001, 100-102.
[7] F. R. Hall, R. D. Fox,.in Pesticide Formulation and Adjuvant Technology. (Eds.: C. J. Foy, D. W. Pritchard),. CRC Press, Boca Raton, FL, USA, 1996, pp. 209-239.
[8] T. F. Tadros, Aspects of Applied Biology, 1987, 14, 1-21.
[9] P. C. H. Miller, M. C. Butler-Ellis, Plant Protection Quarterly, 1997,12, 33-38
[10] R. A. Downer, R. E. Mack, F.R. Hall, A. K. Underwood, in Proceedings 5th International Symposium on Adjuvants for Agrochemicals, (Ed.: P. McMullan), Chemical producers and Distributors Association, Memphis, USA, 1998, pp. 468-474.
[11] B. Richardson, J. Ray in Proceedings 44th New Zealand Weed and Pest Control Conference. (Ed.: A. J. Popay), Tauranga, New Zealand, 1991, pp112-115.
[12] P. J. Holloway, in Chemistry and Technology of Agrochemical Formulations (Ed. D. A. Knowles), Kluwer Academic Publishers, Dordrecht, The Netherlands, 1998, pp. 232-263.
[13] H. Zhu, D. L. Reichard, R. D. Fox, R. D. Brazee, H. E. Ozkan, H. P. Zhu, American Society of Agricultural Engineers, 1994, 94-1069.
[14] D. B. Smith, L. E. Bode, P. D. Gerard, American Society of Agricultural Engineers, 2000, 547-553.
[15] J. W. Ray, B. Richardson, W. Schou, M. E. Teske , A. L. Vanner, G. W. Coker, Canadian Journal of Forestry, 1999, 29, 875-882.
[16] L. Hugo, Air Ag News, 2002, 18, 32-33.
[17] J. J. Spillman, Pestic. Sci., 1984, 15, 97-106.
[18] G. S. Hartley, R. T. Brunskill, in Surface Phenomena in Chemistry and Biology (Eds.: J. F. Danielli, K. G. A. Parkhurst, A. C. Riddiford), Pergamon Press, New York, 1958, pp. 214-223.
[19] D. A. Webb, P. J. Holloway, N. M. Western, Pestic. Sci., 1999, 55, 343-389.

[20] W. A. Forster, M. O. Kimberley, J. A. Zabkiewicz, in Pesticide Formulations and Application Systems. (Eds.: J. C. Mueninghoff, A. K. Viets, R. A. Downer). American Society for Testing and Materials STP 1414, West Conshohocken, PA, USA, 2000, 21, pp.163-174.
[21] W. A. Forster, M. O. Kimberley and J. A. Zabkiewicz, in Proceedings 6th International Symposium on Adjuvants for Agrochemicals, (Ed.: H. De Ruiter). Amsterdam, The Netherlands, 2001, pp. 35-40.
[22] D. H. Bache, in Application and Biology, (Ed. E. S. E. Southcombe). British Crop Protection Council, Croydon, UK, 1985, Monograph no. 28, pp.183-190.
[23] H. Zhu, D. L. Reichard, R. D. Fox, R. D. Brazee, H. E. Ozkan, Transactions of American Society of Agricultural Engineers, 1996, 39, 415-22.
[24] G. A. J. M. J. Akkerhuis, J. A. Axelsen, C. Kjaer, Pestic. Sci.,1998, 53, 252-262.
[25] B. T. Grayson, S. E. Pack, E. Edwards, J. D. Webb, Pestic. Sci., 1993, 37, 133-140.
[26] E. C. Hislop, Aspects of Applied Biology, 1987, 14, 153-172.
[27] R. C. Kirkwood, Pestic. Sci., 1999, 55, 69-77.
[28] R. J. Field, F. Dastgheib, in Pesticide Formulation and Adjuvant technology. (Eds.: C. L. Foy, D. W Pritchard). CRC Press, Boca Raton, Florida, USA, 1996, pp. 241-295.
[29] W. Steurbaut, Pestic. Sci., 1993, 38, 85-91.
[30] H. H. Hull, in Residue Reviews. (Eds.: F. A. Gunther, J. D. Gunther). Springer Verlag, New York, 1970, 31, pp.1-150.
[31] C. E. Price, in The Plant Cuticle. (Eds.: D. F. Cutler, K. L. Alvin, C. E. Price). Academic Press, London, UK. 1982, pp. 237-252.
[32] C. E. Jeffree, in Plant Cuticles – an integrated functional approach. (Ed.: G. Kerstiens). Bios Scientific Publishers, Oxford, UK, 1996, 33-82.
[33] M. Riederer, C. Markstadter, in Plant Cuticles – an integrated functional approach. (Ed.: G. Kerstiens). Bios Scientific Publishers, Oxford, UK, 1996, 189-200.
[34] J. Schönherr, R. Huber, Plant Physiology, 1977, 59, 145-150.
[35] G. J. Hall, C. A. Hart, C. A Jones, Pest Manag Sci, 2000, 56, 351-358.
[36] G. Liakopoulos, S. Stavrianakou, G. Karabourniotis, Annals of Botany, 2001, 87, 641-648.
[37] M. Riederer, J. Schönherr, Arch. Environ. Contam. Toxicol. 1986, 15, 97-105.
[38] J. Schönherr, M. J. Bukovac, Plant Physiology, 1972, 49, 813-819.
[39] P. J. G. Stevens, J. A. Zabkiewicz, in Proceedings 6th EWRS Symposium: Factors Affecting Herbicidal Activity and Selectivity, Wageningen, The Netherlands, 1988, pp. 145-150.
[40] R. E. Gaskin, R. J. Murray, A. P. Singh, J. A. Zabkiewicz, in Proceedings 5th International Symposium on Adjuvants for Agrochemicals, (Ed.: P. McMullan), Chemical Producers and Distributors Association, Memphis, USA, 1998, pp. 61-66.
[41] J. A. Zabkiewicz, W. A. Forster, in Proceedings 6th International Symposium on Adjuvants for Agrochemicals, (Ed.: H. De Ruiter). Amsterdam, The Netherlands, 2001, pp. 19-28.
[42] J. Schönherr, P. Baur, in Plant Cuticles – an integrated functional approach. (Ed.: G. Kerstiens). Bios Scientific Publishers, Oxford, UK, 1996, 135-155.
[43] M. Knoche, P. D. Petracek, M. J. Bukovac, *Pest Manag Sci.* 2000, 56, 1005-1015.
[44] J. Schönherr, H. Bauer, in Adjuvants for Agrichemicals. (Ed. CL Foy). CRC Press, Boca Raton, Florida, USA, 1992, pp. 17-35.
[45] P. Cronfeld, K. Lader, P Baur, in Classification of Adjuvants and Adjuvant Blends by Effects on Cuticular Penetration. (Eds.: A. K. Viets, R. S. Tann, J. C. Mueninghoff). American Society for Testing and Materials STP 1400, West Conshohocken, PA, USA, 2001, pp. 81-94.
[46] M. Knoche, M. J. Bukovac, Pest Manag Sci, 2000, 56, 1016-1022.
[47] R. H. Bromilow, K. Chamberlain, in Plant Contamination: Modelling and Simulation of Organic Chemical Processes. (Eds.: S Trapp, J. C. McFarlane). Lewis Publishers, Boca Raton, Florida, USA, 1995, pp. 37-68.
[48] S. Trapp, M. Matthies, Environ. Sci. Technol., 1995, 29, 2333-2338.

[49] M. Riederer, in Plant Contamination: Modelling and Simulation of Organic Chemical Processes. (Eds.: S Trapp, J. C. McFarlane). Lewis Publishers, Boca Raton, Florida, USA, 1995, pp. 154-190.
[50] L. Boersma, F. T. Lindstrom, C. McFarlane, E. L. McCoy. Soil Science, 1988, 146, 403-416.
[51] K. V. Novozhilova, T. M. Petrova, N. N. Semenova, T. V. Solomina. Arch. Phytopath. Pflanz., 1995, 30, 165-181.
[52] N. M. Satchivi, E. W. Stoller, L. M. Wax, D. P. Briskin. Pesticide Biochemistry and Physiology, 2000, 67-84.
[53] N. M. Satchivi, E. W. Stoller, L. M. Wax, D. P. Briskin. Pesticide Biochemistry and Physiology, 2000, 85-95.
[54] S. Lamb, M. Bean, P. Chapman in Proceedings 6th International Symposium on Adjuvants for Agrochemicals, (Ed.: H. De Ruiter). Amsterdam, The Netherlands, 2001, pp. 106-112.

19 Standards for Sprayers and Cooperation of Test Stations for Mutual Recognition in Europe

Sandro Liberatori

ENAMA – Ente Nazionale della Meccanizzazione Agricola, ENTAM – European Network for Testing of Agricultural Machines, Via Lazzaro Spallanzani, 22/a – 00161 Roma, Email: sandro.liberatori@enama.it

1 Summary

Growing globalisation requires technical standardization in order to guarantee the protection of populations and the environment. Developing internationally approved guidelines for testing application techniques is essential to provide adequate food and protection of the environment. European and international standards ensure state of the art of sprayers and lead to uniform criteria for the evaluation of machines. These standards are developed by the International Organisation for Standardisation (ISO) and the European Committee for Standardisation (CEN) within their technical Committees (TC) 23/Subcommittee (SC) 6 belonging to ISO and TC/Working Group (WG) 3 of CEN. However, standards only set some criteria and only define a minimum level of performance, while there is a growing need for the interpretation of standards to carry out precise and comparable tests for a better evaluation of machines. Besides the continuous development of new standards, improvement of existing standards requires a high degree of specialisation of test facilities. International cooperation among testing stations is imperative. Therefore, in 1999 some European testing stations decided to cooperate and signed an agreement that to date has been accepted by 20 members, proving that this is the way to prepare for the future.

2 Globalisation and Common Standards

At present an important development of world trade is globalisation. Globalisation makes it necessary to establish regulations in order to guarantee a minimum level of quality of the machines being used by the manufacturers, dealers and users. This is very important in the field of chemical applications in agriculture because of the involvement

of operators, environment and final products that will be used by the whole human society. Therefore, internationally approved guidelines have been developed and may be used for testing techniques in skilled testing stations that exist almost in every European country. These testing techniques are the base for the development of specific tests for uniform criteria in the evaluation of machines. This led to the need for cooperation amongst testing stations. The ENTAM Agreement was set up to guarantee standard test conditions and results.
Members foresaw this necessity and a first edition of a common methodology for crop protection machines was made.

3 High Standards for State of the Art Sprayers

It is widely known that standards define the advanced state of the art in a well-defined system. They describe technical rules which have been tested and proven and are accepted practice in the commercial world. Most standards have been studied and developed in European countries (Italy CUNA, Germany DIN, France AFNOR, Great Britain BSI etc.). The importance of internationally approved standards has increased recently because of growing world trade. In 1926, the International Organisation for Standardisation (ISO) was founded to address these concerns. At present, ISO is made up of more than 140 different national standard organisations.
The European Union requires common rules for the clearing of trade barriers. Standardisation has acquired additional value in a European dimension, and the European Committee for Standardisation (CEN) has been established. National and international standardisation bodies participate in the development and upgrade of the standards providing infrastructure and assuring the coherence and the consistency of standards.

3.1 Standardisation and Use of Standards

The use of standards has proven its value whenever exact technical specifications are necessary.
In Europe, the development of a new standard may be proposed by any Member body, any international or European organisation or the EU Commission. The standardisation process is controlled by the CEN Technical Board that assigns the preparation of the first draft to the relevant Technical Committee (TC). Technical work is done by a Working Group (WG) consisting of experts from the Member States. Once a CEN standard has been adopted, the EU Member States have to convert it to a national standard whereby contrary, national standards have to be withdrawn. In Europe, there is a context between legislation and standardisation. Following the "new approach", the EU Council issues directives in order to define essential requirements needed to ensure the free movement of products throughout the Union. These requirements have to be based on relevant technical rules, so called "harmonised standards". Products manufactured in conformity with harmonised standards are presumed conform to the essential requirements. As mentioned before, these standards remain voluntary and the manufacturer may choose

alternate paths. However, the manufacturer must prove that his products are conform to essential requirements.
The legislative procedure for harmonised standardisation in Europe is shown in Figure 1.

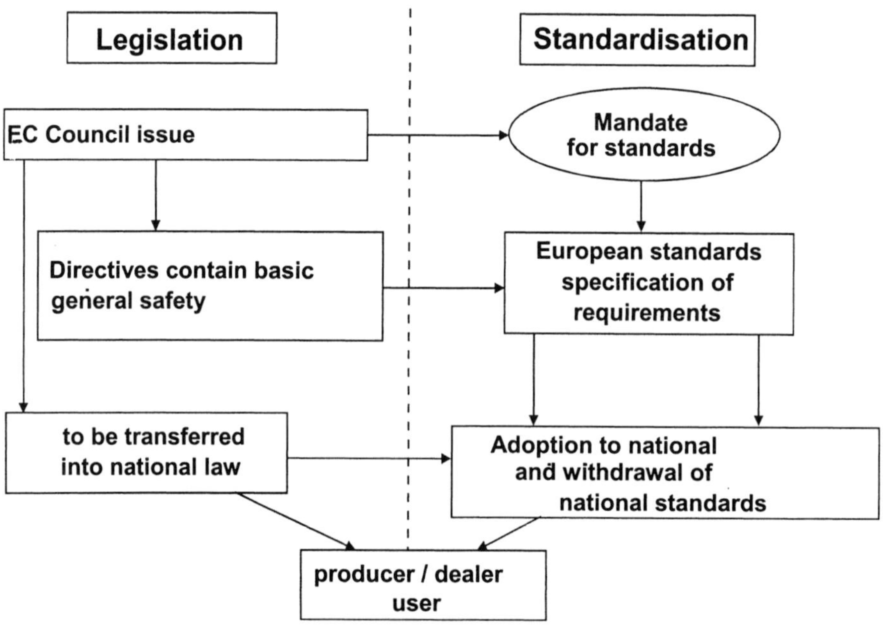

Figure 1. European procedure for legislation and standardisation (Ganzelmeier).

International standards are initiated by the ISO Member Bodies whose organisational structure is similar to CEN. ISO is governed by a Council consisting of Officers and 18 elected Member Bodies. Standardisation is coordinated by the Technical Management Board and standards are developed by WGs subordinated to Subcommittees and TCs.
It is important to emphasise that standardisation is a very transparent process and that it is open to all interested parties, even if general participation is insured by national standardisation bodies. Draft standards have to pass several stages of inquiries and voting. It is interesting to note that approximately 70 % of the time required to develop a new standard is usually spent on getting agreement among participants.

3.2 Standardisation and Crop Protection Machines

Many standards concern crop protection machinery and components because of the numerous possibilities to interchange parts or subassemblies. These standards are for the use of the manufacturer, for the operators' safety, for those dealing with pesticide application and for environmental protection.
The international work on these standards is coordinated by ISO TC23 (tractors and machinery for agriculture and forestry)/SC6 (equipment for crop protection) and at the

European level by CEN TC144 (tractors and machinery for agriculture and forestry)/WG3 (mobile machines and trailers). Table 1 lists all available international and European standards.

Table 1. International and European standards for crop protection machinery

Designation	Type of standardized test system
EN 907	Sprayers and liquid fertilizers - Safety
EN 12761-1	Sprayers and liquid fertilizer distributors Environmental protection General
EN 12761-2	- Low crop sprayers
EN 12761-3	- Air-assisted sprayers for bushes and tree crops
ISO 4102	Connection thread
ISO 4287	Surface texture parameters
ISO 4288	Surface texture assessment
ISO 5681	Vocabulary
ISO 5682-1	Test methods for sprayer nozzles
ISO 5682-2	Test methods for sprayers
ISO 5682-3	Test method for flow control devices
ISO 6686	Anti-drip devices
ISO 8169	Connecting dimensions for nozzles/manometers
ISO 8524	Test of granules applicators
ISO 9357	Nominal tank volume and filling whole diameter
ISO 9898	Test methods for air-assisted sprayers
ISO 10625	Colour coding of nozzles
ISO 10626	Connecting dimensions for bayonet fixing
ISO 10627-1	Data sheets for field crop sprayers - Typical layout
ISO 10627-2	- Technical specifications related to components
ISO 13440	Determination of residues
ISO 13441-1	Data sheets for air-assisted sprayers - Typical layout
ISO 13441-2	- Technical specifications related to components
ISO 14710	Dimensions of nozzle swivel nuts

Analysis of the results of the testing activity in the last few years clearly shows the improvement of the machine performance. The working groups activated by ISO and CEN provide continuous improvement of the standards.

Besides, standards pre-determine the state of the art based on international recognition. Manufacturers, farmers and test authorities consider standards during their decision making process, in order to guarantee the general protection requirements.

4 ENTAM: An International Experience

4.1 What is ENTAM

ENTAM is a network of official testing stations of those European countries that have signed an agreement providing common activities. The aim of ENTAM is to establish and recognise tests carried out on performances, safety and environmental aspects of agricultural machinery based on national, European or international standards or common agreements (or methodologies). The ENTAM gives the manufacturer useful means to improve the machines produced. Publication of test results is the responsibility of the testing station joining ENTAM. Each publication bears the stamp of the testing station.
An ENTAM mark on any report acknowledges that the manufacturer operates in an international market and ensures that all appropriate standards have been met. This gives the farmer an effective tool for the best choice when buying machines.
The first Agreement was signed in 1999 by BLT (Bundesanstalt für Landtechnik), DLG (Deutsche Landwirtschafs Gesellschaft) and CONAMA (named ENTAM since 2000, Ente Nazionale della Meccanizzazione Agricola). Since then, other testing stations have joined the Agreement (Table 2). Today the ENTAM is made up of 19 members. The first steps were difficult because of the different interpretation of the tests and the often incomplete knowledge in the different countries on how to carry out the tests. The wish to coooperate, however, prevailed and some activities started to produce common methodologies and common interpretation of test results.

5 The Development of a Common Methodology

The first draft of an ENTAM methodology for testing crop protection machines based on international standards was defined at a meeting in Germany in 1999. A boom sprayer (Hardi) and an air assisted sprayer (Unigreen) were chosen in a round test that started in 1999 and ended in 2000 giving all testing stations participating in this activity the possibility to test the two machines according to the ENTAM methodology. Subsequently, a special session of the technical ENTAM Meeting in Montpellier was held and results were discussed in order to check the differences caused, for example, by different understanding of the methodology, different ways of carrying out the tests, different equipment, etc.
After much discussion it became clear that most significant differences were caused by a different interpretation of the methodology. In fact, the round test displayed different results with the same test and the same machine. A practical example is given below to help understand this topic.

Table 2. Members of ENTAM

Countries	Institutions
Full Members	
Austria	BLT (Bundesanstalt für Landtechnik)
Belgium	CRA (Centre de Recherches Agronomiques)
Denmark	DIAS (Danish Institute of Animal Science- Research Centre Bygholm)
France	CEMAGRE (Agricultural and Environmental Engineering Research)
Germany	DLG (Deutsche Landwirtschafts-Gesellschaft)
Great Britain	SRI (Silsoe Research Institute)
Greece	IAMC (Institute of Agricultural Machinery and Constructions)
Italy	ENAMA (Ente Nazionale per la Meccanizzazione Agricola)
Spain	EMA (Ministerio de Agricultura, Pesca y Alimentacion - S.G. de Medios de Produccion Agricolas - Unidad Organica: Estacion de Mecanica Agricola)
Switzerland	FAT (Eidgenössische Forschungsanstalt für Agrarwirtschaft und Landtechnik)
Associate Members	
France	ITCF – (Technical Institute for Cereals and Forage)
Germany	BBA (Biologische Bundesanstalt für land- und Forstwirtschaft)
Hungary	HIAE (Hungarian Institute of Agricultural Engineering)
Spain	CMA (Generalitat de Catalunya - Departament d'Agricultura, Ramadeira i Pesca)
Honorary Witness	
FAO	United Nations (Agricultural Engineer Branch - Agricultural Support Systems Division)
New Members	
Argentina	INTA (Instituto Nacional de Tecnologia Agropecuniaria) as Observer Member
Czech Republik	SZZPLS (Government Testing Laboratory of Agricultural, Food Industry and Forestry Machines) as Associate Member
Poland	IBMER (Institute for Building and Electrification of Agriculture) as Associate Member
Poland	PMIR (Chamber of Industry & Commerce of Agricultural & Food Processing Machines) as Associate Member

The volume of residual in the tank on level ground gave different results with a maximum of 28.7 litres and a minimum of 3.45 litres with back flow and with agitation offering a too wide range for the same test (Fig. 2). Considering the fact that it was not a field but a laboratory test it seemed quite clear that the methodology needed to clearly explain the test condition and procedure in order to have a similar output in the different test stations. A special meeting was organised at the BBA testing station in Braunschweig to carry out tests with all the technicians together. This technical meeting agreed on the final version of the methodology containing complete information for all tests.

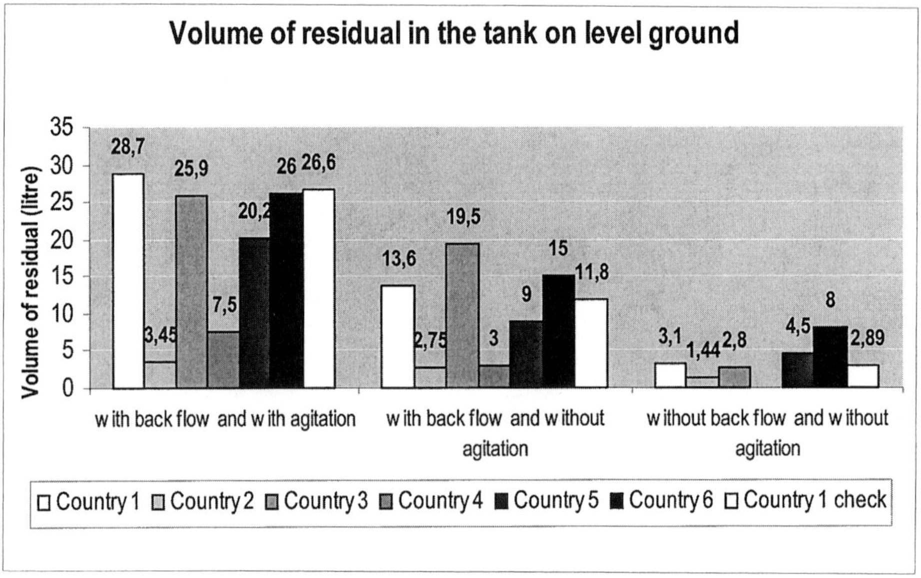

Figure 2. First results of the volume residual in the tank on level ground.

6 The Report: A Printed Document for European Agriculture

An important ENTAM activity is communicating test results. Test reports offer a summary of test results and the possibility to compare them for the best choice of machines. In Italy, in the case of an orchard sprayer, for example, ENAMA publishes test reports which include: technical data (main and auxiliary tank, pump, filters, flow rate adjustment, fan, nozzles, anti-drip devices etc.), description of the machine, test results, comments on the performances, best conditions of use and, also, road regulations (they are different in each country) and safety check.

6.1 Three Aspects: Performance, Environment, Safety

The performances of the machine as well as the environmental aspects are checked according to the methodology. All aspects included in the methodology are taken into account and presented on resume graphics. The safety check includes the reference standards used for the inspections made with the date as validity for the test report. An additional section can give information on the national road regulations and on how to act on public roads. All the results are published in an easy-to-understand compare system, with a brief comment and graphics.

6.2 Guarantee for Farmers, Dealers and Manufacturers

The test report is a guarantee for farmers, dealers and manufacturers because the machine directive and other national and international environment regulations provide for specific requirements and responsibilities.
For the manufacturer, the CE mark is based on self-declaration made after the risk analysis on the machine has been made. A third party check, in this case of an ENTAM member, checking the procedure places, an additional, voluntary mark on the machine. This guarantees quality control giving added value to the product.
With this, the dealer has the guarantee that he is selling products that fulfill existing standards and for the farmer it is important to use machines that fulfill existing standards too. It is well known that despite the EU Directive and national regulations, dealers are, in some cases, selling or renting machines that are not up to standard. In addition, the test report offers useful information for the best choice of machines in relation to his real needs.

7 Machines, Chemicals, Environment, Farmer, Agricultural Products: A Unique Dynamic System

Today, five main aspects must be taken into account: machines, chemicals, environment, farmer, and agricultural products. These five aspects must be considered as a unique system or a dynamic system with close interaction (Fig. 3). Future standards should take these into consideration in the development of better quality in agriculture and environment. Standards should clearly define test procedures to ensure uniformity within all institutions carrying out tests.
A practical example can help to better understand this aspect.
The problems related with container cleaning devices can involve either the engineering of the machines and/or the farmer using the machines and chemicals in the environment. At present, a draft proposal for an ISO standard for devices on the machines is available. During meetings different points of view have been considered. For the farmers' easier use, a cleaning device should be placed on the machine. This raises another concern: that of the environment. Containers will be cleaned on the field with the possibility that the contents of the container may be disposed of in the wrong place. Another solution would be to clean the containers on the farm. This would require special fixed cleaning devices meaning extra costs for the farmer. Where appropriate, the use of hydro soluble containers could be a solution. However, new approaches should be found.
It is important to show that the relation between machines, chemicals, environment, farmers and agricultural products is very close, perhaps more so than it appears to be at first glance.

The ENTAM cooperation represents an important effort towards common approaches and solutions for, as the name implies, a European Union. The ENTAM common methodology is working on finding a unique interpretation. Different test approaches offer different points of view both considering the experience in the testing activities and the experience of farmers and manufacturers in different countries, these, in turn, offer more chances to find the right answer to most problems.

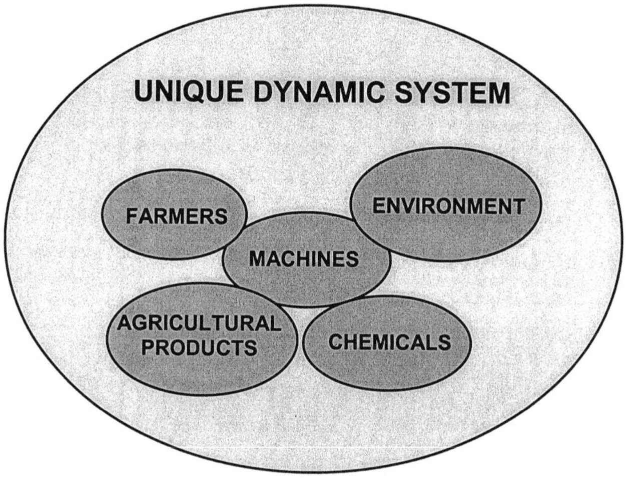

Figure 3. Machine, chemicals, environment, farmer and agricultural products: a unique dynamic system.

8 Conclusions

The application of standards in crop protection machines is an important contribution to the availability of well performing equipment thus improving the safety of the operators, environmental protection and, at the same time, the quality of agricultural products.
Standards offer the highest benefits when there are national regulations such as is the case of the EU Machine Directive. This Directive could be a first important step and may offer protection to farmers from cheap, low quality machinery carrying a CE stamp based only on approximate risk analysis.
The ENTAM network is a unique experience that leads to overcoming most of the existing barriers among European testing stations. A European network working for the European agriculture makes the combination of machine, farmer and chemical a unique dynamic system with a clear formulation and interpretation of standards. Standards are the basic rules for a free and independent market that guarantees the quality of products and the safety of operators and consumers.
The importance of this sector is exemplified by some numbers: more than 4 million crop protection machines are in use in Europe (1998 data). If we consider the European sur-

face of agricultural land to be 75.577.000 ha and 298.286 tons of active ingredients (1999 data), this works out to approximately 4 kg of active ingredient per hectare.
The ENTAM methodology includes the work done by the most important EU testing stations in this field. The next step should be to present the methodology to the EU officials for the preparation of a directive containing guidelines for the testing of crop protection machines.
The new edition of the Machine Directive should provide an additional mark that could be, for example, a *CE-tested* mark in order to give manufacturers as well as farmers the possibility to identify crop protection machines that have been officially tested. This would result in high standard machinery that may be equal to the BBA declaration in Germany or any other compulsory declaration or test in other countries, and would be the base for a high quality mechanisation in crop protection in Europe. This proposal could place all EU countries on the same level concerning the selling and use of new crop protection machines and could be seen as a means of overcoming existing barriers in order to improve quality of the environment and of agricultural products.
A conclusion for this presentation could be the proposal for an *CE-tested* mark based on the efforts made by the ENTAM testing stations that, for the first time, have led to a common and approved methodology for crop protection machines.

Acknowledgements

The work of the ENTAM Members that took part in the round test, as well as the contributions of Prof. Heinz Ganzelmeier (BBA Braunschweig, Germany), Prof. Paolo Balsari (Deiafa Torino, Italy) and Dr. Siegfried Rietz (BBA Braunschweig, Germany) is greatly acknowledged.

20 Suspoemulsions: Key Technology for Tailor-Made Ready-Mix Formulations

Matthias Bratz [1)], Adolf Parg [2)] and Michael Fricke [2)]

[1)] BASF Corporation, 26 Davis Drive, RTP, NC 27709, USA, bratzm@basf-corp.com
[2)] BASF AG, Agricultural Center, D-67114 Limburgerhof, adolf.parg@basf-ag.de, and hans-michael.fricke@basf-ag.de

1 Introduction

When a formulator has to create a new formulation from any given active ingredient, he likes to get an understanding about the nature of that chemical. Solubility in various media, chemical stability in water or organic solvents and the melting point among other data, will give him an idea what types of formulations should be possible to develop. He will keep that in mind and check all new formulation requests against that knowledge and growing his experience with that molecule.

A typical way of thinking would put organic molecules with low water solubility and high melting point into the category of dispersed particle formulations like WP, WG or SC formulations. Molecules that have a low melting point or are even liquids, would be considered for formulations like EC, DC or EW, especially, since many of these molecules will have significant solubility in organic solvents.

When forced to consider ready-mix formulations the above categories become questionable and the creation of the required products becomes more of a challenge. Combinations of different active ingredients frequently make it an objective to combine products with contrasting physical properties, e.g. to combine a water insoluble oil with a high melting solid. Established strategies may fail here and thinking outside the established box becomes necessary.

In this quest suspoemulsions (SE's) offer a valuable and in many cases the only practicable tool to combine a.i.´s with the above properties in one formulation. Thus from the beginning SE´s were used to combine actives with contrasting physical properties, - insoluble in water and oil, high melting on one side and – oil soluble, low melting on the other side.

2 Suspoemulsions in the Market

Besides paying attention to the physical nature and the biological activity of their a.i.´s, formulators have to register and incorporate other trends in the industry, namely in this circumstance, the move to ever more complete products or systems that allow the farmer an easy and convenient one-pass treatment of his crop. These products or systems should be easily adopted and foolproof.
Factors like pack disposal, inventory management and storage requirements are also becoming decisive arguments in the marketplace and are frequently enforced by changes in the regulatory arena (e.g. pressure to reduce organic solvents).
In this environment it takes only some major product combinations that fall in the above categories of physical properties and the development of SE´s is in full swing.

Table 1. Impact of suspoemulsions in major markets

Region	Crop	Solid a.i.	Liquid a.i.	Example
US	Corn	Atrazine	Chloroacetamides	BICEP, Guardsman
Europe	Cereals	Azoles	Morpholines	Opus Team

The corn market in the US and the cereal market in Europe are good examples of how suspoemulsion formulations quickly penetrated a market by combining essential a.i's required in a crop system as a ready mix product.
I will try to further illustrate this by going into more detail about the European cereal fungicide market and exploring challenges and the possible solutions, which can be offered using SE-technology.
The treatment of fungal diseases in the super developed cereal areas has become very complex [1]. In the seventies and early eighties the use of morpholines as mildewicides and azoles as broad-spectrum products were highlighted innovations. To complement each other in the field the products were tank mixed by the farmer. Tank mixing of these products has caused several incidents of incompatibility problems due to the special nature of the morpholines. Aided further by several factors such as the discovery of new active ingredients, biological research, and not to forget the beginning consolidation of the industry, this azole/morpholine-tank mix approach developed into ready mixed formulations. Since many azoles are solids and the morpholines are liquids, SE formulations of these mixtures were designed. Furthermore, the beneficial effect of adjuvants in fungicide treatments was recognized and exploited by coformulating adjuvants [mostly non ionic surfactants (NIS)] as a third component into the SE-products.
In the nineties a new compound class, the strobilurins, was introduced into the market. In Europe azoxystrobin and kresoxim-methyl were heavily used in cereals and the established product concepts and treatment systems had to be adjusted. Namely the importance of anti-resistance management (ARM) was recognized as a key factor in managing crop treatments and product portfolios. Therefore, contrary to first assumptions, product classes like the morpholines were not substituted directly by this new chemistry, but added as a complement to new product combinations to suppress resistance build up.

Due to the physical properties of the morpholines and the need to coformulate with other, mostly solid a.i's, suspoemulsions were again the formulation form of choice found in many sales products.
Carrying the concept of ARM further, more recently the a.i's were not only selected for their fungicidal spectrum but also with respect to their specific mode of action. Importance was especially given to these factors in the mildewicide area. For example, new single site inhibitors with activity against downy mildew like kresoxim-methyl or quinoxyfen were coformulated with morpholines to suppress resistance build-up. (Juwel Top, Fortress Duo). Flexibility in the formulation development process is key for a quick response to these kind of marketing needs. In many cases suspoemulsion technology provided the flexibility and allowed the introduction of new tailor made fungicide products.
With the turn of the century, new products with broad-spectrum activity comparable to the best azoles are entering the market (trifloxystrobin, pyraclostrobin). To find their share in the already highly sophisticated application schemes and developed markets, they need to be presented to the customer as complete, broad-spectrum products, preferably incorporating anti resistance tools. In many instances suspoemulsion formulations will be a cornerstone to combine all required pieces of today's fungicidal armory in one product.
My further presentation will focus on the development of these formulations, possible ways of manufacture, and methods to evaluate their stability and physical properties.

3 Definition and Historical Perspective

Suspoemulsions are defined as a mixture of a dispersed solid phase and an emulsified oil phase in water as the continuous or bulk phase. The solid phase is a solid, water insoluble active ingredient, the oil phase a liquid pesticide in pure form or dissolved in a water insoluble solvent. The oil phase can sometimes also be an oil or adjuvant.
Suspoemulsions are a relatively new formulation type as becomes evident by the fact that a FAO specification was established only as late as 1997. At the same time specific evaluation methods were officially published (CIPAC MT 180) [2].
In the industry the development of SE's started probably as early as the seventies, with several products entering the market in the early eighties. In France the market share for SE's was 1% in 1989 [3]. In the nineties opportunities developed for a more frequent use of suspoemulsion technology for new products [4]. In Europe several ready mix formulations of fungicides containing e.g. azoles and morpholines were commercialized. In the US ready mixes containing atrazine and chloracetamides were gaining market share.

4 Registration of Suspoemulsions

The specification of suspoemulsions is covered by the FAO-Manual (5th Ed. 1998) [1] and most regulatory bodies (specifically in the EU) base their data requirements on the test methods described therein. It is important to recognize that the characterization of SE's is not just the addition of all available methods useful for suspension concentrates (SC) or emulsions (EW).

The most distinctive method to characterize an SE is "Dispersion Stability" (CIPAC MT 180) which is required for registration studies, and has proven to be very valuable in formulation development studies [2].

Multi phase systems like suspoemulsions require special attention in storage stability studies. Therefore, the standard accelerated storage test (CIPAC MT 46) with 14 days at 54°C may sometimes be substituted by using eight weeks at 40°C [5].

Besides CIPAC MT 180, other more general tests like wet sieve residue, pourability and persistent foam are also required.

5 Preparation of Suspoemulsions

In principle, several ways to prepare a suspoemulsion exist and have been described in the literature [6-9]. These include e.g. adding air-milled powder to a preprepared emulsion, emulsifying of an oil phase into an aqueous suspension, or wet-milling a solid in the presence of an emulsion.

The most straightforward method seems to be the simple mixing of preprepared suspensions and emulsions. However, this approach is hampered due to several specific phenomena. In contrast to simple SC or EW formulations, where only one type of dispersed surface is present and needs to be stabilized, at least two distinctive surfaces exist in the SE, -those of the oil droplets and those of the suspended solid particles. In this situation, depletions and displacement of dispersants or emulsifiers take place when EW and SC are mixed.

In particular low molecular weight emulsifiers, which are frequently used for EW or EC's, diffuse onto the solid surfaces, thus displacing dispersants that are stabilizing the solid interfaces. At the same time, oil droplets become depleted of emulsifiers. Whereas the particles become prone to flocculation and sedimentation, the depleted droplets will tend to coalesce. Even worse, destabilization of this kind in SE's will lead to heteroflocculation. Destabilized oil droplets will wet or aggregate with solid particles, thus forming large clusters of unstabilized particles. This normally leads to heavily flocculated systems that are irreversibly damaged. The described phenomena can be observed in the lab straightforwardly by light microscopy.

Generally a stabilization of SE systems can be achieved by selectively choosing surfactants that will not allow migration or transfer between different interfaces, or which will selectively bind to the oil and the solid interface with comparable strength.

These inerts are mostly polymeric or oligomeric in nature and can be both nonionic and anionic. They have multiple binding / interaction sites to the interface, so that a displacement becomes improbable. Several materials have been made available to the industry [6d,e]. The chemistries range from sulfonic condensates (Morwet) or lignin sulphonates (Reaxx) to EO-PO-block copolymers (Pluronics), comb graft polymers, acrylate based copolymers or various other block copolymers.
We have found that the direct emulsification of the oil phase into a suspension can be used as a versatile method in the lab and on production scale to prepare SE's from a wide variety of different active ingredients. In this method an aqueous suspension, which contains all ingredients required by the final formula except the organic phase. This is manufactured to defined specifications The organic phase is then added with gentle stirring to the suspension and is finally emulsified under high shear conditions to give the final product.
For this purpose a fast stirrer with zig-zag blade or a colloid mill such as a rotor-stator mill are useful. After the stirring or the passage through the mill is completed the formulation needs no further processing. Compared with the suspension/emulsion method, fewer vessels are required, low viscous fluids are handled and incorporation of the thickener is more easily achieved.
On a manufacturing scale rotor stator mills have proven to be very efficient. Their use in the preparation of emulsions has been described in a wide body of literature [7]. When preparing suspoemulsions the same principles can be applied.
During the emulsion process the major criteria of process control is the achieved particle size. This is easily measurable, e.g. by laser diffraction or microscopy. The major process parameter to be influenced is the amount of energy used for the emulsification process.
In our experience the fineness of the obtained emulsion is not a virtue in itself. A key for a successful technical process is the stability of the obtained particle distribution at different amounts of energy used for the emulsification step. Thus, achieving a stable particle size distribution is key to control the other physical properties of the formulation (viscosity, dispersibility).
In the following, we will explore some fundamental relations between particle size distribution, energy input, surface tension and viscosities in more detail and will give examples how this relates to product development of SE´s.
In Rotor Stator Mills the dispersed phase is emulsified by shear stress in the conical gap between the rotor and the stator.
The maximum observable particle size (x_{max}) is a function (eq.1) of the surface tension between continuous and disperse phase (γ), the Weber number (We) and inversely proportional to the shear energy (τ) introduced by the colloid mill. Small particles can be expected with high shear and small surface tension values [10-12].

$$x_{max} = {We * 2\gamma}/{\tau} \quad (1)$$

The Weber number is the relation of the viscosities of the dispersed and continuous phase. It should be between 0,01 and 10. Interesting information can be gathered when the changes of the viscosities of dispersed and continuous phase are measured as a function of temperature. Finer droplets are often formed at lower temperatures. All three properties can be controlled; most straightforward is the control of the shear energy.
This equation gives a simplified view however, since at some point more energy will not give automatically smaller particles.
More realistically for a suspoemulsion the correlation between particle size distribution and distribution/separation energy is represented by a curve like as shown in figure 1.

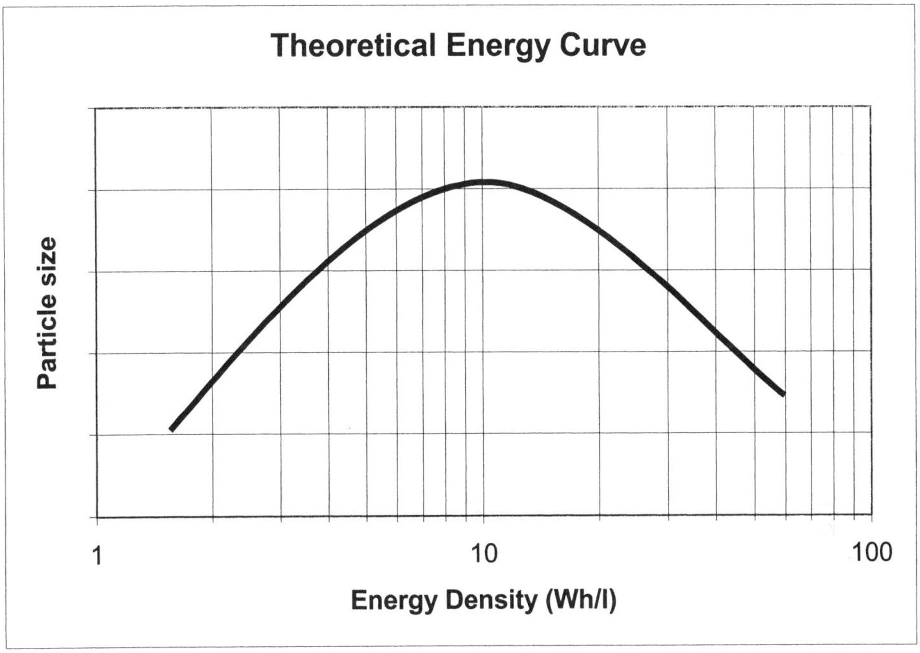

Figure 1. Correlation of particle size and energy density in SE's

At low energy densities, coalescence prevails, while at much higher energies agglomeration and flocculation phenomena will become a problem. The maximum distribution and optimum product quality will be achieved between the two extremes of coalescence and agglomeration.
According to equation 1, the surface tension between the continuous and the disperse phase, the ratio of their respective viscosities (Weber number) and the amount of energy employed play a role. Although, in practice it is sometimes difficult to correlate an observed effect to measurable properties, the principle meaning of these relations is important to be kept in mind.
The optimum particle size distribution cannot be predicted beforehand, since it depends on the used disperse (oil) phase and the emulsifiers and surfactants involved. Therefore

in the course of many formulation projects it was observed that the amount of energy necessary for optimum product quality varied dramatically from formulation to formulation as did the amount of energy required achieving that distribution.
In figure 2 two different systems are shown. On the left side a suspoemulsion is shown, that requires a high amount of energy to be divided and still does not show a maximum in the particle size distribution. When modifying the emulsifier system a drastic change is seen. A spontaneous emulsification occurs with minimum energy input. Further mixing energy does not compromise the system but also does not contribute to further emulsification.

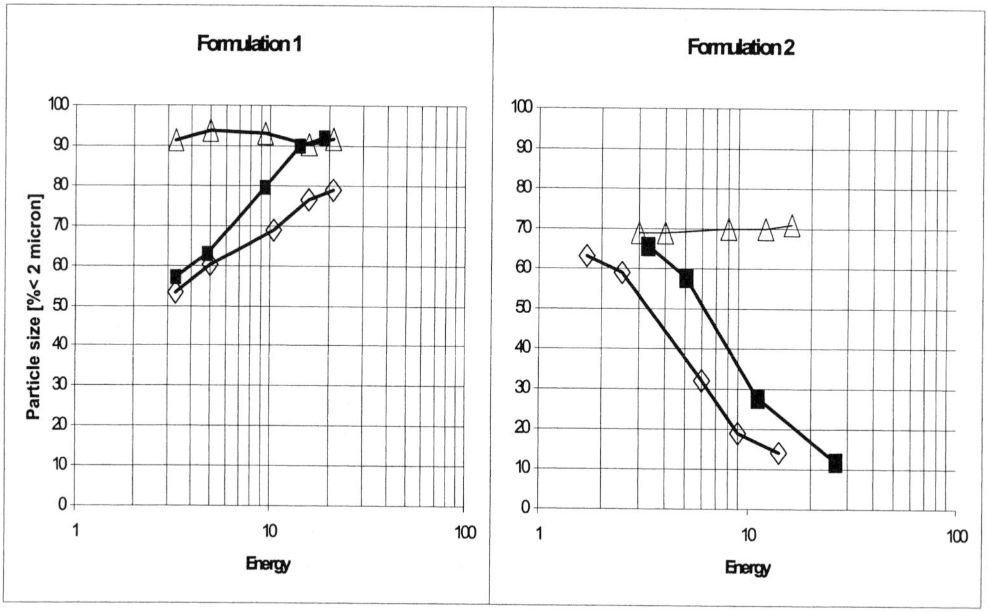

Figure 2. Examples of different formulations

On the other hand, the system shown on the right side is quite different. Beginning with input of relatively small amounts of energy, the particle size distribution does not improve but constantly decreases, showing heavily heteroflocculation in the microscopic picture. Again variation and optimisation of the emulsifiers system yielded a sample that showed nearly spontaneous distribution of the organic phase and resistance to shear induced destruction of the formulation.
In summary the emulsification of oils into aqueous suspensions to give suspoemulsions has proven useful from laboratory to commercial scales.
The process is best understood and controlled by studying the emulsification as a function of energy employed to the system using particle size measurements together with microscopic examinations.

6 Characterization of Suspoemulsions

Characterization of SE´s involves many different aspects. At first there are those methods and protocols that are required to obtain registration of a formulation. These may not be necessarily the same ones that enable the chemist in the lab to quickly differentiate between developmental test candidates. The test methods I will discuss vary in terms of simplicity or depth of insight/understanding they offer. They give quantitative and qualitative results, which may be represented numerically or in a verbal or graphical way.

With complex formulations like SE´s, it is extremely important to have testing protocols that are quickly and easily performed because a lot of different candidates will have to be screened during development.

I would like to distinguish between methods that are performed with undiluted and diluted samples and further on between methods that are performed with the initial or freshly prepared samples on the one hand, and methods or protocols performed with stressed samples on the other hand. Stress for an SE can come from various sources. Most important are temperature, time and shear stress.

Also the purpose to utilize out these methods might be different, ranging from differentiating and screening developmental formulation candidates, assessing storage trials, supporting registration studies or serving as an instrument in production quality control regimes.

6.1 Classical Methods

The most straightforward test for every formulation is the visual inspection of a sample. Especially after storage, observations of hysteresis, serum or supernatant liquid, as well as a check for hard sediment with a spatula can be carried out instantaneously, and doesn't require any fancy instrumentation. Similarly, pourability tests give an indication of viscosity and behaviour in a commercial container. Dilution tests and tests for dispersion in graduated cylinders are equally performed with simple lab equipment. These tests are complemented by redispersability and wet sieve testing. Wet sieve analysis is very valuable, even though it may consume a lot of testing material. It is distinguished from other methods since it yields material that can be isolated, weighed and further examined by other techniques (chemical analysis, HPLC, mass spectrometry).

Even though dispersion stability is a rather crude method, it has a lot of relevance to the actual use of the material in the field by the farmer. Recent refinements are documented in the CIPAC method MT 180 and allow distinction between the stability of the suspended particles and the emulsified oil phase. Unfortunately, only the effects and not the cause of instability can be deduced from these measurements.

6.2 Particle Size Measurements and Microscopy

Much more insight about the diluted sample can be obtained from particle size measurements using light scattering methods [13]. Even though these measurements give nice graphical results or numerical tables, care must be taken in interpreting them. First of all in the context of SE´s it needs to be understood that solid particles and oil droplets will not be distinguished in that measurement. The distribution obtained will represent the sums of all particles. In the illustrative example given, the monomodal distribution around 2μm indicates solid particles and oil droplets of the same size range. In other cases when the size differences between solid and oil particles is big enough a distinction is possible. Bimodal or polymodal distributions will be obtained. The given example illustrates the above. The first picture presents a SE formulation with both the oil phase and the solid particles of comparable size. The second picture shows a distribution obtained from an SE formulation containing epoxiconazol and fenpropimorph (oil) in a distinctive bimodal distribution.

When samples are aged or stressed, an evaluation and interpretation of particle size distributions can become more difficult. While stress in its different expressions (temperature, time, shear) leads almost always to an increase in particle size of both oil droplets and solid particles, the exact nature of the transitions and changes cannot be understood from particle size measurements. In this case microscopic evaluation can clarify the picture. For example, pictures taken from the above monomodal and bimodal distributions confirm the interpretation of the particle size data.

After exposing a sample to stress, a microscopic evaluation becomes mandatory, in order to understand and document the nature of physical changes that might have been occurred. I therefore believe that a microscopic evaluation is one of the most important test methods during development for clarifying and understanding samples of questionable state.

6.3 Other Methods

Another well documented, but less widely used method for formulation development is the measurement of electrokinetic properties [14]. These tests require more distinctive/elaborate sample preparation and are mostly restricted to use during development.

Also widely used and relied on are rheological measurements. The scope of rheological measurements ranges from viscosity measurements to the determination of yield points or oscillatory properties, such as the G' and G''-modules [15]. Since suspoemulsions are not ideal viscous but mostly viscoelastic or dilatant, a wide range of characterization techniques exists. Instrumentation required for this are simple rotary viscosimeters (Brookfield) or more sophisticated stress or shear controlled rotational viscosimeters.

7 Outlook and Summary

Suspoemulsions have proven to be valuable tools to formulate ready mix products of various important active ingredients. Many of these have previously not been accessible due to their contrasting physiochemical characteristics. Examples of their impact can be seen in the US corn herbicide market and the European fungicide market. SE´s have proven to be particularly flexible tools to enable formulators to tailor make a constantly increasing number of difficult new product combinations and adjusting their product ranges to a changing industrial environment.

Processes for the manufacture of SE´s were discussed focusing on the emulsification of oils into aqueous suspensions. Quality control and test methods to accommodate the needs of process control as well as screening tests in the development process were reviewed.

For the future we see a strong continued interest in technologies like the suspoemulsions since product innovations and life cycle management of a.i.´s will rely more and more on new combinations to accommodate changing farming practices and biological challenges.

8 References

[1] Manual on Development and Use of FOA Specifications for Plant Protection Products, 5th Edition, 1999.
[2] CIPAC Handbook, Vol. H, 1998 , p. 310 (Eds.: W. Dobrat, A. Martein).
[3] Seaman, Pest. Sci. 1990, 29, 437-449.
[4] Trends in Pesticide Formulations, PJP Publications, 2001, 53-54.
[5]CIPAC Handbook Vol. F, 1995 , p. 148 (Eds.: W. Dobrat, A. Martein).
[6] A. Wigger, W. Gueckel, Pest. Sci. 1989, 25, 401-409.
[7] P. J. Mulqueen, E.S. Paterson, G.W. Smith, Pest Sci. 1990, 29, 451-465.
[8] J. R. Winkle in Pesticide Formulation and Adjuvant Technology (Eds.: C. L. Foy, D. W. Pritchard), CRC Press, Boca Raton, 1996, 175-185.
[9] T. F. Tadros in Synthesis and Chemistry of Agrochemicals (Eds.: D. R. Baker, J. G. Feynes, G. S. Basarab), American Chemical Society, 76-86.
[10] M. A. Stang, Fortschrittberichte VDI, Reihe 3, Nr. 527.
[11] H. Karbstein, H. Schubert, H. Armbruster, Maschinenmarkt 97, 1991, 50, 20-25.
[12] H. Schubert, H. Armbruster, Chem. Ing. Tech. 1989, 61, 701-711.
[13] CIPAC, Method MT 187, "Particle Size Analysis by Laser Diffraction", accepted as provisional method at the 46th Meeting, Rome, 2002.
[14] R .J. Hunter, Foundations of Colloid Science, Vol. I, II, Clarendon Press, Oxford, 1987.
[15] M. A. Faers, G. R. Kneebone, Pest. Sci. 1999, 55, 312-325.

21 Controlled Release in Crop Protection: Past Experience and Future Opportunities

George B. Beestman

Beestman Formulations Consulting, Madison, Wisconsin, 53717, U.S.A., http://www.beestmanconsulting.com

1 Overwiew

Controlled release is the goal of many formulation processes to produce encapsulated liquid pesticides, coated pesticide crystals, adsorbed and absorbed pesticides into synthetic or natural carriers, guest/host complexes, solid solutions with polymers, or chemical linkages to polymers. Twenty-eight unique processes leading to controlled release of actives are described and guidance for selection is provided [1]. Only a few of these processes can meet the stringent demands of pesticide products for high volume manufacture, particle size, temperature storage extremes, or pumping pressures. Processes to coat pesticide crystals have been reviewed [2]. Comprehensive review of controlled release of pesticides has been published [3]. This short chapter adds more recent developments.

2 Controlled Release

Two distinct controlled release technologies are encapsulation of liquid pesticides and the coating of individual pesticide crystals. Encapsulation of liquid pesticides is an established tool for modern formulators. Commercial microencapsulated pesticide products exist and new developments continue to be made. Coating of individual pesticide crystals without their aggregation is more difficult. While new processes do exist to coat pesticide crystals without aggregation these processes have not yet been utilized to create commercial pesticide products.

3 Microencapsulation of Liquid Pesticides

Two types of microencapsulation are known in the art based upon the shellwall forming chemistry. These are interfacial polymerization and in-situ polymerization. Encapsulating plastic shellwalls are synthesized at the O/W (Oil-in-Water) interface of a pesticide emulsion by reacting oil-soluble monomers dissolved in the pesticide with water-soluble monomers added to the emulsion. This process is referred to as interfacial polymerization.

Novel shellwall chemistry has been developed that produces an encapsulating shellwall around pesticide emulsion droplets utilizing a single monomer or prepolymer dissolved in the pesticide. Heating the emulsion and use of catalyst produces shellwalls. This process is referred to as in-situ polymerization.

3.1 Interfacial Polymerization.

A milestone in application of microencapsulation to herbicides was the development of high concentration formulations. Commercial products containing 480 grams of herbicide in microcapsules per liter volume of formulation were possible. This development was based upon the discovery of dispersing polymers that could emulsify a larger volume of pesticide into a smaller volume of water [4].

Recent adaptations of this technology provide structurally sound shellwalls that can be varied to release in a controlled manner [5]. Conventional methods to alter release from microcapsules include using less monomer and making smaller microcapsules. However, processing in this way to control shellwall thickness could lead to fragile microcapsules. Controlled release with more robust microcapsules was achieved by using specified ratios of unique diisocyanate and triisocyanate monomers dissolved in the pesticide. Diamine monomers added to the emulsion in the typical manner produced robust microcapsules with desired rates of release.

Aqueous solutions can also by microencapsulated in high concentration [6]. To prepare the "reverse phase" W/O (Water-in-Oil) emulsions care must be taken to select monomers that will remain in the dispersed water droplets during the emulsion stage. If the monomers diffuse from suspended droplets into the continuous phase polymerization will happen throughout the emulsion and not at the interface as intended. No microcapsules will be formed. This problem has been addressed utilizing carboxy-functional polymers to associate with amine functional reactive monomers dissolved into the water droplets [7]. Shellwalls are formed at the W/O interface by addition of the oil-soluble monomers to the continuous oil phase. Without the carboxy-functional protective polymers amine monomers would have partitioned out of dispersed water droplets and into the oil phase. Microcapsules would not have been produced.

A recent innovation merges the microencapsulation of pesticides with the use of multiple emulsions W/O/W (Water-in-Oil-in-Water) to isolate incompatible actives [8]. Multiple emulsions are becoming more common tools for the formulator trying to combine multiple active ingredients in a single package, particularly when the active ingredients are not compatible. By microencapsulating one or more of the phases in a W/O/W emulsion the pesticides within each phase are assured to remain apart. Depending upon which phase contains the shellwall forming monomers, it is possible to encapsulate the inner water phase, the oil phase with the inner water phase, or both phases individually. Microencapsulation insures that the dispersed phases remain separate.

3.2 In-Situ Polymerization

Novel shellwall chemistry takes advantage of isocyanate hydrolysis back to amine functionality to produce encapsulating shellwalls "in-situ" at the O/W interface of pesticide droplets emulsified in water [9]. Isocyanate monomers dissolved in the pesticide are the only monomers used. Hydrolysis of some of the isocyanate to form amine groups produces the shellwalls. Self-reaction of the generated amine groups with residual isocyanate groups produces an encapsulating shellwall in-situ. Isocyanate monomers are confined to dispersed pesticide droplets precluding un-reacted water-soluble monomer remaining in the final product. The hydrolysis step evolves CO_2, so that catalysts are used to complete the reaction prior to packaging. This process provides microencapsulated acetochlor with early biological efficacy equivalent to an un-encapsulated herbicide [10]. A wide range of release rates is possible by regulating the degree of crosslinking and affecting droplet size in the emulsion [11].

A recent innovation in in-situ microencapsulation is the development of acid-triggered release of pesticide from the microcapsules [12]. Diols and aldehydes are reacted to form an acid labile acetal moiety. The acetal is then reacted with isocyanate to create a prepolymer. The prepolymer is a polyisocyanate containing the acid labile moiety and suitable for in-situ shellwall polymerization. The prepolymer is dissolved into a pesticide, emulsified into water, and shellwall formed in-situ. Under alkaline or neutral pH conditions in a container, the insecticide is safely contained in the microcapsules. Acid could be added to the spray tank to rapidly release capsule contents prior to application.

Alternate shellwall chemistry for in-situ microencapsulation utilizes etherified urea-formaldehyde prepolymers in the oil phase that are self-condensed with acid catalyst to produce encapsulating aminoplast shellwalls [13]. This process does not have the problem of continuing CO_2 evolution. Water-soluble urea-formaldehyde and melamine-formaldehyde prepolymers can be selected to microencapsulate water or aqueous solutions [14].

Another innovation in in-situ microencapsulation is aminoplast shellwalls containing base-cleavable ester moiety [15]. Polyols reacted with diacids that contain thiol or hydroxy functionality produce crosslinking groups. These crosslinking groups along with urea-formaldehyde prepolymer are dissolved into the pesticide and the in-situ microencapsulation process is completed. The resultant microcapsules may contain an insecticide that is safer to handle under acidic conditions yet will rapidly release the insecticide in the alkaline gut of an insect. Alternately, base could be added to the spray tank to rapidly release capsule contents prior to application.

In an alternate process organic pesticide was microencapsulated in aminoplast shellwalls starting with the urea-formaldehyde or melamine-formaldehyde prepolymers dissolved in the aqueous phase [16]. Growing polymer chains wrap around the pesticide emulsion droplets forming aminoplast shellwalls. In this process the aqueous phase did not have to be heated to the temperature of the molten pesticide. Satisfactory microcapsules were produced with the aqueous phase at temperatures lower than the temperature of the molten pesticide. Excess heat of high melt point molten pesticides could be utilized to catalyze shellwall formation. Use of in-situ microencapsulation was extended to higher melt point pesticides.

3.3 Nanoencapsulation

Formulation of active compounds to the nanometer size range is an active area of development in coatings, pharmacy and other disciplines. Little development has been pursued in Agrochemicals. Yet, there is evidence to suggest that enhanced biological efficacy and other effects may be discovered in sub-micron size-range formulations. Milling crystalline solid pesticides to sub-micron dimensions improved biological efficacy [17]. Microemulsions have provided novel properties compared to conventional emulsions. A new development in production of capsules in the nanometer size-range may open new horizons in pesticides formulations.

Self-emulsifying reactive vinyl monomers were synthesized that emulsify pesticides into water. Polymerization to form encapsulating shellwalls produced capsules in the nanometer size-range [18]. Capsule size can be varied with change in level of self-emulsifying monomer. Higher levels of monomer produce smallest capsules.

The self-emulsifying monomers are alkylethoxylates with a terminal vinyl group and are self-polymerizing. The degree of ethoxylation can be varied to produce O/W or W/O emulsions. Pesticide, monomers, and optionally cross-linkers will self-emulsify to the nanometer size range. Polymerization and cross-linking to form shellwalls produces nanocapsules.

3.4 Controlled Release Developments

Certain pesticidal combinations are antagonistic resulting in loss of efficacy for one or the other of the compounds. To overcome these effects the compounds must be applied separately or the rate adjusted to accomplish full efficacy. These are not attractive solutions to this problem. Controlled release has been a more useful tool. Encapsulating one of the antagonistic pairs allowed full biological efficacy when combined with the un-encapsulated antagonized partner [19]. The problem of antagonism has also been overcome by combining anionic polymers with cationic pesticides [20]. The polymer and pesticide combinations provided improved bioefficacy.

Water Dispersible Granules are a popular type of formulation. Dispersible granule formulations of microcapsules take advantage of high loading with reductions in packaging and transport. Microencapsulated pesticides can be processed to Water Dispersible Granules by spray drying them either separately or in combination with other pesticides [21]. Rapidly dispersing granules of microcapsules were produced using water-soluble polymers in the suspension prior to spray drying [22]. Polyvinyl alcohol was a preferred polymer.

Monomers and prepolymers that are useful to form shellwalls were reacted with interface modifying molecules to create novel surfactants in-situ leading to stable emulsions [23]. The reactions are designed to create polymers with surfactant properties rather than encapsulating shellwalls at the O/W interface. The emulsions were formed without conventional emulsifiers, and were shown to have long term stability without creaming. Absence of classical emulsifiers reduces the tendency toward foaming.

3.5 Coating Pesticide Crystals

Coating of individual pesticide crystals is a technology that is more difficult to accomplish than is microencapsulation of liquid organic pesticides. Discreet crystals in the biologically effective size range of sub-micron to two microns maximum dimension cannot be fluidized in air for the direct application of coatings. Such small sized crystals can be suspended in water but when polymers are insolubilized to form an encapsulating shellwall larger aggregates containing multiples of individual crystals are obtained. Unlike liquid pesticide emulsion droplets in water, individual pesticide crystals suspended in water do not carry monomers to the Organic/Water interface for direct synthesis of microencapsulating shellwalls. For these reasons there are no commercial products composed of individually coated pesticide crystals. New technology was required to microencapsulate individual pesticide crystals.

3.6 Direct Coating

The leading process for direct coating of pesticide crystals in the sub-micron size range is Injection Treatment Coating [24]. Molten coating material is atomized to sonic velocity. Particles to be coated are metered directly into the atomizing stream and become coated on all surfaces. Coated sulfonylurea crystals were stable when formulated with 2,4-D acid while the un-coated sulfonylurea crystals were rapidly decomposed in the same composition. While the process is very efficient and effective, it is not suited for research purposes. Less than one kilogram of powder per minute is not efficiently coated. Much higher rates of powder flow coat most effectively.

Sub-micron to micron-size particles cannot be fluidized in air for direct application of coatings. A new process that is capable of suspending sub-micron particles and provides for coating of the particles is Mechanically Assisted Impaction Coating [2]. Particles to be coated are suspended in a bed of magnetic discs. The magnetic discs are in rapid spinning motion responding to an alternating magnetic field that surrounds the coating device. Particles to be coated along with solid coating materials are in constant collisional contact with the magnetic discs. The small coating particles become shot-peened onto the surface of the larger particles to be coated. Particles of coating material must be smaller than one-tenth of the size of particles to be coated. This dry-on-dry process is well suited to water-sensitive compounds. Small laboratory experimental units are available. Production units capable of hundreds of kilograms per hour are being used.

Pesticide crystals coated by impaction of small coating particles onto larger core particles may not be fully coated on all surfaces. Controlled release of the pesticide may not be obtained. Therefore the technology still needs to be proven with pesticides. However, the process does offer volume and particle size to meet the demands of commercial pesticide products.

Numerous other solid particle coating techniques are reviewed [2]. None can produce individually coated particles in large volume and in the biological size range of pesticide crystals.

3.7 Suspension Coating Of Particles

Pesticide crystals can be coated by the process of coacervation [1]. Water-soluble polymers are used to suspend the crystals. The suspension is then manipulated to cause the polymers to insolubilize and harden to encapsulating shellwalls. While effective, these processes do not produce sub-micron particles. Usually matrix particles from thirty microns to fifty microns maximum dimension are obtained. Each matrix particle contains many of the primary particles. Full biological efficacy in foliar applications would be lost.

A new process from the paint industry may have application for the suspension coating of pesticide crystals [25]. Low molecular weight water-soluble vinyl polymer chains are synthesized and the process stopped with a terminal vinyl group on each polymer chain. Hydrophobic acrylic monomers are then added to create the hydrophobic strongly adsorbing backbone polymer. As each water-soluble polymer terminal vinyl group reacts with the growing hydrophobic backbone polymer it becomes inserted like a tooth on a comb. Indeed, these polymer structures are referred to as "comb" polymers.

A special class of comb polymers each containing a reactive functionality on the adsorbing backbone was synthesized for suspension coating of pigments. The pigments are wet milled to sub-micron size using the comb polymers as suspension stabilizers. To the suspension of pigments is then added a monomer that is reactive with the functionality built into the backbone of the stabilizing polymers. The resultant reaction chemically cross-links the stabilizing polymers to the pigment surface. Extremely stable pigment suspensions with sub-micron size range particles can be produced. This process provides for the suspension coating of individual crystals in the volume and particle size requirements of pesticide products. This process should find utility with pesticides.

4 Future Prospects

Processes are now available to provide controlled release to liquid pesticides, molten lower melting point pesticides, and solid higher melting point pesticide crystals in the size range and volume demanded of pesticide products. These processes will continue to develop into more robust and simpler systems. New processes will be developed.

End-use properties of release on demand, stabilization of compounds, enhanced efficacy, reduced environmental loss, successful combination of incompatible compounds, reduced handling toxicity, reduced phytotoxicity to crops, and universal formulation compositions will continue to improve pesticide use. Controlled release technologies, just beginning to demonstrate commercially viable options, will grow to become the most utilized of formulator's tools.

5 References

[1] R. E. Sparks, I. C. Jacobs, Selection of Coating And Microencapsulation Processes, Controlled Release Delivery Systems, Marcel Dekker, Inc., USA, 1999, pp. 3-29.

[2] G. B. Beestman, Microencapsulation of Solid Particles, Controlled Release Delivery Systems, Marcel Dekker, Inc., USA, 1999, pp.31-54.

[3] H. B. Scher, Controlled Release Delivery Systems, Marcel Dekker, Inc., USA, 1999.

[4] G. B. Beestman and J. M. Deming, Encapsulation by Interfacial Polycondensation and Aqueous Herbicidal Composition Containing Microcapsules Produced Thereby. United States Patent 4,280,833, 1981.

[5] M. E. Seitz, R. J. Brinker, and J. N. Travers, Microcapsules with Readily Adjustable Release Rates. United States Patent 5,925,595, 1999.

[6] G. B. Beestman, High Concentration Encapsulation of Water-Soluble Materials, United States Patent 4,534,783, 1985.

[7] M. Lykke, K. K. Mistry, and K. C. Symes, Particles Having a Polymeric Shell and their Production, United States Patent 6,359,031, 2002.

[8] D. K. Rodham, G. Ramsey, D. J. Brown, and T. F. Tadros, Formulation, United States Patent Application 25,986, 2002.

[9] H. B. Scher, Encapsulation Process and Capsules Produced Thereby, United StatesPatent 4,285,720, 1981.

[10] H. B. Scher and M. Rodson, Microencapsulated Acetochlor Herbicidal Composition, United States Patent 6,340,653, 2002.

[11] J. L. Chen, M. Rodson, H.B. Scher, and K. S. Lee, Microencapsulated Compositions, United States Patent 6,133,197, 2000.

[12] J. E. Van Koppenhagen, H. B. Scher, K. S. Lee, I. M. Shirley, P. Wade, and R. Follows, Acid-Triggered Microcapsules. United States Patent Application 68,169, 2002.

[13] H. B. Scher and M. Rodson, Microencapsulation Process, United States Patent 4,956,129, 1990.

[14] M. Rodson and H. B. Scher, Water-in-oil Microencapsulation Process and Microcapsules produced thereby, United States Patent 6,113,935, 2002.

[15] J. E. Van Koppenhagen, H. B. Scher, K. S. Lee, I. M. Shirley, P. Wade, and R. Follows, Base-Triggered Release Microcapsules, United States Patent Application 004,059, 2002.

[16] R. Nastke and E. Neueunschwander, Process for the Preparation of Spherical Microparticles Containing Biologically Active Compounds, United States Patent 5,908,632, 1999.

[17] R. M. Strom, D. C. Price, and S. D. Lubetkin, Aqueous Dispersions of Agricultural Chemicals, United States Patent Application 51,175, 2001.

[18] D. Herault, Encapsulation of Emulsions, PCT Patent WO 051,536 2002.

[19] H. P. Krause, G. Schnabel, G. Frisch, J. Wurtz, U. Bickers, E. Hacker, T. Auler, A. Melendez, and D. Haase, Combinations of Crop Protection Agents with Organic or Inorganic Carrier Materials, United States Patent Application 55,436, 2002.

[20] G. Frisch, U. Bickers, K. A. Young, E. Hacker, and G. Schnabel, Combinations of Crop Protection Agents with Anionic Polymers,United States Patent Application 10,099, 2002.

[21] J. M. Surgant and J. M. Deming, Formulation of Water-Dispersible Granules and Process for Preparation Thereof, United States Patent 4,936,901, 1990.

[22] R. J. Lo, J. L. Chen, and H. B. Scher, Dry Water-Dispersible Compositions of Microencapsulated Pesticides, United States Patent 6,358,520, 2002.
[23] H. B. Scher, P. J. Mulqueen, N. D. Green, and C. J. Piper, Novel Emulsions, United States Patent Application 40,065, 2002.
[24] R. D. Wysong, G. B. Beestman, and G. A. Schurr, Crop Protection Composition Comprising a Crop Protection Solid Particle Coated With Water-insoluble Coating Material and a Crop Protection Mixture Comprising the Same, United States Patent 6,015,773, 2000.
[25] M. Fryd, and K. B. Visscher, Particles Dispersed W/Polymer Dispersant Having Liquid Soluble and Cross-Linkable Insoluble Segments, United States Patent 6,262,152, 2001.

Environmental Fate and Safety

22 Advances in Validation of Environmental Exposure Model Predictions for Regulatory Purposes

Paul Hendley

Syngenta Crop Protection Inc.,410 Swing Road, Greensboro, NC 27455, USA, E-mail: paul.hendley@syngenta.com

1 Introduction

The validation of exposure estimates for pesticide risk assessments is a fascinating topic and draws together many complex issues. It is a growing discipline and this paper is not intended to review the field since others [e.g. 1,11] have published comprehensive literature surveys. Instead, this publication seeks to raise awareness of some conceptual issues surrounding the validation of aspects of the exposure component of the risk assessment. Exposure estimation technology has made many advances recently but it can be argued that we have yet to "connect the dots" between practical experience and model estimates in a way that can be readily explained to the public; this paper will challenge some of the current thinking and suggest ways that regulators, industry and academia can work together to lay the foundations for understanding and explaining exposure estimation approaches in order to engender the confidence of the public and government agencies.

In the past, complex exposure assessments generated via more sophisticated models were the preserve of a small elite international group of experts who understood the strengths, weaknesses and underlying assumptions associated with the exposure models. However, the realization of the need for more sophisticated assessments and recent developments in model "shells" have resulted in more complex exposure assessments being conducted by a much broader group of scientists, few of whom would class themselves as model experts. It is therefore essential that we take steps to ensure that the implications of the routine regulatory use of these modeling tools and scenarios are transparent to this new model-user community. To add to the confusion, the various approaches used by different regulatory bodies (e.g. US Environmental Protection Agency's (USEPA), the European Commission Food Safety Office and the Pest Management Regulatory Agency (PMRA) in Canada) to estimate exposure in aquatic and terrestrial ecosystems differ significantly and this contributes to apparent international complexity.

This paper will show that relatively simple approaches could be taken now to greatly improve the transparency of exposure estimates made to support risk assessments and to more readily identify similarities and differences between the various approaches (and their underlying assumptions) used worldwide. This will provide "validation" in terms of improved confidence with impacts much greater than can be achieved in the short term by more rigorous formal scientific validation exercises.

2 Are Regulatory Models an Issue?

It is important to understand why regulatory models need more than purely technical validation. Firstly, they serve the needs of a range of stakeholders in addition to exposure scientists; for example, risk assessors from other disciplines such as ecology and human health, as well as risk managers who are charged with implementing solutions that meet socio-economic goals based on the outcome of the risk assessment. To meet these needs, models used for regulatory purposes need to have clearly understood purposes and deliverables. Additionally, because model output is used to help decide issues with economic and political implications where the protagonists often hold highly polarized opinions, the model output is frequently the topic of scrutiny and criticism by one party or another. These debates have been exacerbated by the use of emotionally charged terminology such as "worst-case" scenarios.

The continuing trend from simple risk assessments involving Toxicity/Exposure ratios (TER's) or Risk Quotients (RQ's) to more complex assessments based on the probability of exposure has raised the profile of exposure modeling without providing underlying support for non-modeler stakeholders to feel comfortable with the information being provided. As mentioned above, the lack of apparent international standardization has added to this lack of confidence. To add to the questions in the minds of some key influencers, the new and potentially confounding element of "understanding the uncertainty associated with exposure estimates" has entered the debate recently as a result of discussion on probabilistic exposure assessment e.g. [2] and [3]. A Pellston workshop has recently been held on the topic [4] but unfortunately, discussion on the subject has so far tended to focus on ways of measuring it rather than on how to interpret its significance. Thus to some regulators and to the general public, it may appear that the estimation tools used by scientists are prone to much uncertainty and, perhaps, unsuited to clear decision making. Similar concerns have also sprung from the need to make decisions at the tails of estimated distributions (e.g. the use of 99.9^{th} percentile estimates in human drinking water exposure estimation).

The last point to be made is that most regulatory exposure transport models were developed at the field scale and that is where historic validation efforts have largely focused. However, increasingly, the public and regulators want to understand how the current exposure model predictions relate to watershed scale monitoring data with a focus on human drinking water. Currently the models tend to grossly over-predict at these scales [5,6]; while there are scientific reasons for this, clear explanations and plans to address the problem in a fashion that the public will find credible are needed soon. From a purely science-based approach, the stock answer of "needs more research" would, in time, resolve these issues; however, regulatory decisions normally have a fixed timeline and decisions have to be made "now" with available information and thus urgent efforts are needed to improve validation approaches as well as to build a bridge between the science of model validation and public and regulatory confidence.

3 Regulatory Models Redefined

When asked to consider validation of pesticide exposure models, specialists tend to jump straight into a discussion of the step 1 and 2 leaching, runoff and drainage models required in various regulatory schemes. (Note: throughout this chapter, the terms tier 1 & 2 and Step 1 & 2 are used interchangeably and reflect the different terminology used in EU (FOCUS) and USA for the protective preliminary risk assessment and more predictive modeling with "canned scenarios".) The roles of these mathematical tools are to predict the degradation and transport of pesticides in the environment to estimate concentrations in water bodies at or near the "edge-of-field" or in soil pore water. These models allow the estimation of sensitivities to various chemical, weather and agronomic parameters and, if validated, can extend measured data sets to better understand the probability of various levels of exposure in the field. While these are very important and provide the end-step for pesticide risk assessments, it can be argued that a broader view should be taken to include all the "assumptions" inherent in our approaches since many of these involve underlying (albeit simple) models.

Under this broader definition, we would define models as "frameworks to help interpret or extend data". These frameworks are *always* characterized by some form of conceptual model and some underlying models/assumptions and they *may* have predictive capability. Examples fitting this definition include:

- "Observations of "xxx" are normally distributed"
- "Soil dissipation rate will be modeled as first order"
- "Wash-off is proportional to rainfall intensity"
- Darcy's law: Flow rate in a saturated homogeneous porous media is directly proportional to the hydraulic gradient
- "Chemical degradation rate, soil permeability and rainfall are primary drivers for deciding leaching scenarios"
- PRZM, EXAMS, PEARL, MACRO, FOCUS shells, GENEEC, FIRST....

The key issue is that ALL these "models" have inherent conceptual models and assumptions and unless their impacts are understood then the credibility of the risk assessment may be in doubt.

4 Traditional Validation Approaches

Classical exposure model validation approaches have been based around *validation* – "determining that adequate agreement exists between the quantity being modeled and the model output" [e.g. 7] and *verification* – "ensuring that the model behaves (runs) as intended" [e.g. 7]. Alongside this, approaches have considered the strengths and weaknesses of hypothesis testing and the risks related to making Type 1 and Type 2 errors [8] along with consideration of accuracy and precision. Increasingly, [9,14-17] Monte Carlo approaches have been used to take some account of the information on the range of input variables. The goal of many model validation exercises is a statement such as:

"Predicted values from the model should be within a "factor of X" of reality M% of the time and within a "factor of Y" of reality N% of the time".

However, the regulatory relevance of these classical validation approaches has been subject to question for a number of reasons:

- Applicability of extrapolating from a validation conducted with field specific information to broader scales (especial with site specific calibration).
- Lack of convincing "weight of evidence" supporting model validity via consistent validation approaches.
- Lack of observed data to validate extremes used for regulatory decisions.
- Tendency of authors to fixate on details of the validation rather than addressing the broader brush regulatory picture.

Hence in addition to the technical description of a model validation italicized above, in order to address regulatory questions, it appears that additional goals must be included:

- Build confidence and credibility in a particular model application by explaining the implications of findings from the validation for the specific regulatory PURPOSES of the particular model.
- Agreement and use of consistent approaches for expressing the findings of model validation.
- Develop concepts and approaches for "ground truthing" model output with field data where strict validation may not be possible in order to develop "weight of evidence" support from multiple locations and data sources.
- Provide clear and agreed expositions of the conceptual models behind ALL regulatory approaches in addition to agreed lists of strengths, weaknesses, assumptions and potential sensitivities in order to help users of model output feel confident about the "suitability for purpose" of each regulatory model.
- Provide long term studies to develop field data to support model development focused on pressing regulatory needs.

5 Monitoring and Modeling - Strengths and Weaknesses

The validation goal italicized above mentions that model output should simulate "reality" with some specified degree of precision. Increasing frustration comes from this implicit and extraordinary assumption that data measured in the field reflects "reality" more closely than model output. This issue became apparent during the FEMVTF process [10] and has also been mentioned by other authors. It should be sufficient to "validate" a tier 1 model designed to be protective in risk assessments by demonstrating that it is consistently more protective than the accepted tier 2 model that follows it in the stepwise process. However, in all other validation approaches, some measured field data is needed. Questions that must be asked about the generation of the field data used for validation include:

- How well did the sample integrate across space and time (e.g. depth and width distributions of residues in a stream and the likelihood of even mixing at the sampling point)?
- Do the analytical measurements reflect "total" chemical or separate sediment bound from dissolved components?
- Is the Level of Detection (LOD) appropriate for the regulatory goals?
- Did the sampling strategy "miss" peaks in time and/or space?
- Can the extent and timing of chemical use in the watershed relative to weather conditions be quantified?

Table 1 lists strengths and weaknesses of monitoring and modeling approaches that are relevant when balancing relative values of model output and matching field data for a validation study.

From a validation perspective, an important point for consideration is exemplified in Figure 1. Typically, pesticide exposure model estimates are driven using data from laboratory studies where the number of variables is controlled; the resulting model output is most often validated by comparison with data from managed field transport studies (e.g. prospective ground water or simulated runoff studies) performed at plot to field scale. Traditional validation can be effective because these studies tend to have more intensive measurement of several variables (e.g. hydrology, tracers and pesticide) across time and space. However, *regulatory* interest increasingly focuses on the relationship between model output and monitoring datasets which, as described above, may be characterized by less detail, higher measurement uncertainty and considerable watershed variability (but may have the advantage of having data on multiple analytes). A comprehensive model validation strategy needs to build a "weight of evidence" approach to allow an assessment of the predictive power of a model at various scales. Clearly, different criteria are needed to address the degree of validation achieved for either type of dataset.

Table 1: Comparison of Strengths and Weaknesses of Modeling and Monitoring Data Sets

Modeling		Monitoring	
PRO's	CON's	PRO's	CON's
Relatively quick - days to months Can start irrespective of season	Model algorithms may oversimplify / misrepresent compound behavior	"Real World" measurements • residues reflect actual agronomy / Ag landscape	Requires months to several years • Start date depends on season
Cost effective	Reality "poor" • May use conservative assumptions /scenarios • Output may focus on weather extremes	Reality "rich" • Real world hydrology • Accounts for actual pesticide use	A few unique combinations of circumstances - no "context" • Missing pesticide use data • Location, weather, timing
Can evaluate "what-if" scenarios and sensitivities	Generates lots of "data"!	Data can calibrate/ validate models	Costly and needs time to set up
Can incorporate effectiveness of possible mitigation alternatives	General public reluctance to accept predictions/low credibility	Avoids use of conservative modeling assumptions /scenarios	Difficult/time-consuming to plan
Predicts across space and time	Calibration/validation essential	Accounts for heterogeneity • Integrates data across watershed	Often mis-used • More public confidence/credibility!!
Not constrained by analytical LOD	Building in uncertainty in inputs not obvious	More public confidence/credibility	Unusual weather may distort data
Can quantify relative to benchmarks	Data on watershed scale pesticide use unavailable	Does not require algorithm/ model development & validation	Probabilistic interpretation difficult
Generates lots of "data"	Useful watershed and flowing water scenarios not currently available	Provides "benchmark" data for comparison with new chemistry	Difficult to express uncertainties
Data easy to put into context		Can address diverse stakeholder needs • Good tool for State & Federal Gov	Cause & effect difficult to cross-correlate with biological monitoring Highly skewed distribution of data - "non-detects" an issue – LOD key Comparisons difficult due to lack of "control watersheds" Not much "data" provided

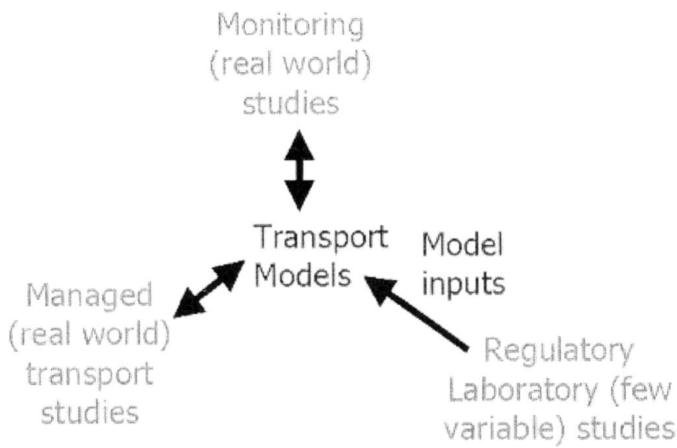

Figure 1. Relationships between model output, managed study and monitoring results

Consideration of Table 1 and Figure 1 shows that it is critical to understand the implications of the field sampling strategy, analytical scope, study scope etc. that generated the field data used for validation. The obvious conclusion is that where programs are set up to help validate models, some sort of coordinated multi-scale nested design would be ideal to replace the currently haphazard coordination of regulatory studies. This might start by including soils from the region in laboratory studies and would include conducting any necessary managed regulatory studies *within* a monitored watershed. The monitoring program itself might perhaps use nested monitoring designs such as a few more intensely sampled sub-areas within larger, less intensely monitored watersheds (as in the United States Geological Survey's (USGS) National Water Quality (NAWQA) program [24]). From an ecological risk assessor's point of view, this type of design might also ultimately allow for the inclusion of effects monitoring.

6 FIFRA Environmental Model Validation Task Force (FEMVTF) – a Classic Example

The FEMVTF program [1] proved to be a very useful learning exercise for all involved in the collaboration. In the interest of brevity, the results will not be repeated here since they have been very recently reported [10 – 17]. Key findings may be summarized as follows:

- 18 sets of chemical parameters and field data (leaching and runoff) were made available for use by the public.
- Detailed literature survey was published.
- Novel approaches for quantifying validation performance were investigated.

- Sensitivity analyses showed expected patterns with soil bulk density being an often problematic but significant field measurement.
- Modeler "individuality" needs management – detailed input guidance and model shells appear to be key elements for improvement.
- Leaching predictions were satisfactory for bromide tracer in soil and soil pore water (factor of 2) and, after site specific calibrations, estimates of pesticide soil pore water concentrations were also within a factor of 2.
- Runoff pesticide transport was predicted within an order of magnitude and, after site specific calibration, within a factor of 3.
- The extent of agreement was generally considered better than expected.
- Practical tools were developed to assist sensitivity analysis and Monte Carlo validation approaches.

Important non-technical learning came from the realization of the difficulty of quantifying the "quality" of a validation in terms that were meaningful to regulators as well as the importance of understanding the deficiencies in the field data set used for comparison with model output.

7 Idealized Validation Process

Figure 2 is a flow chart linking key aspects of exposure model conduct and validation which will be discussed in the following sections. It represents one pathway for ensuring that the key elements of a validation process have been addressed.

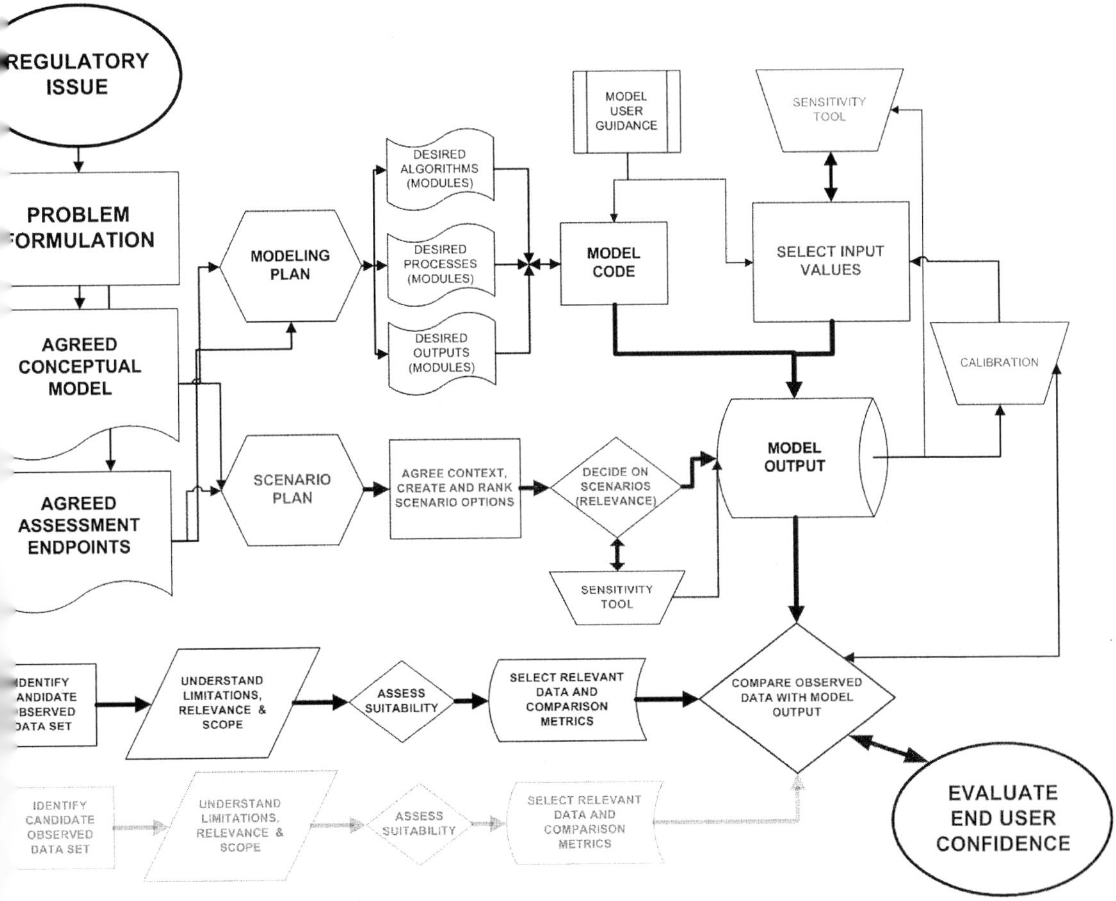

Figure 2. Idealized diagram for model validation

7.1 Problem Formulation

This critical step in all risk assessments has increasingly been highlighted in reviews and one seminal reference is from USEPA itself [18]; more recently, a particularly helpful review was prepared for the Pellston Workshop on Uncertainty [19]. Regulatory risk assessors too often miss or minimize this step in the mistaken belief that they are addressing a standard problem with a routine approach. While this may be true for studies where a step 1 or 2 modeling approach satisfactorily addresses a concern; once additional investigations become necessary, a problem formulation analysis is obligatory.

Problem formulations typically clarify the precise regulatory concern to be addressed and integrate the available information which is then used to develop a conceptual model describing the hypothesis(es) for exposure. The problem formulation will also identify the exposure information needed to address the regulatory concern and sources of potential variability and uncertainty. This leads to an understanding of the scale of the analysis and of the smallest units which need to be considered (e.g. for ecological concerns, the modeled entity might be daily concentrations in a small pond (USA) or ditch (EU) whereas, for a human drinking water issue, the modeling target might be annual or quarterly estimates of reservoir concentrations). This, in turn, indicates the type of model validation/confidence building required and the types of measured data needed to compare with model output. Depending on the complexity of the problem, probabilistic distributions of concentrations across the entire chemical use area in terms of magnitude, spatial or temporal distribution might be needed to best address the regulatory concern.

7.2 Conceptual Models

ECOFRAM [2] identified the importance of the conceptual model as a tool to represent the relationships hypothesized to be critical to a pesticide exposure scenario. Given the long history of interest in pesticide exposure, it is perhaps surprising that internationally agreed diagrams describing the underlying factors contributing to the most common exposure processes are not commonly available. To be successful, a conceptual model should be a multi-level construct. For example, a soil dissipation scenario may describe bulk transport processes such as volatility, wind and water erosion but, at the next level, a "runoff module" might describe the factors leading to sheet and rill flow while beneath that an even smaller-scale diagram would consider inter-relationships between raindrop influenced particle detachment, soil saturation and mixing/ desorption processes in the water-soil interface. Conceptual models for key regulatory exposure scenarios (e.g. Non-target plant and arthropods, terrestrial soil dissipation, runoff and spray drift entry into water bodies) need to be agreed internationally and refined as science develops. This is an achievable goal since conceptual models merely reflect science issues and would not commit authorities to include all elements in their regulatory frameworks.

Given that model developers almost always used such diagrams in the early stages of planning their models, it is also surprising that conceptual models are not readily available describing the processes addressed in major regulatory models. This discrepancy probably arose because, until recently, exposure modeling was a tool for a small pool of experts who were intimately familiar with the models. The renewed importance of exposure models to a broader group of stakeholders means that such diagrams would now be invaluable to provide simple visual tools to help understand why a particular model might fail to address key aspects of a particular problem.

It is strongly recommended that the development of such conceptual models for both exposure processes and models would provide agreed unifying frameworks for regulators, risk assessors and modelers to work within.

7.3 Model Validation Planning

As stated above, effective problem formulation will identify the scope of the exposure estimation exercise and the type of model validation/regulatory confidence building required. It should be realized that the validation exercise indicated may not be a formal validation in the sense of a scientific rigor. It may merely be a simple comparison showing that model predictions are conservative relative to monitoring measurements to help regulators feel comfortable that the assessment is suitably conservative.

At this point, it is instructive to consider the temporal and physical scales of the key sensitive variables and how they may impact planning. Figure 3 shows where some key chemical, field and agronomic factors fit when examined in a temporal and spatial framework to help understand which factors may be particularly sensitive for a given regulatory need. Moreover, broader consideration tends to group the factors that drive typical ecological risk assessments into a "sooner and smaller" region than those typically important for human drinking water issues and, within these "spaces", acute and chronic aspects may even be differentiated. The diagram suggests why validation may be more difficult for exposure assessments targeted at human drinking water because these are associated with larger temporal and spatial scales typified by more innate heterogeneity.

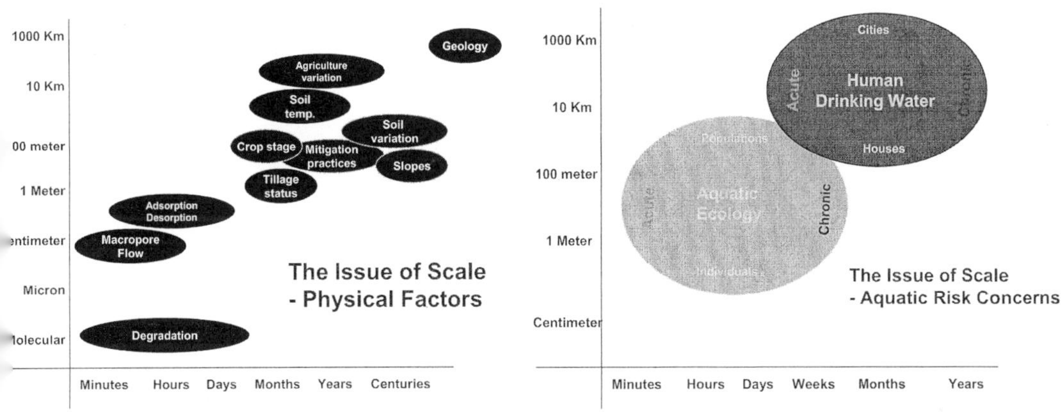

Figure 3. Scale issues to consider when identifying key model processes

Figure 4 illustrates an anomaly in regulatory attitudes to "model" validation. While there has been long standing concern about the accuracy of predictions made using transport models in the EU and USA, the expectation that soil dissipation data should be fitted as a first order process is widespread, irrespective of the extent to which field or laboratory dissipation data supports the assumption. The examples show the greatly improved "quality of fit" throughout the dissipation period resulting from use of a reversible binding adjustment and yet the regulatory models and many regulators require a first order rate constant without consideration of the quality of fit of this "model". It is critical to realize that there is *nothing wrong with making simplifying assumptions* as long as their impact on the exposure assessment and subsequent modeling is understood. For example, figure 5 shows the effect that changing the half life (from a value determined via a first-order fit to one based on a non-linear fit) on the results of soil dissipation data can have in an evaluation of pore water concentration at 1 m. Clearly, understanding the extent to which the simple first-order model assumption might impact the model output is critical to model validation exercises. However, the broader message is that we need to be careful to apply uniform standards to the desire for model validation among all the "models" identified by the problem formulation.

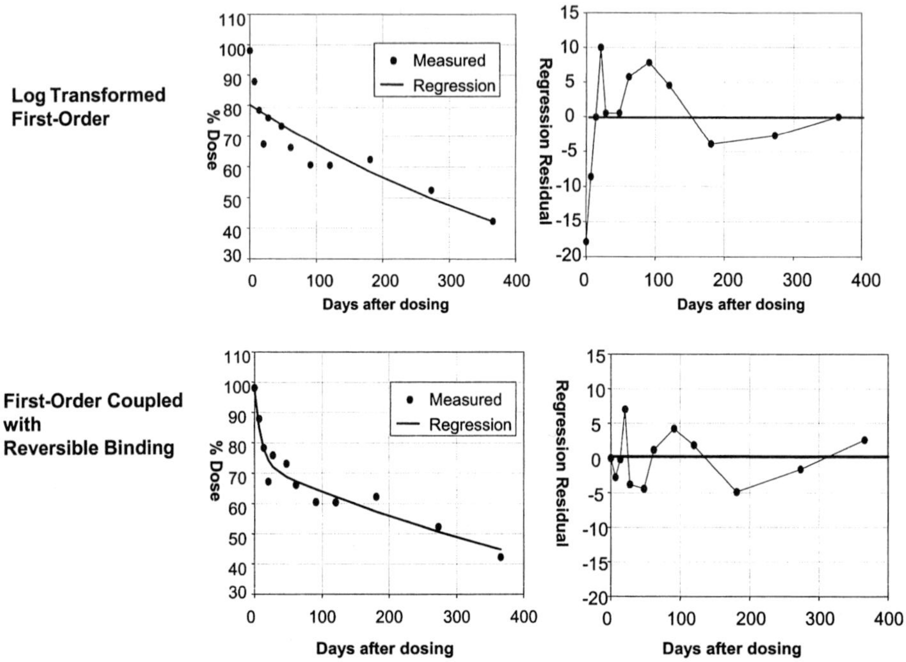

Figure 4. Curve fits obtained by selecting different models for soil degradation kinetics

7.4 Selecting Input Values

FEMVTF [1, 10 -17] identified that modeler individuality could be one of the more significant sources of model output disparity and identified developing model "shells" such as those used under FOCUS [20] as well as very detailed guidance on input selection as the best ways for reducing this source of validation discrepancy.

Regarding chemical inputs, there has been much discussion around the desire of some regulatory bodies to incorporate some form of uncertainty factor when data for a parameter is sparse. It is hard to understand why this topic has been so hotly debated since a simple sensitivity analysis should show whether the parameter is significant (and therefore perhaps meriting more detailed experimental work) or, if insensitive, the uncertainty factor is probably not significant. A more significant issue is the handling of correlated input variables in validations and regulatory modeling. For example, half life and adsorption constant are often codependent and therefore it is inappropriate to select extreme values (typically longest half life and lowest adsorption) if the combination cannot occur in the field. This is equally important when using Monte Carlo approaches in the more recent model validation exercises which build combinations of input variables into multiple model runs [9,15,16].

Agronomy and soil data is one of the weaker areas for exposure modeling. While current databases may have average values or ranges of parameters such as organic matter and clay content for a given soil, it is unusual to find regional databases including information from real-world single samples that provide correlated information. This information is frequently available for the managed study sites used in many validation exercises and this lack of correlated variables when regional databases are used may contribute to apparently decreased model performance at these scales.

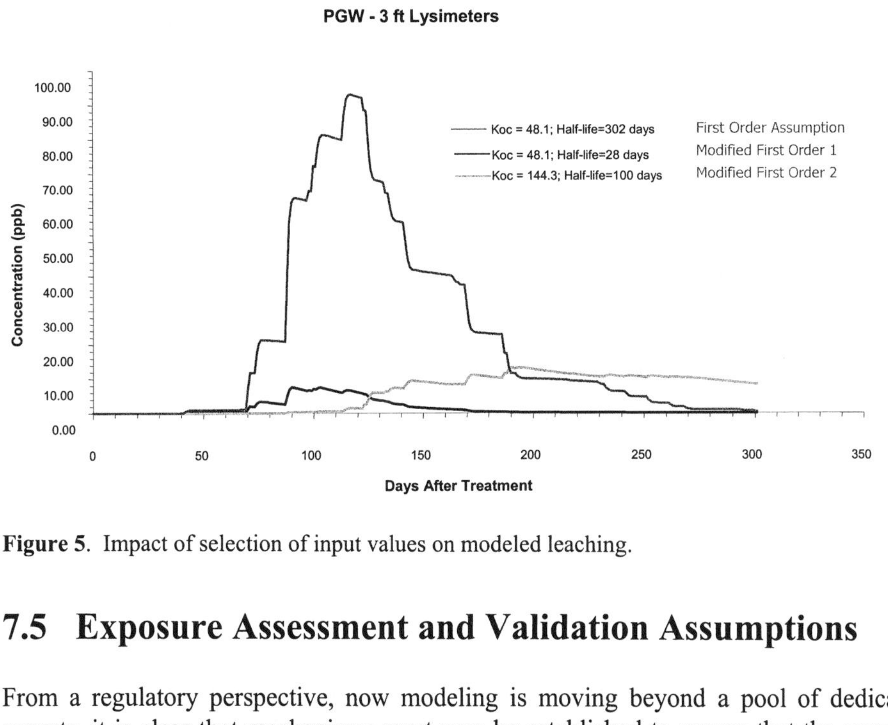

Figure 5. Impact of selection of input values on modeled leaching.

7.5 Exposure Assessment and Validation Assumptions

From a regulatory perspective, now modeling is moving beyond a pool of dedicated experts, it is clear that mechanisms must now be established to ensure that the assumptions underlying models and their implications are clearly understood. For each model required for regulatory decision making, a list of assumptions inherent in the conceptual model, code, input selection and output processing should be agreed and made publicly available. The listing should describe the potential impact of the assumption on exposure estimation – for example, the potential impact of slow mixing of a drift or runoff "pulse" entering a water body. This documentation could be done once and revised as needed using an "open source" model overseen by a model-owner group (as in the computer software industry).

This approach needs to be aligned with a revision of perceived Good Modeling Practice (GMP) which would require comments on the impact of model assumptions in a report on a particular exposure assessment or validation. Considering validation, the issue becomes a little more complex with the need to understand the potential impact of discrepancies between the potentially different sets of assumptions associated with the problem formulation, the exposure modeling and the measured data used for the validation comparison.

7.6 Scenarios, Models and Validation

A fundamental point that is not well understood is that almost all scenarios are "valid" as long as their underlying assumptions and implications are understood. While they may not closely reflect reality, they may be perfectly suitable for a specific regulatory purpose. Protective scenarios will have different characteristics from predictive ones while higher-tier scenarios should incorporate more realism. Generally, this process is being followed in USA and Europe where there is a move to generate increasing numbers of local detailed scenarios to deal with particular crop/region issues.

However, the relative importance of scenarios has been yet been recognized in what may be termed the validation of generic exposure assessments such as examining the leaching potential of a compound in the EU using the FOCUS groundwater scenarios [20] or running US EPA's standard scenarios for runoff and drift [25].

Table 2 compares the assumptions inherent in a US EPA Tier 2 scenario with reality and it becomes clear that some of the key factors in the scenario can have an enormous impact on the modeled concentration because they are directly proportional to the expected environmental concentration (EEC). The Percent Crop Area (PCA) (which indicates how much of the watershed grows the crop of interest) used is the maximum known in the USA and therefore tends to be very conservative. In reality, the percent of the crop present that is treated with the chemical of interest (% market share) is a further major reduction factor not included in standard assessments. Finally, the extent to which direct adjacency of treated crops to the entire perimeter of water bodies occurs becomes critical for drift entry estimations since the impact of increasing intervening distance is pseudo exponential. When these direct impacts on potential exposure are compared with the factors more usually considered when examining sources of uncertainty and validation, it becomes apparent that scenario parameterization may be one of the most important factors impacting validation *and regulatory confidence building* between exposure model output and monitoring data at a watershed scale [e.g. 5]. It is therefore particularly concerning that one of the least readily available pesticide datasets is local use on a watershed basis or even the percentage of each crop treated with different chemical.

Table 2: Comparison of Tier 2 model scenario assumptions with reality

Standard Scenario	Reality
10 Ha watershed delivery	Variable area watershed
1 ha pond	Variable area ponds
2 m depth	Variable depths
No flow/overflow	Flow and overflow
Max PCA% cropped	<<100% cropped
100% drift adjacency	<100% drift adjacency
• Despite max PCA	• Non-linear impact
100% market share	<<100% market share
Maximum App. rate	Range of rates
Applications simultaneous	Variable timing
Uniform (high) slope	Non-uniform slopes
One soil	Multiple soils

In addition to the factors described above, an understanding of the *relevance* of a particular scenario – is it 99th or 50th percentile in terms of some selected assessment endpoint (e.g. leaching potential at 1m) is equally critical. Without understanding this information, it is almost impossible to provide useful comparisons and "validatory confidence building" between scenario-based modeled exposure and field data. It is important to consider agronomic correlations when developing scenarios; for example, it has been shown that for a HUC8 watershed in Yazoo County MS, slopes greater than 3% account for 16% of the land. However, when the slopes specifically associated with cotton cropping are considered, only 3% of the cotton area is cropped on slopes in excess of 3%. Such differences may be important for a higher tier exposure assessment and indicate an area where Remote Sensing and GIS become important adjuncts to understanding and parameterizing scenarios.

7.7 Model Sensitivity

When considering validation approaches, it is important to understand the model's inherent sensitivities to input parameters. Again, different strategies are needed to address different types of regulatory exposure modeling. For site specific validations such as those performed by FEMVTF, the agronomy, actual rainfall, exact application dates, rates and times are known and the key variables become the range of chemical and soil property variation expected. However, where standard scenarios are being examined across multiple weather years, an additional range of sensitivities needs to be considered. The importance of physical scenarios was dealt with above but another agronomic parameter has been shown to be important for estimating runoff potential. In an evaluation [21] of four compounds selected to exemplify a wide range of adsorption and half life characteristics, it was shown that varying the application date within a 90 day window had between 2 and 700 fold greater impact than varying either soil half life or adsorption constant over actual measured ranges. This is because changing application dates ensures

that by using 36 weather years, a "worst case" of a major storm occurring very shortly after application will always be obtained.
This shows that, for some regulatory model uses, the implications of sensitivity associated with agronomic factors such as application date need to be considered. The approach used in the experiment described above [21] and by FOCUS [26] also suggests one mechanism for standardizing validation approaches by using a suite of test chemicals and associated properties covering agreed extensive ranges of key parameters. In keeping with this and the suggestions for agreeing documentation of the conceptual models and assumptions underlying each model, it is recommended that generic sensitivity information (some of which is already available [e.g. 14]) for each regulatory model be made readily available.

7.8 Tiered Model Validation and Calibration.

Calibration, specific fine tuning of parameters to refine model performance, is a feature of most sophisticated validation exercises but generates considerable debate. For some, it is the outcome of careful science controlling certain variables so that others may be manipulated with certainty. To others, it is "fiddling with the controls until the desired answer is obtained". Generally, satisfactory calibration is very data intensive and it is this author's opinion that ideally for a regulator to feel confident about extrapolating a site specific calibration to a wider area some form of coordinated plan of nested field studies and monitoring would be needed.
Various authors [1,9,12,22,23] have suggested stepwise model validation and calibration procedures and the example from FOCUS [23] has been used successfully [22]; the underlying concern is that attempts to calibrate chemical transport when simpler outputs (e.g. the hydrology) may not have been adequately simulated can result in empirical rather than logical corrections that will therefore not be extrapolable to other situations. Stepwise approaches typically start by ensuring that water movement is adequately predicted, then movement of any other conserved parameter (e.g. sediment erosion or bromide ion) before attempting to validate and/or calibrate the pesticide transport element of the model. It might also be assumed that, without having a reasonably accurate prediction of the amount of chemical present in the soil column each day and the rate with which it may be degrading at each depth, it is unlikely that predictions of transport will be meaningful. Moreover, depending on whether the chemical concerned exhibits irreversible binding, not only the mass present but the amount *available* for transport should be understood before considering calibration of the transport element. This raises the interesting proposition that the traditional terrestrial field dissipation field study may find utility as a means of validating the "mass present" predictions of an exposure model.

7.9 Goodness of Fit

The FEMVTF group investigated various statistical approaches for expressing the results of a formal validation [15-17] and found the size of the dataset insufficient for classical statistics (the FEMVTF dataset was relatively large!) However, they had some success with a Monte Carlo approach as recommended by Haan [9]. They found that care was needed to set a lower bound threshold to remove low predicted residues from further consideration and to ensure that effort was directed at comparing *significant* predicted concentrations with field values. Nevertheless, a number of model experts still believed that a "by eye" examination was critical; clearly there remains more work to be done to learn to express validation results in ways that will be meaningful to regulators, casual users and modeling experts. To achieve this, internationally agreed standard methods for reporting validation results need to be developed.

7.10 Uncertainties and Validation

As mentioned in the introduction, the recent trend towards probabilistic exposure modeling has raised the level of attention directed at the understanding of uncertainty and recent validation exercises [15-17] have incorporated certain input parameter uncertainties into validation. Unfortunately, developments so far on incorporating uncertainty into exposure assessments have focused more on tools for incorporating this rather than on rationales for understanding the impacts of uncertainty to be included in an assessment. As a result, it can appear that complex models are "worse" than simple models because they combine more sources of uncertainty and therefore apparently have more uncertain output. This impression is probably misleading in many cases, but until we have better mechanisms for considering appropriate sets of uncertainties, this will remain a deficiency in risk assessment. For exposure assessment validations, the situation is more complex since there is not yet an agreed approach for describing the outcome of the model validation let alone the associated uncertainty.
At the recent Pellston Workshop on Uncertainty in Probabilistic Risk Assessment [4], there were discussions about uncertainty in aquatic exposure assessments. A few interesting points [21] arose including the finding that the current "black box" regulatory transport models are not readily amenable to convenient uncertainty analysis. An example Bayesian erosion "module' was created as an experiment and this could be easily linked to uncertainty analysis software. This raises the long-standing wish of expert modelers for an international unified modular modeling system from which desired modules could be matched to needs identified during problem formulation. Given standard components (probably parameterized according to local regulatory and physical needs) validation of modules might truly become a developing weight of evidence approach and international findings might be mutually supportive.

8 "Moditoring" – the Way Forward?

From the foregoing text, it is clear that modeling can develop huge amounts of data but the underlying models, assumptions, scenarios and lack of standardization mean that there is doubt about the value of the output – especially where the desired endpoint is an understanding of potential water concentrations in human drinking water taken from a larger reservoir or river. Nevertheless, although monitoring data may presently be given more credence by the public than model output, in reality, it is merely a snapshot of the universe of possible outcomes reflecting a limited number of seasons, one or more sites and one combination of weather /agronomy/chemical variables. As separate datasets, neither model output nor monitoring data satisfies regulatory needs.

Hence the author proposes the term "moditoring" as the specific combination of modeling and monitoring. Measured field data needs to be extended using multi-year, multi-scenario spatial modeling to extrapolate to the larger chemical use area. Success will require multiple developments including agreement on a watershed scale model framework. More importantly, it is likely that it will require a broader approach to validation that will meet the needs of the broader group of users for today's regulatory approaches. This approach will probably be more empirical than current site specific models and will focus more on weight of evidence approaches to support desired levels of regulatory confidence.

The author recommends that future government and industry monitoring initiatives and field study programs have model validation as the primary goal and develop coordinated nested approaches of various levels of complexity.

9 Conclusions

Regulatory exposure modeling has changed from being the preserve of a small connected group of international experts to a diffuse group of regulatory users, few of whom might consider themselves model "experts". Urgent efforts are needed to improve validation approaches as well as to build a bridge between the science of exposure model validation and public-regulatory confidence in the "validity" of regulatory decisions based on model output. In some cases, this might be as simple as demonstrating a realistic degree of conservatism in model output relative to monitoring data or constructing "weight of evidence" arguments. While continued refinement of internationally consistent technical model validation approaches remains important, more pragmatic steps could be taken now to greatly improve the transparency of exposure estimates made to support risk assessments and to more readily identify similarities and differences between the various approaches (and their underlying assumptions) used worldwide. These steps would provide much greater improvements in regulatory confidence in the short term than can be achieved by more rigorous formal scientific validation exercises.

The importance of rigorous regulatory problem formulation for a model validation cannot be overstressed as the foundation of estimates of the *regulatory validity* of model output. All exposure models have inherent conceptual models and assumptions and unless their impacts on the regulatory endpoint are understood, then the credibility of resulting risk assessments may be in doubt. There is nothing "wrong" with simplifying assumptions or model scenarios as long as their impact on the assessment and subsequent risk assessment are documented. It is strongly recommended that international regulators should stimulate development and documentation of conceptual models for both exposure processes and models; this would provide agreed unifying frameworks for regulators, risk assessors and modelers to work within. Similarly, we need to update

Good Modeling Practice requirements for modeling and validation studies to include discussion of the implications of any discrepancies between the assumptions underlying a particular problem formulation, the models used and the field data used for validation. The importance of model scenario/agronomy conceptualization, parameterization and relevance for successful multi-year regulatory exposure modeling needs be more widely understood and documented. In short, intensive efforts are needed to fully document assumptions, sensitivities and related validation efforts for all regulatory models; without this, it will remain difficult to build bridges between model output and field monitoring data via formal or non-traditional means.

As we increase awareness of exposure model validity, it is critical to realize that monitoring data is typically a snapshot of a particular set of circumstances; unless the assumptions inherent in the experimental design are appreciated, monitoring data may be even more misleading than model output. Increasingly, regulatory interest focuses on how model output relates to regional and watershed scale monitoring data while traditional model validation focuses on field scale studies; however, because of the nature of broader scale monitoring data, different criteria will be needed to best express the degree of "validation" achieved. Measured field monitoring data ideally needs to be extended via multi-year spatially discrete modeling to put the results into a meaningful regulatory context. It is proposed that future watershed scale monitoring programs should be developed hand in hand with model development and validation experts ("moditoring") using a coordinated nested design strategy to provide various temporal and spatial data densities. The development of an international modular model would allow specific validations of modules to help build confidence across many international regulatory communities. In addition, key datasets such as pesticide usage for watershed scale models should be made an international research priority.

Perhaps Aristotle [27] had the best advice to exposure modelers and regulators interested in validating and building confidence in model output:

"It is the mark of an educated mind to rest satisfied with the degree of precision which the nature of the subject admits and not to seek exactness where only an approximation is possible."

10 References

[1] FIFRA Environmental Model Validation Task Force. FIFRA Environmental Model Validation Task Force. Final Report. April 27, 2001. Obtainable via www.FEMVTF.com or via EPA citing MRID 45433201.

[2] Ecological Committee on FIFRA Risk Assessment Methods (ECOFRAM), "ECOFRAM: ECOFRAM Aquatic Report. Guidance on Methods for Aquatic Risk Assessment Methods for Aquatic Ecological risk Assessment", June 2000.

[3] EUPRA, "Probabilistic Risk Assessment for Pesticides in Europe: Implementation and Research needs. A Report from the European Workshop on Probabilistic Risk Assessment for the Environmental Impacts Of Plant Protection Products, The Netherlands, June 2001", Editor: Andy Hart, Central Science Laboratory, York, UK, 2001.

[4] "Quantifying Uncertainty in Risk Assessment: Practical Approaches and their Application to Estimating Ecological Risks of Pesticides", report on the findings of a workshop held in Pensacola, FL, USA., SETAC, February 24 to March 1, 2002 (in preparation).

[5] P. Hertl, W. Phelps, D. I. Gustafson, S. H. Jackson, R. L. Jones, M. H. Russell, M. J. Schocken, " A Comparison of US EPA's Tier 1 and 2 Index Reservoir Model Estimates to Drinking Water Reservoir Monitoring Results in Selected US Systems in 1999/2000", Poster presented at IUPAC 2002, Book of Abstracts, Part 2, page 125.

[6] R .L. Jones, "Comparison of Modeling and Monitoring results for Residues of Aldicarb, Bromoxynil, Carbaryl and Fipronil in Drinking Water from Surface Water", poster presented at IUPAC 2002, book of abstracts, part 2, page 126.

[7] ASTM, "Standard practice for Evaluating mathematical Models for the Environmental Fate of Chemicals", Designation E 978- 92, Committee E-47 on Biological Effects and Environmental Fate, ASTM Committee on Standards, West Conshohocken, PA, USA, Approved Feb. 15, 1992.

[8] L. A. Burns, "Performance Standards and Prediction Uncertainty", manuscript in preparation.

[9] C. T. Haan, B. Allred, D. E. Storm, G. .J. Sabbagh and S. Prabhu, "Statistical Procedure for Evaluating Hydrologic/Water Quality Models", Trans. ASAE, 1995, 38 (3), 725-733.

[10] J. P. Carbone, "Predicting Fate and Transport: the Pesticide Root Zone Model", Env. Tox. Chem. 2002, 21 (8), 1533-1534.

[11] R. L. Jones and G. Mangels, "Review Of The Validation Of Models Used In Federal Insecticide, Fungicide, and Rodenticide Act Environmental Exposure Assessments", Env. Tox. Chem. 2002, 21 (8), 1535-1544.

[12] P. Singh and R. L. Jones, "Comparison of Pesticide Root Zone Model 3.12: Runoff Predictions With Field Data", Env. Tox. Chem. 2002, 21 (8), 1545-1551.

[13] M.H. Russell and R. L. Jones, "Comparison of Pesticide Root Zone Model 3.12: Leaching Predictions With Field Data", Env. Tox. Chem. 2002, 21 (8), 1552-1557.

[14] . J. Wolt, P. Singh, S. Cryer, and J. Lin, "Sensitivity Analysis for Validating Expert Opinion As To Ideal Data Set Criteria For Transport Modeling", Env. Tox. Chem. 2002, 21 (8), 1558-1565.

[15] P. L. Havens, J. P. Carbone, W. Warren-Hicks, and M A. Fouch, "Development of a Monte Carlo Sampling Shell for the Pesticide Root Zone Model and its Application by the Federal Insecticide, Fungicide, and Rodenticide Act Environmental Modeling Validation Task Force", Env. Tox. Chem. 2002, 21 (8), 1566-1569.

[16] W. Warren-Hicks, J. P. Carbone, and P. L. Havens, "Using Monte Carlo Techniques to Judge Model Prediction Accuracy: Validation of the Pesticide Root Zone Model 3.12", Env. Tox. Chem. 2002, 21 (8), 1570-1577.

[17] J. P. Carbone, P. L. Havens, and W. Warren-Hicks, "Validation of Pesticide Root Zone Model 3.12: Employing Uncertainty Analysis", Env. Tox. Chem. 2002, 21 (8), 1578-1590.

[18] EPA, "Guidelines for Ecological Risk Assessment", Washington D.C. , 1998. US. EPA.

[19] A. Hart, S. Ferson, J. Shaw, "Problem Formulation for Probabilistic Ecological Risk Assessments", "Quantifying Uncertainty in Risk Assessment: Practical Approaches and their Application to Estimating Ecological Risks of Pesticides", report on the findings of a workshop held in Pensacola, FL, USA., SETAC, February 24 to March 1, 2002 (in preparation).

[20] "FOCUS groundwater scenarios in the EU review of active substances" - The report of the work of the Groundwater Scenarios Workgroup of FOCUS (FOrum for the Co-ordination of pesticide fate models and their USe), Version 1 of November 2000. EC Document Reference Sanco/321/2000 rev.2, 202pp.

[21] P. Hendley, T. L. Estes, „ Uncertainty in Aquatic Aquatic Exposure Assessment", "Quantifying Uncertainty in Risk Assessment: Practical Approaches and their Application to Estimating Ecological Risks of Pesticides", report on the findings of a workshop held in Pensacola, FL, USA., SETAC, February 24 to March 1, 2002 (in preparation).

[22] M. Thorsen, P. R. Jorgensen, G. Felding, O. H. Jacobsen, N. H. Spliid, J. C. Refsgaard, "Evaluation of a Stepwise Procedure for Comparative Validation of Pesticide Leaching Models", J. Env. Qual., 1998, 27, 1183-1193.

[23] FOCUS, "Leaching models and EU registration. Final report of the work group – Forum for the coordination of pesticide fate models and their use" funded by the European Commission, The European Crop Protection Association and COST Action 66, 1995, Dco. 4952/VI/95.

[24] USGS National Water Quality Assessment (NAWQA) Program, 1991 to present available via internet at http://water.usgs.gov/nawqa/nawqa_home.html.

[25] e.g. "Drinking Water Screening Level Assessment; Part A: Guidance for Use of the Index Reservoir : in Drinking Water Exposure Assessments", USEPA Office of Pesticide Programs. USEPA, 2001 (available via www.epa.gov/pesticides/trac/science/reservoir.pdf).

[26] FOCUS, "Evaluation of FOCUS Surface Water Step 1, 2 and 3 Scenarios using representative Test Substances", Appendix G of "Focus Surface Water Scenarios In The EU Evaluation Process Under 91/414/EEC. Report prepared by the FOCUS Working Group on Surface Water Scenarios", EC Document Reference SANCO/4802/2001-rev.1. 221 pp., 2001. (Available via www.iuct.fhg.de/FOCUS/swreport/).

[27] Aristotle, Nicomachean Ethics, (e.g. translation by W.D. Ross at www.knuden.liu.se/~bjoch509/works/aristotle/nico_ethics.txt), approx 350 BC.

23 Environmental Fate and Impact Considerations Related to the Use of Transgenic Crops

Gijs A. Kleter and Harry A. Kuiper

RIKILT (National Institute for Quality Control of Agricultural Products), P.O. Box 230, NL 6700 AE Wageningen, Netherlands, E-mails: g.a.kleter@rikilt.wag-ur.nl and h.a.kuiper@rikilt.wag-ur.nl

1 Introduction

The area of commercially cultivated genetically modified (GM) crops has grown rapidly from their introduction in 1996 onwards, totalling approximately 55 million hectares in 2001 [1]. The major commercial GM crops are soybean, maize, cotton, and canola (Figure 1), and most of the GM crops cultivation takes place in United States, Argentina, Canada, and China. Most of these GM crops have been modified to express traits of agronomic importance, such as herbicide tolerance and insect resistance. Herbicide tolerance is acquired by the introduction of enzymes that either degrade the herbicide or that are herbicide-insensitive forms of the native target enzyme in the crop. Herbicide tolerant crops facilitate weed management by enabling application of post-emergence herbicides over the top of crops. Insect resistance has been conveyed to GM crops by the introduction of, in most cases, an insecticidal protein from the soil bacterium *Bacillus thuringiensis*. Preparations of this bacterium have been used as biological pesticides for decades. Other agronomic traits in commercial and experimental GM crops include resistance to viruses, bacteria, moulds, and nematodes.

Switching from conventional to GM crops expressing agronomic traits likely affects pesticide usage, because these traits serve to improve yields, insect pest control, and weed management. Herbicide tolerance traits, for example, are linked to the use of a specific herbicide, which may substitute for alternative herbicides that are used for weed management in conventional crops. Shifts in pesticide usage by adoption of GM crops will likely have an impact on the environment, but data in this respect are scarcely available and sometimes even contradictory.

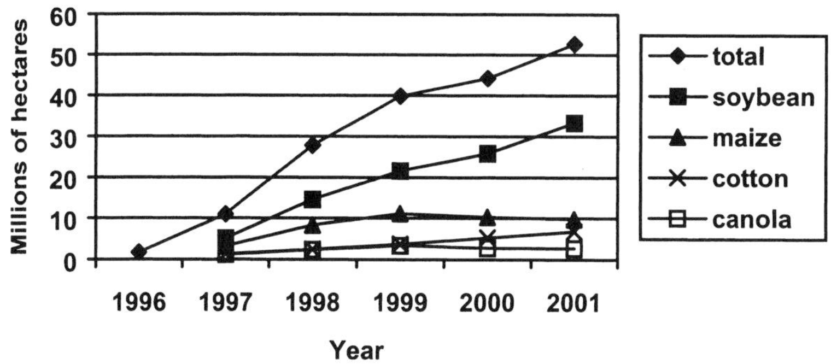

Figure 1. Global cultivation of GM crops [1-3]

Besides analysis of alterations in usage patterns of agrochemicals, the intrinsic chemical and (eco)toxicological properties of the applied chemicals should be taken into account, in order to complete a full quantitative assessment of the impact of GM crop cultivation on the use of agrochemicals and its environmental consequences. This is an important endeavour given the broad public concern on the cultivation of GM crop. In many cases it may not be easy to compare data on pesticide use obtained from GM and traditional crops. Pesticide usage will be affected by other factors besides GM crop adoption. The changes in pesticide usage in GM crops should therefore be put into perspective against a continuously altering background of agricultural crop management, such as integrated pest management and the introduction of more effective and less toxic chemicals.
A three-year project "Impact of transgenic crop cultivation on the use of agrochemicals and the environment" was recently initiated under the umbrella of IUPAC. Within this project, the environmental impact of altered pesticide usage patterns on genetically modified crops will be assessed. To this end, data on the actual use of pesticides on GM crops will be collected. Most of these data, which are available from literature and the Internet, pertain to the American and Canadian areas of cultivation. Where possible, data will also be collected that have been obtained in other nations where GM crops are grown. Other factors than GM crops that may have contributed to altered use of pesticides (*e.g.* new registrations) will be taken into account and their impact compared with that of GM crop adoption.
From these data on pesticide usage, the environmental effects associated with adoption of GM crops will be assessed both qualitatively and quantitatively. Qualitatively, the impact will be predicted based on, for example, properties of pesticides, quantities of applied pesticides, times of application, pest characteristics, and geographic conditions. Subsequently, the quantitative assessment will employ methods such as environmental indicators that allow for a comparison of the environmental impacts of different pesticide regimes. The environmental impact may also cover other important parameters associated with altered agronomic practices under GM crop cultivation, such as tillage practices and farmland biodiversity.

Finally, following this risk assessment, the sorted scientific data will be subject to a risk benefit analysis. In risk benefit analysis, negative and positive aspects are weighed against each other, which may lead to, for example, measures with the aim to minimise risks and to maximise benefits. The project activities are summarised in Figure 2.

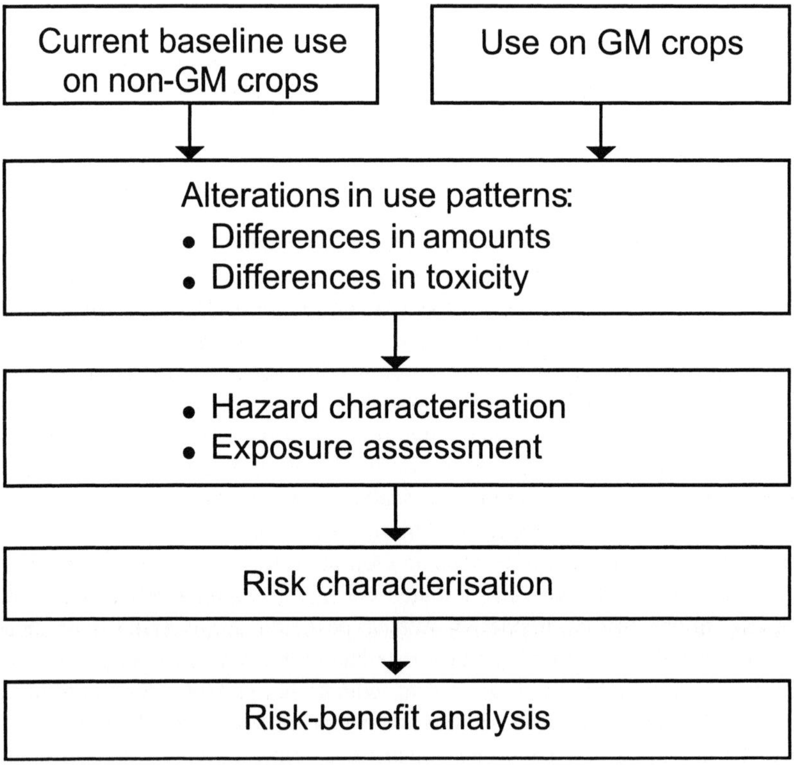

Figure 2. Environmental impact assessment of pesticide use on transgenic crops: Activities within the IUPAC project "Impact of transgenic crop cultivation on the use of agrochemicals and the environment"

2 Surveys on Pesticide Usage in GM Crops

2.1 Methods of Carrying out Surveys

Various types of studies exist on the different pesticide regimes applied to genetically modified crops, including field trials and farmer surveys. Field trials represent model conditions and model practices, and would render results for our study that are not fully representative for the impact of GM crops on pesticide usage in practice. Therefore, preference is given to studies on the real-practice use of pesticides in GM crops, including farmer surveys.

Most of these farmer surveys have been carried out in the United States of America, for example by the National Center for Food and Agricultural Policy (NCFAP, http://www.ncfap.org) and the U.S. Department of Agriculture's Economic Research Service (http://www.ers.usda.gov). Such survey results were recently reviewed [4,5,6].

Recently, a comprehensive NCFAP survey was released on known and anticipated environmental effects of the adoption of 40 GM crops, including the effect of GM crops on pesticide use [7]. These crops include eight GM crops that are commercially cultivated in the United States, while the remaining crops either are under development or have been approved, but have not been adopted by farmers yet. The commercially cultivated crops include: herbicide tolerant canola, herbicide tolerant cotton, insect resistant cotton, herbicide tolerant maize, insect resistant maize, virus resistant papaya, herbicide tolerant soybean, and virus resistant squash. The survey considers, among others, pesticide use.

With regard to the applied methodology, Gianessi et al. [7] noted that preferably the comparison should not be carried out between the year preceding GM crop adoption and the years following adoption, but between GM crop cultivation and alternative treatments within the same year. The impacts that each crop may have on agriculture are also specified for those U.S. states where these crops are grown.

Marra et al. [4] review strategies for collecting data on farm practices, including:

- Interviews with operators of randomly selected crop fields.
- Farm-level surveys in which farmers are interviewed about their practices and performance, which differs from the previous strategy as it may concern , for example, multiple fields and different practices applied to each field.

These authors acknowledge that within-farm comparison of GM- versus non GM-crop cultivation correct for some background factors that influence the outcome of field-level surveys, such as between-farm differences in yields, pest pressure, and education. Different results were generated from surveys of farmers who grew both GM- and non GM-varieties on the same farm compared to surveys that sampled GM data from adopters' farms and non GM data from non adopters' farms.

Farmers' impacts by adoption of genetically modified crops have also been studied by the Economic Research Service (ERS) of the U.S. Department of Agriculture (USDA). For these studies, data have been employed from farms that were randomly selected from groups of farms with similar attributes. The raw data have been further processed in order to adjust for the systematic differences between adopters and non adopters that would influence the outcome of the comparison. This is because the individual farmer's choice to adopt a biotech crop is not random and is driven by, for example, farm size, farmer education, pest levels, and access to financial resources. These driving factors have been incorporated into a statistical model that has been applied by the ERS to the farm survey data to determine the impact of biotech crop adoption [8]. The impact of the adoption of herbicide tolerant soybean, herbicide tolerant cotton, and insect resistant cotton on pesticide use was thus calculated for the year 1997 (see also [9]).

2.2 Results from the Surveys

The findings of Gianessi et al. [7] on the impact of the major GM crops (soybean, maize, cotton, and canola) on pesticide use are summarised in Table 1. Data from additional recent studies and reviews are highlighted per crop in the following sections.

2.2.1 Herbicide Tolerant Canola

Herbicide tolerant canola varieties constitute a major part of the canola grown in Canada [10]. During 1999, for example, these varieties accounted for 70% of the cultivated Canadian canola. More precisely, transgenic glyphosate tolerant-, transgenic glufosinate tolerant-, and non transgenic imidazolinone tolerant- varieties accounted for 38%, 17%, and 15% of total canola. Stephenson [10] compared the herbicide regimes applied to the herbicide tolerant varieties and the conventional varieties. Measured in amounts of active ingredient, the amounts of ethalfluralin or trifluralin (1400 grams per hectare) applied pre-emergence to conventional canola was higher than the highest amount of herbicide applied to herbicide tolerant canola (glyphosate, 900 g/ha). The adoption of herbicide tolerant canola was estimated to have decreased herbicide use on Canadian canola from 6.226 million kilograms (100% conventional) to the actual use of 4.381 million kilograms (70% herbicide tolerant).

Table 1. Impact of GM crops on pesticide use in the United States (summarised from [7])

Crop	Trait	Impact [1]	Method of calculation
Soybean	Herbicide tolerant	27.8 million pounds less herbicide active ingredients in 2001 (50.016 million acres)	Comparison of glyphosate use with a theoretical alternative herbicide mix for each US state based on experts' views
Maize	Herbicide tolerant	5.805 million pounds less herbicide active ingredients in 2001 (5.809 million acres)	Comparison of glyphosate/glufosinate treatment with a theoretical alternative mix of three post-emergence herbicides, assuming a 25% reduction in pre-emergence applications on herbicide tolerant maize
Maize	Insect resistant	5.575 million pounds potentially less insecticide active ingredients in 2001 (14.927 million acres)	Calculation of pesticide use in a simulation of high insect pest infestation based on historical data, assuming 100% pest control by insect resistant Bt maize. Rationale for the simulation were the low actual infestation levels in recent years that would have obscured the potential change in insecticide use under insect pressure.
Cotton	Herbicide tolerant	6.2 million pounds less herbicide active ingredients in 2000 (15.402 million acres)	Comparison of herbicide use per acre in 2000 with herbicide use in same US state in year preceding introduction of herbicide tolerant cotton
Cotton	Insect resistant	1.870 million pounds less insecticide active ingredients in various years (1998-2000) (5.144 million acres)	Estimates from university experts and Monsanto on impact of Bt cotton in specific U.S. states
Canola	Herbicide tolerant	0.531 million pounds less of active ingredient per year (0.871 million acres)	Comparison of estimated glyphosate use per acre with that of theoretical alternative mixes of three herbicides

[1] Impact calculated for all assayed states of the United States, except for Canola, where only North Dakota, where most of the U.S. canola is located, was surveyed.

2.2.2 Herbicide Tolerant Maize

Weed management in conventional maize commonly involves the use of soil herbicides in combination with atrazine. Herbicides are applied both pre- and post-emergence, the latter being only effective on weeds below a certain height limit. Herbicide tolerant maize renders weed management more flexible, for instance glyphosate can be used for post-emergence control on weeds up to a larger height limit for weed plants [7].
Carpenter et al. [5] note that, while glyphosate use in U.S. maize production has increased in the years following adoption of herbicide tolerant maize, total herbicide use has decreased, much to the expense of metolachlor. In addition, they highlight some findings previously published by Wauchope et al. [11] that cultivation of herbicide tolerant maize in watershed areas may improve water quality by a shift in the herbicide residue pattern of surface waters of the soil herbicides alachlor and atrazine towards low or nil residues of glufosinate and glyphosate, which have a higher threshold level than the soil herbicides.
Phipps and Parks [12] predicted a substantial reduction of herbicide use on maize in the EU, if herbicide tolerant maize were adopted. These authors based their prediction on a comparison between standard herbicide programmes for maize in the EU (1.74-1.90 kg active ingredient per hectare; PRE atrazine & POST bromoxynil, or PRE flufenacet + terbutilazine & POST nicosulfuron + sulcotrione) with glyphosate /glufosinate programs (0.60/1.08 kg active ingredient per hectare).

2.2.3 Insect Resistant Maize

Maize is grown in most U.S. states, where European corn borer and Southwestern corn borer are major lepidopteran insect pests. The transgenic Cry1Ab protein expressed in Bt maize can provide protection against these two pests. Control of these insect pests depends among others on the timing of pesticide applications, because the larvae spent part of their life inside maize stalks where pesticides may not be effective at all.
The study by Gianessi et al. [7] gives an account of previous studies that show that Bt maize adoption was not economically rewarding during 1998-2000. The authors argue, however, that pest populations were unusually low in these years due to several conditions, such as premature development of the overwintering pest generation concurrent with delayed planting of maize in late spring. Therefore the authors simulated the pesticide use as it would have been in 2001 on the acreage of Bt maize if it had experienced high-infestation levels of corn borers (Table 1). This example underscores the need for careful consideration of background conditions during the interpretation of factual data on pesticide use.

2.2.4 Herbicide Tolerant Soybean

For herbicide (glyphosate) tolerant soybean, it is noted that adoption concurs with a background trend towards post-emergence herbicide applications from 1990-1995 [7]. Glyphosate (a.o. sold as Roundup) can be used post emergence over the top of the herbicide tolerant soybean crops. As a broad spectrum herbicide it affords protection against a wide variety of annual and perennial weeds, including the difficult-to-control water-

hemp. Alternatives of glyphosate require mixing of three or more herbicides to achieve the same broad range of action against weeds [7].
Gianessi et al. [7] therefore compared the amount of glyphosate that had been used on herbicide tolerant soybeans in 2001 to the amount of mixed herbicides that would be required to achieve the same degree of weed control. For each U.S. state, an alternative mix of herbicides was composed for comparison, based on the outcomes of a survey among experts, *e.g.* a mix of Boundary (metribuzin & s-metolachlor), Flexstar (fomesafen), and Select (clethodim) in Iowa (Table 1).
Farm-level data sampled in 1997 and 1998 by the U.S. Department of Agriculture (USDA) also indicate that adoption of herbicide tolerant soybean is associated with an an effective overall decrease in herbicide applications. In more detail, an increase was observed in the applications (acre treatments) of glyphosate and a decrease in other herbicides (except acetamide). A calculation model was used for the interpretation of these data to single out other effects that may have influenced this outcome, such as systematic differences between adopters and non-adopters of genetically engineered crops [8]. Similarly, Carpenter et al. [5] conclude that the number of herbicide applications has been largely reduced. This trend can be explained by the increased flexibility of weed control provided by glyphosate tolerant soybeans. For the amount of active ingredients applied per acre, however, the USDA data were less clear, showing mixed trends for the various regions, and in total a three-percent increase [13]. This increase differs from the decrease in herbicide use that has been calculated for 2001 (see above). It cannot be concluded, however, whether this is due to different methodologies or choice of year.

2.2.5 Herbicide Tolerant Cotton

Weeds in cotton can cause sizeable yield reductions and account for extensive weed control measures in cotton farming, including hand-weeding that is still carried out at a large scale. Combinations of pre- and post-emergence herbicide applications are commonly employed in cotton. Herbicide tolerant GM cottons express tolerance to either bromoxynil or glyphosate, the latter rapidly gaining adoption in the U.S, and allow for post-emergence, over-the-top application of these herbicides. In addition, the recently introduced herbicide pyrithiobac can also be applied over-the-top of conventional cotton. None of these herbicides can provide control of all weeds, and combinations with more persistent soil herbicides are recommended [7].
Gianessi et al. [7] observed that while total herbicide use on herbicide tolerant cotton in the U.S. had decreased by 6.2. million pounds of active ingredient, glyphosate use had increased by 8.8 million pounds.
Based on a farm-level survey by the U.S. Department of Agriculture in the Southeastern cotton growing states of the U.S. in 1997, it was concluded that herbicide tolerant cotton did not cause a significant decrease in the number of pesticide applications [8,9].

2.2.6 Insect Resistant Cotton

Genetically modified, insect resistant cotton expressing the Cry1Ac protein provides protection against several of the major insect pests in U.S. cotton: the pink bollworm, cotton bollworm, and tobacco budworm. Given their strongly resembling appearance, the cotton bollworm and tobacco budworm are referred to as the combined "bollworm/budworm" complex.

According to Gianessi et al. [7], insect resistant Bt cotton has been introduced in many U.S. states to combat the bollworms that have acquired resistant against the broad-range pyrethroid insecticides. In all states, the amount of active ingredient and the number of pesticide applications have subsequently decreased (Table 1).

However, other pests are not controlled by the transgenic cotton, such as boll weevil and whitefly. For the boll weevil, successfull eradication programs are carried out in cotton growing states of the U.S.. Combined strategies are deployed in these eradication programs. Weevil populations, for example, are treated with a powerful insecticide (*e.g.* malathion) during diapause (hibernation) of the weevil and during the season at sites of weevil appearance [14]. Insect resistant cotton can also contribute to the management of these pests as it may obviate the need for insecticide applications that would not only destroy the target pests but also predators and natural enemies. On the other hand it is argued that the use of pesticides to eradicate boll weevil also eliminates predators of the bollworm, which would necessitate the use of Bt cotton in eradication programs.

Cotton farmers in the Southeast U.S. who were surveyed the by the U.S. Department of Agriculture in 1997 applied most insecticides less frequently on Bt cotton than on conventional cotton. No less or more frequently did these farmers apply organophosphate and pyrethroid insecticides to Bt cotton. In addition, adoption of Bt cotton correlated with yield increases, *i.e.* for each 10 % of adoption, yields increased 2.1% [9].

In a farm-level survey of 360 farms in the Southeast U.S. in 1996, the impact of Bt cotton adoption on number of insecticide sprays was studied. The direction in which the number of insecticide sprays changed depended on the selection and background-correction of data employed for calculation. A decrease and an increase in the number of insecticide sprays were found if the comparisons of Bt cotton *versus* conventional cotton were within-farm (both Bt and conventional cotton on the same farm) and between all acres (also on different farms), respectively [4].

Carpenter et al. [5] conclude in their review that insecticide use in cotton has decreased by two million pounds from 1995 to 1998 after adoption of Bt cotton in the U.S. Similar reports of decreased insecticide use also come from other countries, including China and Australia, where Bt cotton has been adopted. It has been suggested that the decrease in broad spectrum insecticides is due to the beneficial insect populations that are sustained in Bt cotton.

The Australian cotton industry has experienced reductions in the insecticide sprays applied to Bt cotton. The number of sprays against cotton bollworms were consistently decreased (by 38-52 %) during four seasons (1996/97 - 1999/2000; *e.g.* 9.7 sprays on conventional cotton *vs.* 4.2 sprays on Bt cotton in 1996/1997), as were the number of total pesticide sprays on Bt cotton. In addition to protection against bollworm, it is also mentioned that Bt cotton needed less sprays against tipworm in a year of heavy infestation [15,16]. With regard to the yields, Bt cotton showed variable results comparative to conventional cotton. This variability is attributed to the environmentally influenced, variating expression of Bt proteins, with expression being high in warm seasons with high irradiation [15].

In China, Bt cotton scientists have developed and commercialised their own lines of Bt cotton and cotton with stacked genes (Bt and cowpea trypsin inhibitor) besides Monsanto's approved Bt cotton. Chinese farmers in the Shandong and Hebei provinces of mainland China, who cultivated Bt cotton, used less pesticides than farmers cultivating non Bt cotton (10.3 and 57.8 kg/ha, respectively). Concomitantly, farmers often reduced the number of pesticide sprays from twelve to three or four sprays [17]. In addition, it has been observed that in these provinces in 1999, Bt cotton adoption led to approximately 10% higher yields. This yield effect was geographically dependent, because in other regions the observed benefit was not equally high [18].

3 Assessing the Environmental Impact of Changes in Pesticide Use

The pesticide reduction data from the surveys mentioned above are expressed as amounts of active ingredients that have been saved, such as pounds per acre or kilogram per hectare. Such comparisons of amounts of active ingredients do not take into account the dose-effect relationships that these substances may have, not only on the target pests but also on the environment and human health. Glyphosate and glufosinate, for example, are applied at high doses to combat weeds. This may seem environmentally disadvantageous compared to the application of other types of herbicides, such as acetolactate synthase inhibitors, which are applied at lower doses. If dose-response relationships would be taken into account, however, the comparison need not be disadvantageous for glyphosate and glufosinate, given their comparatively low toxicity for humans and animals, high soil binding activity (low run-off) and rapid degradation by soil microorganisms.

To allow for comparison of the environmental effects of pesticides, various tools have been developed to express these effect(s) in quantifiable terms. One example is the Environmental Impact Quotient (EIQ) developed by Kovach and co-workers [19]. The aim of the EIQ is to transform information on the toxicological and environmental impacts of pesticides into a usable format to facilitate the choice which pesticide to apply in practice. For each pesticide, the EIQ value is based on an equation which brings together mode of action, environmental behavior, and toxicity to humans, animals, and wildlife. To calculate the environmental impact of the application of a certain pesticide, its EIQ is multiplied with the application dose (applied amount of active ingredient). To date, EIQs are available for more than 200 pesticides, including chemical and organic pesticides (see website under reference [19]).

Besides the EIQ, many other environmental indicators have been developed. The value of the EIQ for this project resides in its coverage of all types of pesticides and the assessment of their possible impact in both environmental and toxicological (human, animal, and insect) terms. A review and a website compiled by Dr. Levitan give an overview of indicators, including those employed by government agencies and those used in decision support systems for advice to applicators (http://www.cfe.cornell.edu/erap/PRI/PRI.cfm) [20]. Another review is provided by Reus and co-workers [21] who compared the following eight environmental indicators employed in the EU: Environmental

Yardstick, Hasse Diagram, SYNOPS_2, p-EMA, Ipest, EPRIP, SyPEP, and PERI. They noted, among others, that these indicators may serve different purposes, including support for farmers' decisions, monitoring for policy makers, and environmental labelling of products. Differences between the indicators pertained to the manner in which data were incorporated with respect to, for example, toxicity, environmental compartments, and organisms. Recommendations by the authors included, among others, the harmonisation of indicators.

An example of how the EIQ can be applied to assess the environmental impact of altered pesticide use patterns is provided by Gallivan and co-workers [22]. The authors used historical figures of pesticides used on crops grown in the Canadian province Ontario to calculate the reduction in usage of pesticides, stratified into the various crops on which they are used or into their active ingredients. In addition, the EIQ was employed to determine the reduction in pesticide risk associated with the quantitative and qualitative (active substance) changes in pesticide use. Results show that from 1983 till 1998, reductions in pesticide usage and pesticide risks were comparable: 38.5% and 39.5%, respectively. Major reductions have been achieved, which cover a substantial part of crop area and requires comparatively few treatments per acre. Further reductions in pesticide usage and risks may be achieved in crops such as fruits and tobacco, which account for a relatively large part of the aggregate environmental impact of pesticides because of the high pesticide use in these crops. Furthermore, the authors conclude that GM herbicide tolerant soybean may account for the 90% increased use of glyphosate in Ontario from 1993 to 1998 [22]. The authors also mention newer herbicides with lower EIQs and lower application rates than glyphosate.

In a similar fashion, the EIQ has been applied to Canadian canola cultivation in order to estimate the environmental impact of the adoption of herbicide tolerant canola, which includes transgenic varieties. In recent years, herbicide tolerant canola varieties (glyphosate-, glufosinate-, and imidazolinone-tolerant) have been adopted by Canadian farmers at a larges scale, accounting for 80% of the national canola acreage in 2000. The adoption of these herbicide tolerant varieties has been associated with a concomitant decrease in herbicide use on canola. The environmental impact of herbicide use on canola determined by the EIQ method decreased in correlation with the decreased amounts of herbicide applied to the crops, rather than this would be due to different intrinsic properties (EIQ) of the herbicides involved [23].

To estimate the environmental impact of large-scale biotech crop cultivation, we applied the EIQ methodology to the data on pesticide use in biotech crops form Gianessi et al. [7]. The outcomes of the calculations are given in Table 2. Besides the total environmental impact per acre (EI/A), its farm worker-, consumer-, and ecology- components are shown (the total EI/A is the average of these three components). The EI/A is calculated by multiplying the EIQ with the pesticide application rate, *i.e.* the amount of active pesticide ingredients applied per acre (lbs ai/A). In our calculations, the EI/A for each separate active ingredient was calculated based on Gianessi's data on its rate (lbs ai/A) prior to summation of the EI/As for all the applied ingredients. For some pesticides, no EIQ has been established and for these pesticides, the average EIQ value for either herbicides or pesticides was employed.

As previously noted in Table 1, the methodology of impact assessment was different for each crop. Pesticide reduction in transgenic herbicide tolerant soybean, for example, was predicted by a scenario in which glyphosate was substituted by mixes of alternative herbicides, while the reduction in herbicide tolerant cotton was calculated by comparing the pesticide use in the year 2000 with that of the year preceding cotton adoption.

A consistent difference of -78% for pesticide use and environmental impacts between non transgenic- and transgenic, insect resistant- cotton is observed. This consistency relates to the fact that the same insecticides were assumed to have been applied to both non transgenic and transgenic cotton, the impact being caused by different numbers of insecticide sprays that had been applied to these crops. The consistent difference of -100% observed for the impacts of insect resistant maize relates to the fact that Gianessi et al. [7] assumed that this transgenic maize would provide 100% control of European corn borer.

In general, the extent of risk reduction for humans and the environment of pesticide use on transgenic crops is dominated by a reduced exposure, and not by a shift towards toxicologically more benign pesticides. In addition, in the case of herbicide tolerant cotton and -soybean, results show that the environmental benefit of transgenic crop adoption is comparatively higher for farmworkers and consumers than for the ecological component. Such calculated differences in benefits allow for a detailed analysis and “weighing” of the separate components of risks and benefits in a following stage of the risk evaluation process.

Table 2. Environmental impact (EI) of pesticide use in GM crops (based on pesticide use data [7])

Item	Non transgenic	Transgenic	Difference	Difference %
Herbicide tolerant canola [1]				
Pesticide use, lbs ai/A	1.1	0.5	-0.6	-55
Total impact, EI/A	30.9	16.2	-14.7	-48
Farmworker impact, EI/A	17.0	8.0	-9.0	-53
Consumer impact, EI/A	9.3	3.5	-5.8	-62
Ecology impact, EI/A	66.5	37.2	-29.3	-44
Herbicide tolerant cotton [2]				
Pesticide use, lbs ai/A	2.3	1.9	-0.4	-18
Total impact, EI/A	61.0	54.0	-7.0	-11
Farmworker impact, EI/A	42.2	33.2	-9.0	-21
Consumer impact, EI/A	18.4	14.6	-3.8	-20
Ecology impact, EI/A	122.4	114.3	-8.1	-7
Insect resistant cotton [3]				
Pesticide use, lbs ai/A	0.7	0.1	-0.5	-78
Total impact, EI/A	21.6	4.7	-16.9	-78
Farmworker impact, EI/A	10.6	2.3	-8.3	-78
Consumer impact, EI/A	4.0	0.9	-3.1	-78
Ecology impact, EI/A	43.6	9.5	-34.2	-78
Herbicide tolerant maize [4]				
Pesticide use, lbs ai/A	3.4	2.4	-1.0	-30
Total impact, EI/A	102.5	69.9	-32.9	-32
Farmworker impact, EI/A	52.2	39.9	-12.3	-24
Consumer impact, EI/A	26.7	19.0	-7.7	-29
Ecology impact, EI/A	224.3	150.0	-74.2	-33
Insect resistant maize [5]				
Pesticide use, lbs ai/A	0.4	0.0	-0.4	-100
Total impact, EI/A	18.8	0.0	-18.8	-100
Farmworker impact, EI/A	20.4	0.0	-20.4	-100
Consumer impact, EI/A	5.3	0.0	-5.3	-100
Ecology impact, EI/A	30.8	0.0	-30.8	-100
Herbicide tolerant soybean [6]				
Pesticide use, lbs ai/A	1.5	1.0	-0.6	-38
Total impact, EI/A	40.9	30.8	-10.1	-25
Farmworker impact, EI/A	29.9	15.2	-14.7	-49
Consumer impact, EI/A	12.8	6.7	-6.1	-48
Ecology impact, EI/A	80.1	70.6	-9.5	-12

[1] comparison of alternative spray programs

[2] total use in 2000 ("transgenic") versus base year ("non transgenic")

[3] statewise comparsion of number of sprays to control PBW/BW, impact based on average use (lbs) and EIQ for these sprays; Arkansas and Mississippi not included due to lack of data on # of PBW/BW sprays

[4] calculation base on substitution scenario of spray programs

[5] calculation based on average use and EIQ of pesticide sprays applied to control European corn borer

[6] calculation based on statewise differentiated programs to control weeds as effective as glyphosate

4 Bt Cotton: the Learning Case of Arizona

Several reviews on the topic of pesticide usage before and after the introduction of genetically modified Bt cotton crops report a substantial decrease in numbers of sprayings and quantities of insecticides used by cotton growers, for example in the U.S. state of Arizona. Closer scrutiny of such reviews on the history of pest management in cotton in Arizona, however, learns that this decrease may not be solely related to the adoption of Bt cotton. Rather, the adoption of Bt cotton coincides with the education of farmers on pest management, the introduction of new, effective insecticides for cotton pests and an insect plague in the year preceding Bt cotton introduction. The following data on Arizona are taken from two reviews [24,25] as an illustration of the difficulties to single out specific changes in pesticide usage on Bt crops.

The three top pests of cotton in Arizona are silverleaf whitefly, Lygus bug, and pink bollworm. In 1992, 1994, and 1995, heavy outbreaks of silverleaf whitefly occurred and contributed considerably to total pesticide applications in those years. Lygus is another pest, for which still "old" insecticides, such as acephate and endosulfan, are used. The largest plague of pink bollworms occurred in 1979, when on average 6.8 pesticide applications were applied to control this pest.

The year 1996 and the years thereafter have witnessed a trend towards decreased pesticide use in Arizona cotton. Several factors can be associated with this decrease. Bt cotton was introduced in 1996 as part of a renewed integrated pest management (IPM) program. This program also included whitefly control and farmer education about pest management, Bt-resistance management, and threshold levels for spraying. In the same year (1996), two insect growth regulators, Knack (pyriproxyfen) and Applaud (buprofezin), that are used in the management of whitefly were approved.

Some features need to be considered in the interpretation of pesticide use data borrowed from the pesticide use reporting systems. These features include the type of data entered into the system and the factors that cause variability of the data. The reporting systems of various American states (California, New York, and Arizona) are not identical and are separate from the annual average statistics of the National Agricultural Statistics (NAS). In the case of the Arizona reporting system, for example, data are entered by professional pesticide applicators, for whom reporting is mandatory. The voluntary pesticide use reporting by other groups, such as agricultural producers, may vary, depending, for example, on their awareness. In addition, not all pesticides are on the reporting list [26]. Data from the Arizona system also show that geographical variations occur, for example variation in insecticide use between counties, which may reflect different insect infestation levels. Insect infestation, in turn, is influenced by climatological conditions, such as dry winters causing low levels of Lygus infestation in the following cotton growing season.

From the data available on the Internet, it may not be possible to "decouple" Bt cotton from other factors that may have contributed to general changes in the use of cotton pesticides following Bt cotton adoption. Changes in specific substances, however, can be linked to Bt cotton adoption, such as the decrease in the use of Gossyplure, a pheromone for the control of pink bollworm, in Arizona. Similarly, the increase in Roundup usage in cotton has been linked to the adoption of Roundup Ready varieties [25]. In addition, the important role of Bt cotton in pest programs has been acknowledged. Integrated Pest Management in cotton, for example, is facilitated by the use of Bt cotton as it obviated the use of broad-range insecticide sprays directed against certain early insect pests. In Australian cotton industry, for example, the most pronounced comparative decrease of insecticide applications in Bt cotton occurred early in the growing season [15]. This allows for predator and other natural pest enemy populations to sustain, which help to suppress development of pest populations, eventually with the help of additional feed sprays [16]. Surveys on IPM in cotton carried out before introduction of Bt cotton show contradictory results for its effect on pesticide use (reviewed by Fernandez Cornejo et al. [27]). For example, IPM practices among American cotton producers in 1994 were associated with high pesticide use and high yields [28]. Bt cotton would therefore provide an avenue for IPM programs to cut pesticide use while maintaining high yields. In addition, Bt cotton has been incorporated into the U.S. Boll Weevil Eradication Program based on similar considerations.

Other factors that may also have an effect on the pesticide use on cotton are, among others:

- Substitution of broad-range insecticides with low-toxicity, selective insecticides against insects that are not controlled by Bt cotton. As noted above, for example, two insect growth regulators were introduced as part of an IPM program in Arizona in 1996 that substituted for pyrethroid and other insecticides used for control of the whitefly *Bemisia* [29].
- New transgenic insect resistant cotton varieties. New varieties with "stacked" insecticidal genes are currently entering the market, that would provide less variable control of cotton bollworm and other insect pests due to the variation of Bt gene expression in single Bt gene varieties. Examples are Monsanto's Bollgard II expressing two Bt proteins (Cry1Ac and Cry2Ab) and the Chinese SGK321 cotton expressing a Bt protein and cowpea trypsin inhibitor. Gianessi et al. [7] predict that Bollgard II will reduce pesticide use on U.S. cotton by 986,000 pounds, thanks to more consistent control of bollworm and additional control of beet army worm and soybean looper.
- Non transgenic, (partially) insect resistant cotton varieties. Examples are:
  - -Lines with okra-shaped leaves, which allow more light to protrude into the canopy and thereby leaves less room to shade-preferring insects.
  - -"Nectariless" lines that do not produce nectar on which the insects can feed.
  - -Early maturing lines, with which the damage by insects infestations in the late season can be avoided (note: okra-leafed lines also mature early).

5 Impact on Pesticide Residues in Food Products

It may come to mind that the introduction of crops that impact pesticide use will inevitably lead to shifts in pesticide residue patterns in foods derived from these crops. In insect-resistant crops, for example, reductions in insecticide sprays against the target insects can be anticipated and concomitant reductions in crop residues. For herbicide tolerant crops, on the other hand, the ability to apply herbicides over the top of these crops (rather than directed sprays in herbicide sensitive crops) can be anticipated to lead to increased residue levels. Besides this anticipated increase, also the metabolism of the herbicides in tolerant crops may be altered from that in sensitive crops, leading to different metabolite patterns. In glyphosate tolerant maize, for example, an increase of the level of glyphosate metabolite AMPA has been observed comparative to the AMPA level in glyphosate sensitive maize [30]

It should be borne in mind that herbicides like glyphosate and glufosinate, that are used on herbicide tolerant crops, are also used on conventional crops. These herbicides are applied to crops shortly before harvest to "desiccate" these crops or to kill the tops of root crops in order to facilitate harvest. This use of herbicides will contribute to consumers' exposure to pesticides. It has been estimated by the FAO/WHO Joint Meeting on Pesticide Residues, for example, that the largest part of glufosinate ingested through food by Europeans is from potatoes, on which glufosinate is used as a "leaf killer" [31].

6 Conclusions

The general impression from the data reported so far is that overall pesticide use on GM crops is associated with a decrease in pesticide use. The reduction in pesticide use is expressed as significant quantities of pesticides (lbs or kg). Some critical remarks have to be made about these figures. First, other developments that concurred with the adoption of GM crops may have contributed to a decrease in pesticide use. Many governments have committed themselves to reducing pesticide use, and specific national programs have been initiated to achieve this goal, including programs for integrated pest management (IPM). Secondly, changes in quantities of aggregate pesticide use do not reflect the environmental impact that these changes in pesticide use may have, due to different properties of these pesticides, such as environmental persistency, carry-over to other compartments, and altered dose-relationships for environmental- and health effects. Therefore, interpretation of the pesticide usage data with regard to their implications for environment and human health requires an assessment of the environmental and human risks associated with the use of the specific pesticides involved.

It is one of the aims of the IUPAC project to qualitatively and quantitatively assess the impact of altered pesticide use on GM crops. Furthermore, other impacts will also be taken into account, which may include weed ecology and soil conservation. Finally a risk-benefit analysis of these impacts will be made by evaluating the various outcomes of our research, which may provide a tool for further policies in the area of GM crop cultivation and pest management. This is an important endeavour given the public concerns regarding large scale cultivation of GM crops and its environmental consequences.

7 Acknowledgement

The authors gratefully acknowledge financial support from the Netherlands Ministry of Agriculture Nature Management and Fisheries (program no. 390 'Safety of food and animal feed'). In addition, the authors thank T. A. Brimner, G. R. Stephenson, and the IUPAC project team for their input of scientific data.

8 References

[1] C. James, Global Review of Commercialized Transgenic crops: 2001, ISAAA Briefs No. 24-2001, International Service for the Acquisition of Agri-Biotech Applications, Ithaca., 2001 http://www.isaaa.org/publications/briefs/isaaa_briefs.htm.

[2] C. James, Global Review of Commercialized Transgenic Crops: 2000, ISAAA Briefs No. 23-2001, International Service for the Acquisition of Agri-Biotech Applications, Ithaca, 2001, abstract can be found under. http://www.isaaa.org/publications/briefs/isaaa_briefs.htm.

[3] C. James, Global Review of Commercialized Transgenic Crops: 1998, ISAAA Briefs No. 8-1998, International Service for the Acquisition of Agri-Biotech Applications, Ithaca., 1998, abstract can be found under http://www.isaaa.org/publications/briefs/isaaa_briefs.htm.

[4] M. C. Marra, P. G. Pardey, J. M. Alston, The Payoffs to Agricultural Biotechnology: An Assessment of the Evidence. EPTD Discussion Paper No. 87. International Food Policy Research Institute, Washington D.C., 2002.
see http://www.ifpri.cgiar.org/divs/eptd/eptpubs.htm.

[5] J. Carpenter, A. Felsot, T. Goode, M. Hammig, D. Onstad, S. Sankula, Comparative Environmental Impacts of Biotechnology-Derived and Traditional Soybean, Corn, and Cotton Crops. Council for Agricultural Science and Technology, Ames, 2002..

[6] H. A. Kuiper, G. A. Kleter in Proceedings of the IUPAC-TACTRI/COA International Workshop on Pesticides 2000, Taiwan, 3-6 October, 2000, pp. 57-71.

[7] L. P. Gianessi, C.S . Silvers, S. Sankula, J. E. Carpenter, Plant Biotechnology: Current and Potential Impact for Improving Pest Management in U.S. Agriculture: an Analysis of 40 Case Studies, National Center for Food and Agricultural Policy, Washington D.C., 2002, can be found under http://www.ncfap.org/40CaseStudies.htm.

[8] J. Fernandez Cornejo, W. McBride, Adoption of Bioengineered Crops, Agricultural Economics Report No. 810, U.S. Department of Agriculture, Economic Research Service, Washington D.C., 2002, can be found under http://www.ers.usda.gov/publications/aer810/.

[9] J. Fernandez Cornejo, W. McBride, Genetically Engineered Crops for Pest Management in U.S. Agriculture, Agricultural Economics Report No. 786, U.S. Department of Agriculture, Economic Research Service, Washington D.C., 2000.
abstracts can be found under http://www.ers.usda.gov/publications/aer786/.

[10] G. R. Stephenson, Impact of the Introduction of GM Crops on Pesticide Use in Canada. Unpublished paper, 1999.

[11] R. D. Wauchope, T. L. Estes, R., Allen, J. L. Baker, A. G. Hornsby, R .L. Jones, R. P. Richards, D. J. Gustafson, Pest Manag. Sci. 2001, 58, 146-160.

[12] R. H. Phipps, J. R. Park, J. Anim. Feed Sci. 2002, 11, 1-18.

[13] W. Lin, G. K. Price, J. Fernandeze Cornejo, "Estimating Farm-Level Effects of Adopting Herbicide Tolerant Soybeans", U.S. Department of Agriculture, Economic Research Service, Washington D.C., 2001.
see http://www.ers.usda.gov/Briefing/biotechnology/herbicide-tolerent_soybean(new).pdf
[14] J.W. Smith, Ann. Entomol. Soc. Am. 1998, 91, 239-247.
[15] D. Anthony, "Genetically Modified Cottons: Costs and Expectations," Paper presented at the technical seminar of the 59th plenary meeting held in Cairns, 5-10 November 2000, International Cotton Advisory Committee, 2000.
see http://www.icac.org/icac/CottonInfo/tis/biotech/documents/techsem/english.html.
[16] G.P. Fitt, Crop Prot. 2000, 19, 793-800.
[17] C. Pray, D., Ma, J., Huang, F., Qiao, World Dev. 2001, 29, 813-825.
[18] J. Huang, N. Li, H. Van Meijl, F. Van Tongeren, "Global Trade and Biotechnology Boosts to Non-Food Crop Productivity in China", Paper presented at the 5th Annual Conference on Global Economic Analysis, Taipei, Taiwan, 2002.
see http://www.gtap.agecon.purdue.edu/resources/res_display.asp?RecordID=995.
[19] J. Kovach, C. Petzoldt, J. Degni, J. Tette, N. Y. Food Life Sci. Bull. 1992, 139, updated version can be found under http://www.nysipm.cornell.edu/publications/EIQ.html.
[20] L. Levitan, I. Merwin, J. Kovach, Agric. Ecosyst. Environ. 1995, 55, 153-168.
[21] J. Reus, P. Leenderse, C. Bockstaller, I. Fomsgaard, V. Gutschke, K. Lewis, C. Nilsson, L. Pussemier, M. Trevisan, H. Van der Werf, F. Alfarroba, S. Blümel, J. Isart, D. McGrath, T. Seppälä, Comparing Environmental Risk Indicators for Pesticides, Center for Agriculture and Environment, Utrecht, 1999, http://www.clm.nl/en/caper.html.
[22] G .J. Gallivan, G. A. Surgeoner, J. Kovach, J. Environ. Qual. 2001, 30, 798-813.
[23] T. A. Brimner, G. R. Stephenson. "Influence of Herbicide-Tolerant Canola on Herbicide Use Patterns and the Environmental Impact of Weed Management", Poster presented at the IUPAC International Congress on the Chemistry of Crop Protection, Basel Switzerland, August 4-9, 2002.
[24] P. C. Ellsworth, J. S. Jones, "Cotton IPM in Arizona: a Decade of Research, Implementation and Education", Cotton Magazine, Western Agricultural Publishing Company Online, Fresno, 2001, also see http://www.westagpubco.com/Archive/Articles/cotton_IPM_in_Arizona.htm.
[25] G. K. Agnew, P. B. Baker, "Pesticide Use in Arizona Cotton: Long-Term Trends and 1999 Data", University of Arizona, Pesticide Information and Training Office, Tucson., 2000a, can be found under http://ag.arizona.edu/pito/articles/cottonrep2000.htm.
[26] G. K. Agnew, P. B. Baker, "Arizona Cotton Pesticide Use Data: Opportunities and Pitfalls, University of Arizona, Pesticide Information and Training Office, Tucson, 2000, can be found under http://ag.arizona.edu/pito/articles/beltwide2000.htm.
[27] J. Fernandez Cornejo, S. Jans, M. Smith, Rev. Agric. Econ. 2000, 20, 462-488.
[28] J. Yee, W. Ferguson, J. Prod. Agric. 1999, 12, 618-623..
[29] P. C. Ellsworth, J. L. Martinez Carillo, Crop Prot. 2000, 20, 853-869.
[30] JMPR in Pesticide Residues in Food - 1997, FAO Plant Production Paper 146, Joint Meeting on Pesticide Residues, Food and Agriculture Organisation of the United Nations, Rome, 1997, pp. 509-534.
[31] JMPR in Pesticide Residues in Food - 1999, FAO Plant Production Paper 157, Joint Meeting on Pesticide Residues, Food and Agriculture Organisation of the United Nations, Rome, 1999, pp. 121-129, 217-218, 224, 258-259.

24 Development and Registration of Pesticides With Reduced Risk Characteristics

Kenneth D. Racke

Dow AgroSciences, 9330 Zionsville Road, Bldg 308/2B, Indianapolis, Indiana, 46268, USA, E-mail: KRacke@dow.com

1 Summary

During the past decade a reduced risk pesticide registration program has been in operation in the United States. The primary objective of this program is to give registration priority and accelerated approval to products with more favorable regulatory characteristics than currently available alternatives. Since 1994 a total of 28 such reduced risk pesticides have been approved within shortened regulatory review periods of 14 to 22 months versus the traditional 24 to 38 months or longer. This includes 10 insecticides, 9 herbicides, and 9 fungicides. The Environmental Protection Agency (EPA), which administers the program, defines a reduced risk pesticide as one that reduces pesticide risks to human health and non-target organisms, reduces the potential for contamination of valued environmental resources, and broadens adoption or increases the effectiveness of integrated pest management programs. The reduced risk program has been supported by both industry and environmental organizations alike. Industry discovery R&D programs have grown more highly attuned to the early market entry benefits of selecting those products for development and commercialization which compare favorably from a registration perspective with those products already available. From a government regulatory standpoint the reduced risk program offers a more proactive, positive approach to pesticide regulation than traditional initiatives which tend to focus inordinate levels of resources on the reevaluation and restriction of older products. The success of this program in accelerating the introduction of new products with highly favorable characteristics into the marketplace serves as an attractive model for other national/regional regulatory authorities and international advisory bodies to consider. A fast-track process for reduced risk pesticides has now been adopted by the pesticide coordination program of the North American Free Trade Act (NAFTA) countries (Canada, Mexico, U.S.) and incorporated into recently promulgated pesticide registration process legislation in Brazil. The Codex Committee on Pesticide Residues is also considering reduced risk pesticide classification as one of several criteria for prioritization of technical evaluations with the FAO/WHO Joint Meeting on Pesticide Residues (JMPR).

2 Pesticide Development

The discovery, development, registration, and commercial introduction of new pesticides for control of weed, insect, and disease pests is a difficult and expensive process. In addition to targeted molecular design approaches, a significant amount of screening occurs to identify patentable chemistries with novel modes of action. Once a decision is taken to develop a new active ingredient, a major testing phase focused on field performance, health and environmental safety, manufacturing optimization, and commercial planning is initiated. This may last from 3 to 5 years, after which time a regulatory review period of anywhere from 1.5 to 4 or more years occurs.

For every new active ingredient that enters development, there may be 20,000 or more candidate compounds and analogues that will have been screened for biological activity. The discovery and development phases are collectively both costly and time-consuming, and it may take 5-10 years after entry into the marketplace for the original investment to be recovered. The cost of bringing a single product to the marketplace from discovery through launch may be USD 50 to 60 million or more. Thus, even a single year delay in reaching the marketplace can be very costly in terms of value recovery and patent protection. By using discounted cash flow analysis, it has been estimated that a one-year delay in registration approval and product launch may reduce the value of a new pesticide product by USD 5 million [1].

3 Regulatory Trends

There are several regulatory trends which, during the past decade or more, have focused an inordinate degree of attention and an increasing level of resources on reevaluation of older pesticides. During this period, it has also become more difficult, costly, and time-consuming to obtain registration approvals for newer pesticide products.

One trend has been for ever-increasing levels of regulatory scrutiny and increased testing requirements. This scrutiny has been driven by both technical and societal issues, and some recent examples would include the interest in potential endocrine effects, acute dietary intake assessment, and ground and surface water entry. In addition, there have been political factors which have yielded increased demands on the pesticide registration process. An example of this in the U.S. would be the Clinton Administration's "Children's Health Initiative", which drove highly conservative and restrictive approaches for pesticide evaluation despite a lack of compelling evidence that children in the U.S. were at any undue risk of adverse effects from the use of pesticides [2]. The result of such increased scrutiny, which often has been generated by issues related to older products, is that it has become more difficult and expensive to register new products.

A second trend has been for regulatory agencies to focus on reducing pesticide use as a national priority. Various countries have set goals for percentage reduction in quantities of pesticides used or area treated with pesticides. Examples would include Sweden and the U.S., which during the early-to-mid 1990's set 10-year targets to reduce pesticide use by half [3,4]. Although the concept is a popular one, use restrictions for older, less active, higher-application-rate pesticides may not necessarily lead to any reduction in potential pesticide human health or environmental impacts if the market merely shifts to lower-application-rate products with higher potency. Indeed, a reduction in pesticide use in some of the developed and often temperate countries with such programs may lead to an increased use of pesticides for export crop production in developing and often subtropical or tropical countries.

Finally, an increasing number of regulatory authorities have been involved in comprehensive reevaluation or reregistration programs focused on older chemicals. These programs are designed to review the safety aspects of existing pesticides in light of current data requirements and more stringent assessment procedures. Examples of such programs would include the EU Review under Directive 91/414, U.S. EPA Reregistration, and the Australia Existing Chemicals Review Program. In general, these efforts have turned out to be more complicated, costly, and lengthy affairs than originally envisioned. Although restrictions or use cancellations of some older products have resulted from these programs, the allocation of regulatory resources for a focus on older product review has decreased the capacity of regulatory bodies to evaluate and approve newer products.

4 Reduced Risk Initiative

Although all of the above trends have certain merits, there can be a tendency for the programs they generate to focus primarily on creating additional restrictions for older products and lead secondarily to increased barriers for the introduction of new ones. These dual emphases can also result in allocation of so much regulatory resource on older products that review and approval of newer and more desirable alternatives may be slowed. One recent regulatory initiative in the United States, the reduced risk pesticide program, has departed somewhat from these approaches and created a more proactive and positive approach to pesticide regulation.

The general principle of the reduced risk pesticide registration program in the U.S. is to give registration priority and accelerated approval to products with the most favorable characteristics as compared to currently available alternatives. The overall objective is to accelerate the introduction of these type of products so that marketplace choices rather than increased regulatory restrictions can lead to replacement of older products and technologies with newer ones. Under the program, pesticides classified as "reduced risk" products must meet several or all of the following criteria as compared with currently available alternatives [5]:

- reduced risks to human health
- reduced risks to non-target organisms
- reduced potential for contamination of environmental resources (water, air, soil)
- broadened adoption of IPM or makes it more effective

The reduced risk pesticide program was first envisioned during the late 1980's and early 1990's, and by 1992 public comments were invited on a proposed program for providing regulatory incentives for development of safer pesticide products. During 1993 a voluntary reduced risk pesticide initiative was introduced which described interim criteria for "reduced risk" active ingredients, guidelines for submission of reduced risk rationale by registrants, and a streamlined registration process. During 1994 the first pesticide active ingredient, hexaflumuron, was registered under the reduced risk program with an elapsed regulatory review of only 14 months versus an average of more than 24 months for pesticides in general. By 1997 more detailed guidelines were promulgated for expedited regulatory review under the reduced risk program [5], including adoption of formalized reduced risk criteria, a standardized submission format, and establishment of a Reduced Risk Committee at EPA. During 1998 EPA established a priority system of review for all registration submissions, with several categories of reduced risk products adopted by the Agency as top priorities for registration review. In addition to first-time registrations of new active ingredients, during 1996 the program was expanded to include accelerated registration approval for additional crops and uses for active ingredients already classified as reduced risk.

With the current system in place, a manufacturer assembles a detailed justification for classification of a new active ingredient (or new use for an existing active ingredient) which must accompany the application for registration. This document makes explicit comparisons between the new pesticide and those products currently in use for the same pests and/or crops with respect to human safety, environmental impacts, and pest management practices. Soon after submission to EPA, the Reduced Risk Committee thoroughly evaluates the proposal and reaches a decision as to classification of the new pesticide or use as a reduced risk priority. For those few situations in which the reduced criteria are convincingly met, the product is given a high priority for registration review and advanced toward accelerated approval.

Since the introduction of the reduced risk initiative at EPA in 1994, a total of 28 new active ingredients have been classified as reduced risk products and approved in an expedited fashion. These include 9 herbicides, 9 fungicides, and 10 insecticide products. During this period a total of 88 new chemical active ingredients have received registration approval in the U.S., thus making that proportion of new products classified as reduced risk approximately one-third. Figure 1 compares, on a fiscal year basis (Oct 1 to Sep 30), the total number of new active ingredients registered versus the number classified as reduced risk and given expedited approval. The real benefit of the program to both pesticide manufacturers and pesticide users has been the shortened time for regulatory approval and market introduction. Pesticides eligible for fast-track review as reduced risk products have experienced review periods of 14 to 22 months whereas pesticides which did not quality as reduced risk products took generally more than twice as long for registration approval (24 to 38 months or more). Thus, reduced risk pesticide products have reached the market 1 to 2 seasons in advance of other pesticides.

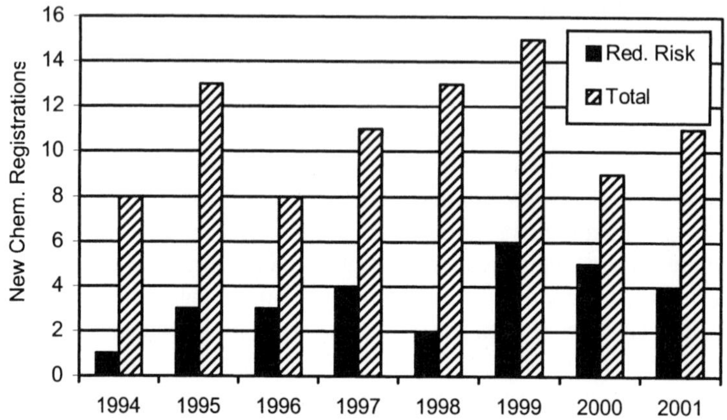

Figure 1. U.S. EPA New Chemical Registrations for Insecticides, Herbicides and Fungicides (not including biopesticides, repellents, or rodenticides)

5 Reduced Risk Products

A listing of pesticide active ingredients approved under the reduced risk pesticide initiative appears in Table 1 along with identification of the first registered use(s). Since their introduction, some of these products have achieved outstanding commercial success and displaced the use of significant quantities of older products in both agricultural and urban pest management programs. Others have not enjoyed such clear success in the marketplace and have instead become highly specialized niche products serving a select clientele. To provide a practical context of the reduced risk paradigm and highlight the market impacts which may result, four case studies of reduced risk pesticide products will be briefly presented (Table 2). These will include examples which impact pest management programs in corn, cotton, turfgrass, and the urban environment.

5.1 Hexaflumuron

Conventional control of termites is heavily reliant upon prophylactic application of liquid insecticides to form a soil barrier around and beneath a structure to termite entry. These barriers, designed to repel or kill termites, degrade over time and must be reapplied every 5-10 years. A barrier termiticide application requires significant quantities of active ingredient, 5-10 kg, carried in 300 to 600 liters of water. Conventional products used as soil termiticides include organophosphates such as chlorpyrifos and isofenphos, pyrethroids such as bifenthrin and cypermethrin, and more recent products such as imidacloprid and fipronil. Issues associated with use of the existing products include the introduction of large quantities of pesticide into the environment whether termites are present or not, potential contamination of water systems (wells, ponds) during the application process due to unforeseen drainage, worker exposure concerns, and odor complaints by residents due primarily to carrier solvents in the pesticide formulations.

Table 1. New Chemicals Registered as Reduced Risk Pesticides by U.S. EPA

Year	Product	Type	First Registered Uses
1994	Hexaflumuron	Insecticide	Termite control
1995	Flumiclorac-pentyl	Herbicide	Corn, soybeans
1995	Hymexazol	Fungicide	Seed treatment
1995	Tebufenozide	Insecticide	Walnuts
1996	Fludioxonil	Fungicide	Corn seed treatment
1996	Cadre	Herbicide	Peanuts
1996	Mefenoxim	Fungicide	Fruits, nuts
1997	α-Metolachlor	Herbicide	Row crops
1997	Azoxystrobin	Fungicide	Turfgrass, fruits, vegetables
1997	Imazamox	Herbicide	Soybeans
1997	Spinosad	Insecticide	Cotton, turfgrass
1998	Carfentrazone-ethyl	Herbicide	Corn, soybeans, wheat
1998	Cyprodinil	Fungicide	Stone fruits, nuts
1999	Bifenazate	Insecticide	Ornamentals
1999	Diflufenzopyr	Herbicide	Corn
1999	Pymetrozine	Insecticide	Potatoes, tobacco, ornamentals
1999	s-Dimethenamid	Herbicide	Corn, soybeans, peanuts
1999	Fenhexamid	Fungicide	Grapes, strawberries
1999	Trifloxystrobin	Fungicide	Vegetables, pome fruit, grapes
2000	Methoxyfenozide	Insecticide	Cotton, pome fruit
2000	Indoxacarb	Insecticide	Vegetables, pome fruit, sweet corn
2000	Flucarbazone-sodium	Herbicide	Wheat
2000	Fenpyroximate	Insecticide	Ornamentals
2000	Buprofezin	Insecticide	Vegetables
2001	Zoxamide	Fungicide	Grapes, potato, tomato
2001	Fluazinam	Fungicide	Peanuts, potato
2001	Novaluron	Insecticide	Ornamentals
2001	Mesotrione	Herbicide	Corn

The first pesticide approved under the reduced risk program by EPA was the insecticide hexaflumuron for structural protection from termite damage. Hexaflumuron is a slow-acting, chitin synthesis inhibitor, with low mammalian and fish toxicity [6]. The product, although of low mobility, is moderately persistent in the soil environment and is of high toxicity to nontarget aquatic arthropods such as daphnids. How then was hexaflumuron determined to be of reduced risk caliber? It was the mode of action and method of delivery (baiting) that distinguished hexaflumuron as a reduced risk alternative to conventional liquid barrier termiticide treatments. Instead of a soil barrier, below-ground stations initially containing wood blocks are placed in the soil around a structure to be protected. Once termites locate the wood blocks during this monitoring phase, a bait tube containing 0.5% hexaflumuron impregnated on a paper matrix is inserted. After several months during which foraging termites feed on the bait matrix and share food with their nestmates, the entire termite colony is eliminated by the slow-acting toxicant. Instead of kilograms of active ingredient serving as a temporary barrier to termites with the conventional approach, only 2-5 g of hexaflumuron is required to eliminate an entire termite colony. The wood-block monitoring mode is then resumed until detection of future termite activity. This IPM approach has represented a new paradigm in termite protection, and at present more than 750,000 homes and other structures in the U.S. employ the Sentricon* Colony Elimination System (*Trademark of Dow AgroSciences).

Table 2. Characteristics of Case Study Reduced Risk Products

Compound	Hexaflumuron	Azoxystrobin	Spinosad	Diflufenzopyr
First registered use(s)	Termite control	Turfgrass	Cotton, turfgrass	Corn
Application method	Enclosed bait station	Ground spray	Ground or aerial spray	Ground spray
Application rate	2-3 g ai/structure	250-600 g ai/ha	50-150 g ai/ha	6-11 g ai/ha
Mode of action	Chitin synthesis inhibitor	Electron transport inhibitor	Rapid excitation of nervous system by persistent activation of nicotinic acetylcholine receptors	Auxin transport inhibitor
Rat oral LD_{50}	> 5000 mg/kg	> 5000 mg/kg	> 5000 mg/kg	> 5000 mg/kg
ADI	0.02 mg/kg/d	0.18 mg/kg/d	0.027 mg/kg/d	0.26 mg/kg/d
Field soil DT_{50}	73 days	8 - 34 days	< 1 day	4 days
Soil sorption, Kd	147 - 1326 low mobility	2 - 23 mod. mobility	5 - 323 low mobility	0.2 – 1.6 very mobile
Bluegill LC_{50}	> 100 mg/L	1.1 mg/L	5.9 mg/L	135 mg/L
Daphnid LC_{50}	0.0001 mg/L	0.2 mg/L	1.5 - 14 mg/L	15 mg/L
Quail LD50	> 2000 mg/kg	> 2000 mg/kg	> 1333 mg/kg	> 2250 mg/kg
Reg. Approval	1994	1997	1997	1999
Review Period	14 months	17 months	18 months	13 months

5.2 Azoxystrobin

For intensively managed turfgrass areas such as golf courses, parks, and lawns fungicides are the most heavily relied upon products used to deal with pest situations. Conventional products, which generally act by contact or systemic action, often must be used repeatedly and in combination to deal with the variety of turfgrass diseases which may occur during the season. These include contact fungicides such as captan and chlorothalonil and systemic products such as benomyl and metalaxyl. Issues such as pest resistance, surface and ground water quality, and reentry exposure have generated regulatory concerns during recent years with the use of fungicides in turfgrass areas. Several of the fungicides currently in use were considered by EPA to be probable or possible human carcinogens.

Azoxystrobin was the first reduced risk pesticide approved by EPA specifically for turfgrass pest control, and its broad-spectrum activity and unique mode of action provided important support for both resistance management and IPM programs. Registration of azoxystrobin as a reduced risk product occurred during 1997 [7]. Azoxystrobin poses low acute and chronic mammalian toxicity, is non-carcinogenic, and exhibits low toxicity to fish and birds. In addition, it is expected to have low environmental impact due to its moderate persistence and mobility. Within a short time after its introduction, azoxystrobin became a leader in the turf disease management market. Following this successful introduction for turfgrass pest control, subsequent approvals have occurred for use on variety of fruit and vegetable crops.

5.3 Spinosad

Cotton is one of the crops most heavily reliant upon insecticide use. Although Bt cotton has been introduced in some areas, insecticide sprays remain as important components of cotton pest management programs. In some areas of the cotton belt, the majority of applications are made by aircraft. Issues which have been of regulatory concern in recent years include pest resistance associated with the frequent sprays, beneficial insect effects, worker exposure, and spray drift impacts on surface water quality. Products currently in use include organophosphates such as azinphos-methyl, methyl parathion, acephate, and disulfoton, pyrethroids such as lambda-cyhalothrin and carbamates such as methomyl.

Spinosad was the first reduced risk cotton insecticide approved in the U.S., and registration occurred during 1997 [8]. It is a biologically produced natural product of the bacterium Saccharopolyspora spinosa, first isolated from soil during the early 1980's. Secondary metabolites produced during fermentation of S. spinosa include the insecticidally-active spinosyns A and D, which comprise the product spinosad. Spinosad exhibits a novel mode of insecticidal action, and thus offers relief from the resistance which plagues cotton pest control. It is also highly specific in its insecticidal activity, and is not toxic toward mammals, birds, fish, aquatic invertebrates, or beneficial predators and parastioids. EPA classified spinosad as reduced risk product based on its low mammalian toxicity, low environmental impacts, and compatibility with IPM programs. Since its introduction, spinosad has became one of the leading insecticidal products used on cotton in the U.S., and it has also been approved for use on more than 150 additional row, fruit, and vegetable crops as well as for turfgrass pest control. Spinosad has also recently been approved for use in certified organic agriculture by the U.S. Department of Agriculture.

5.4 Diflufenzopyr

Corn is one of the most heavily herbicide-reliant crops, and intense weed pressure results in the use of a number of pre-emergent and post-emergent products. Regulatory concerns have focused largely on potential groundwater and surface water quality impacts, although concerns with worker exposure and chronic toxicity of some herbicides have also arisen.

Diflufenzopyr was approved by U.S. EPA as a reduced risk herbicide for corn during 1999 [9]. It controls a range of broadleaf weeds and also suppresses grasses, and is highly compatible with IPM programs. It has low acute toxicity toward mammals and other nontarget organisms, and is not a carcinogen or developmental toxicant. Diflufenzopyr also was one of the first reduced risk pesticide products jointly reviewed by U.S. EPA and the Canada Pest Management Regulatory Agency under the North American Free Trade Agreement.

6 Benefits and Issues

There are a number of benefits that have been realized through implementation of the reduced risk pesticide program in the U.S. First, it has been a positive step toward the accelerated introduction into the market of new products with highly favorable safety profiles. Rather than being faced only with increased restrictions on existing products, pesticide users increasingly wider choices among both new and existing products for a variety of crop pest management scenarios. Second, the possibility of accelerated approval has provided an additional incentive for agrochemical companies to focus early during new product discovery and development efforts on those products with reduced risk attributes. For a number of companies, the value associated with potential reduced risk product classification is being used as one of the criteria by which selection for those compounds to advance through the development process is occurring. Third, the reduced risk program has been a uniting factor in the agrochemical arena. It is one of the very few initiatives which has enjoyed the support of government regulators, agrochemical manufacturers, farmers and other pesticide users, and environmental activists alike.

The reduced risk program is by no means a panacea. Not all reduced risk pesticides, having been afforded regulatory priority of effort, have become successful commercial products. In addition, not all reduced risk products will necessarily pose less risk than current alternatives for all use patterns and in all circumstances. The concept of reduced risk involves a relative comparison; an active ingredient may be considered reduced risk for some uses but not for others. Finally, the fact that a new or existing product has not achieved reduced risk product status should not be construed as an indication that these types of products necessarily pose greater risks. The actual risk posed by the use of any pesticide product, whether it has been approved as a reduced risk product or not, is a function of many factors including formulation type, method of application, use of protective clothing, and environmental variables. In addition to the accelerated regulatory review and market introduction afforded by the reduced risk program, there are other mechanisms by which the advancement of reduced risk products may be encouraged. Since the mid-1990's, several pesticide products have been publicly recognized at the highest levels of government in the U.S. with the Presidential Green Chemistry Challenge Award. Past award winners include tebufenozide (1998), spinosad (1999), and hexaflumuron (2000), all of which also had previously been recognized as reduced risk products by U.S. EPA. Future ideas for consideration would include reduction of registration fees, increased period of data protection, extended patent protection, and increased period of reregistration renewal for reduced risk products.

7 International Applicability

The success of the reduced risk pesticide program in the U.S. raises the question of its applicability for other national and regional registration authorities as well as international advisory bodies. It would appear that the concept of an accelerated review and approval system for reduced risk products would be compatible with the regulatory processes of many national and regional regulatory authorities. Certainly many national authorities have developed quite similar and very extensive efforts for the reregistration evaluation of older pesticides.What should reasonably prevent a similar widespread adoption of the reduced risk pesticide approach described in this paper? Few authorities, however, appear to have developed a recognized process for fast-track approval of the most desirable pesticide products. Thus, the dossiers of many newer products with highly favorable characteristics may be caught in the queue of lengthy review and approval processes while regulatory authorities focus an inordinate amount of effort on the review and restriction of older products. One can only speculate as to the advantages which are being lost by ignoring the selective power of the marketplace for choosing the most appropriate and safest products from a widened portfolio of options. One encouraging development was the fast-track process for reduced risk pesticides incorporated into recently promulgated pesticide legislation in Brazil.

It is not necessary that regulatory authorities change their review and approval process for new products so as to focus such a large amount of effort on reduced risk products that the approval of other new products is slowed. As mentioned earlier, products which may not necessarily qualify for reduced risk classification may still offer substantial benefits and merit timely approval and market introduction. However, by placing some priority of effort on accelerated review of reduced risk products and perhaps diverting resources from seemingly endless and time-consuming reevaluation programs, regulatory authorities should be able to adopt a reduced risk program approach without interrupting the regular evaluation of new products in general.

One possibility for advancement of reduced risk programs would be mutual recognition or review of reduced risk pesticide products by cooperating national regulatory authorities. For example, a product approved as a reduced risk pesticide in one participating country might then be advanced to fast-track review and approval in other participating countries. Such an approach for joint review of reduced risk chemicals is currently being piloted by the North American Free Trade Agreement (NAFTA) Technical Working Group on Pesticides. Under this plan, U.S. and Canadian regulatory authorities reach a joint decision on classification of a new active ingredient as a reduced risk product, and then cooperatively conduct a joint, accelerated regulatory review which results in simultaneous approval in both countries.

The possibility for accelerated review of reduced risk alternatives exists also with international advisory bodies. For example, the establishment of Codex MRLs is a lengthy and time-consuming process and it may take 6-8 years or more between the time a new compound is nominated for review and final approval of CXLs by the Codex Alimentarius Commission. Part of this time is spent with the proposal in the queue behind other chemicals awaiting technical review by the FAO/WHO Joint Meeting on Pesticide Residues (JMPR). The Codex Committee on Pesticide Residues recently adopted reduced risk pesticide classification as one of several criteria for prioritization of JMPR scheduling for new chemical evaluations. Other international advisory bodies and processes also might benefit from a reduced risk priority approach. These include the FAO Specification Process and World Health Organization Pesticide Evaluation Scheme (WHOPES).

8 Conclusion

We live in a world of choices when it comes to establishing the priorities of individual regulatory processes. As has generally been the case, massive investment of resources can continue to be devoted to cycles of data evaluation, regulatory reevaluation and reregistration, and imposition of increased restrictions for the many older products currently employed in pest management programs. Some measure of attention to these older products is obviously warranted. However, with the continued limitation of resources in both the government and industrial sectors, an undue emphasis on the old may not allow sufficient resources to be directed to support the timely approval and market availability of the next generations of pesticides. The accelerated review and approval of new products with enhanced safety and environmental profiles is a proactive and positive step which unleashes marketplace forces which can effectively accelerate the replacement of older pesticide products. It is also an endeavor that places regulatory agencies and manufacturers in a collaboration which is of benefit to both pesticide users and detractors. Where we place our emphasis and energy in coming days, on the old or the new, will determine to a great extent how soon the world may have unfettered access to the safest and most desirable pest management tools.

9 References

[1] M. L. Leng, Chemtech, 1991, July, 408-413.
[2] S. Huebner, K. Chilton, Issues in Science and Technology, 1998, Winter, 35-38.
[3] V. Bernson, G. Ekström, Pesticide Outlook, 1991, 2, 33-36.
[4] L. R. Goldman, K. J. Garvey, C. C. de Nava, K. Looye, in Proceedings of the 8th International Congress of Pesticide Chemistry: Options 2000 (Eds.: N. N. Ragsdale, P. C. Kearney, P.C., J. R. Plimmer). American Chemical Society, Washington, DC, 1995, pp. 372-381.
[5] U.S. EPA Guidelines for Expedited Review of Conventional Pesticides Under the Reduced-Risk Initiative and for Biological Pesticides, PRN 97-3, September 4, 1997.
[6] A. S. Robetson, N.-Y. Yu, Down to Earth, 1995, 50, 1-7.
[7] U.S. EPA Pesticide Fact Sheet: Azoxystrobin, Office of Pesticide Programs, February 7, 1997.
[8] G. D. Thompson, R. Dutton, T. C. Sparks, Pest Management Sci. 2000, 56, 696-702.
[9] U.S. EPA Pesticide Fact Sheet: Diflufenzopyr. Office of Pesticide Programs, January 28, 1999.

25 Environmental Fate Considerations for Pesticides in Tropical Ecosystems

Elizabeth Carazo

Centro de Investigación en Contaminación Ambiental (CICA) Universidad de Costa Rica. San José, Costa Rica; E-mail:ecarazo@cariari.ucr.ac.cr.

1 Introduction

The tropics are the region between 23.5° north and south of the Equator. All of Central America, extended parts of South America, Asia and Australia and a major portion of Africa are within this region. Half of the world's population live in the tropics, mostly in developing countries with a substantial portion subsisting at or below the poverty level. This situation intensifies the need to resolve problems related to human intervention on natural ecosystems.

The mean daily solar radiation reaching the tropics is twice that received in temperate zones. The length of the days in the tropics remains nearly unchanged the whole year [1]. Rainfall is the principal parameter used to classify tropical climates due to the small variation in temperature. At the equatorial zone, rain is continuous, while away from the equator near the tropical parallels the climate gradually changes to a bimodal pattern of a rainy and a dry season. Desert conditions prevail at higher latitudes. Low lands constituting 87% of the area are characterized by year-round high temperatures, the absence of temperature limitations gives the tropics a great agriculture potential [2].

Tropical ecosystems have a prominent role in the conservation of genetic diversity, as these habitats contain the majority of the diverse species of the world [3]. The World Conservation Strategy [4] emphasized the need for conserving the genetic diversity by sustainable maintenance and management of natural ecosystems. The tropics provide challenges for the generation of reliable data on the consequences of human intervention on natural ecosystems. One such a challenge is agricultural production, which is the mainstay of most tropical countries. Agriculture involves an extensive use of pesticides, in order to minimize losses caused by pests. For example in Latin America, Brazil, Colombia and Mexico rank among the 20 leading world agrochemical consumers and even a small country like Costa Rica uses considerable quantities of pesticides per cultivated hectare for some crops [5]. The use of pesticides accomplishes a noticeable increase in agricultural productivity but also a dependence on chemical control of pests. Pests destroy up to one-third of the world's food crops during growth, harvesting and storage and in developing countries crop losses are even higher [6]. Hence, now and in the near future, the continued use of pesticides is inevitable. It should be noted that substantial quantities of pesticides are still used for disease vectors control (e.g. dengue and malaria) and the control of urban and animal pests. Since, conventional chemical pesticides will remain with us, for a wide range of uses, for the foreseeable future and we need to ensure that management practices are substantially improved to accommodate their safe and efficient use [7]. Problems with pesticides in developing countries are exacerbated due to the limitations in resources necessary to confront them.

In order to provide a strategy to reduce or mitigate risks to human health and the environment, it is imperative that the fate and behavior of pesticides are adequately understood [8]. This necessitates that environmental fate studies must be scientifically sound, broadly applicable and unbiased [9].

2 Pesticides Fate in Tropical Soils

What is unique about tropical soils is their behavior and management in the tropical environment [2]. The soil climate is the only common attribute that tropical soils share and what differentiate them from soils in temperate regions [1]. Soil climate is important, first, in soil formation and soil-plant relationships that involve biological processes activated by living organisms specific to the tropical environment, and second, to soil management practices that are essentially technological packages, adapted to particular ecological conditions. These packages cannot be transferred as they adapt to the environments to which they are introduced [1]. Thus, extrapolation of data on the fate of pesticides in temperate soils to tropical soils has proved to be unsatisfactory. Subtropical and tropical soils differ from temperate soils mainly in the quality of the clay minerals, and in the amount and quality of iron oxides and organic matter [10]. Tropical soils are not uniform due to vegetation, origin of material and geomorphology. The basic common property is the absence of seasonal variation in soil temperature [2], which is called isotemperature regimes.

In discussing the fate of pesticides in soils, the key processes controlling it should be addressed, e.g. sorption/desorption, degradation (biological and chemical), volatilization, leaching and runoff. How each mechanism contributes to the overall process depends on the physicochemical properties of the pesticide (vapor pressure, water solubility, partition coefficient), agricultural practices (frequency and rate of application) and climatic conditions. When pesticides are used properly they are very effective and adverse effects are minimized but trace amounts are often found in environmental compartments even far from the application sites. With the agricultural practices used in developing countries, pesticides residues may be expected in soil, since it acts as a sink for these chemicals but the processes mentioned previously play an important role in their dissipation from soils. In general, herbicides and insecticides are used mainly in the tropics but in some countries, e.g. Central America, fungicides are also extensively used to control fungal diseases in crops such as bananas.

A comprehensive review article about the fate of pesticides in tropical soils, reported several interesting facts about pesticide research in tropical soils [11]. For example, in a 10 year period, literature citations on research in tropical soils as opposed to temperate soils were disproportionate in a ratio of 1:3. Literature citations on the tropical soils were mainly from the Asia/Pacific-India region with only a few from Latin America with the exception of Brazil. Also, few references related to studies on the fate of a given pesticide in both tropical and temperate soils. In fact, not only are publications on pesticide fate in tropical soils scarce but are often difficult to obtain and may contain inadequate experimental details for a complete interpretation. This situation still continues, with many studies not being published or reported in publications difficult to access.

2.1 Pesticides Residues in Tropical Soils

Organochlorine pesticides were banned in the tropics during the 1970-80's but in some countries it did not take effect until the early 1990's, e.g. DDT and toxaphene in Nicaragua (1993) [12]. In addition to the use of organochlorines in vector control programs there are exceptions to the ban for agricultural purposes, e.g. endosulfan. Residues of organochlorines are detected in the tropical agricultural soils at almost every site sampled. In the Chinandega district, Nicaragua, high residues levels of toxaphene (17 to 44 $mg \cdot kg^{-1}$) and DDT's (0.19 to 0.98 $mg \cdot kg^{-1}$) were reported [13] at levels 40 and 4 times respectively, higher than found in lagoon sediments near the mouth of the Atoya river, that irrigates this area. The amounts of organochlorine residues accumulated in agricultural soils will continue to be the source of pesticide contamination of the lagoon systems for many years. Endosulfan was found in the same soils at concentrations of 0.65 $mg \cdot kg^{-1}$ but in lagoon sediments, the residues were below 1 $\mu g \cdot kg^{-1}$. Pesticides residues were monitored in soils from Guanacaste, Costa Rica for several years to evaluate the pesticide impact on an irrigation project [14]. The organochlorine compounds found in samples were α-HCB and β-HCB (45%), lindane (84%), endosulfan (16%), heptachlor (37%), heptachlor epoxide (11%), chlordane (39%), o,p-DDE, p,p-DDT (42 %), p,p-DDE (82%) and o,p-DDT (61 %) . The levels of the particular organochlorine compounds varied between 0.1 to 35 $\mu g \cdot kg^{-1}$.

Reports of other pesticide residues are less frequent. Propanil was present in 42% of the soil samples analyzed as well as ethyl parathion in 29%, methyl parathion in 17% and chlorpyrifos in 45% of the samples [14]. The concentrations of all three individual compounds were low (0.5-14 $\mu g \cdot kg^{-1}$). Residues of 2,4-D were found only in a few samples but MCPB was present in the 60% of the samples, the levels of these two compounds varied between < 0.01 to 0.12 $mg \cdot kg^{-1}$. In El Salvador, where paraquat is extensively used, different regions of the country were monitored. Residues in Sonsonate soil samples averaged 7.14 $mg \cdot kg^{-1}$, while in Ahuachapán, one sample contained as much as 16.47 $mg \cdot kg^{-1}$ paraquat [15]. Tatuì soils from Sao Paulo State, Brazil, contained trifluraline residues of 1.2 $mg \cdot kg^{-1}$ six weeks after application also methyl parathion residues up to 2 $mg \cdot kg^{-1}$ [16]. Reports on residues of newer pesticides in soil are rare.

2.2 Pesticide Dissipation from Tropical Soils

Fipronil is a relatively new insecticide used in many tropical countries. It degrades by oxidation/reduction and hydrolysis mechanisms, Pesticides are known to be dissipated from tropical soils by various mechanisms[11]. In Panama, the dissipation of propanil, atrazine, maneb and carbofuran has been reported in sandy loam and clay soils from Cerro Punta, El Coco and El Ejido [17, 18]. All these experiments were conducted in field lysimeters under the same climatic conditions; the dissipation rates during the initial phase were related to the soil properties. Soils with a high organic content and low amounts of clay seemed to greatly facilitate dissipation of the compounds, whereas a high clay content prolonged the rate of dissipation. A high rate of dissipation of the pesticides was observed at different field sites shortly after application and was attributed to changes in temperature, humidity and atmospheric pressures, volatilization, water co-distillation and evapotranspiration from the soil. In addition, leaching could result in significant dissipation, in the case of atrazine 21.6%, of the original applied concentration had moved to the underground layers and with carbofuran, the leachate accounted for 50% of the applied concentration. In another study with Panamanian soils from Cerro Punta and El Ejido in field lysimeters, volatilization and microbial degradation appeared to be the most important mechanisms involved in the dissipation of α-endosulfan [19]. Endosulfan was also observed to undergo rapid dissipation mainly due to volatization, photolysis and hydrolysis [16, 20]. Although on steep slopes lacking vegetative cover, runoff was the major route for residue transport, even when rainfall was moderate. When endosulfan was applied to field soil columns in a Philippine rice field, the dissipation rate was slower as it passed down the columns [21]. The same behavior has been reported for metalochlor and in both cases it was attributed to reduced microbial degradation in the subsurface soil layers [22]

The dissipation of the herbicides, hexazinona, tebuthiuron, imazapir and glyphosate was studied under tropical conditions in Hawaii, Peru, and Panama [23]. Dissipation occurred much faster under these tropical than temperate conditions. The persistence of herbicides in Hawaii under field controlled conditions was close to that found in Peru and Panama. For hexazinona and tebuthiuron, leaching and persistence could increase the risk of ground water contamination, although leaching of hexazinona when applied at normal rates appeared to be slight. Imazapir dissipated quickly from the hawaiian soil plots, but not from leaching, since the herbicide remained in the upper 25 cm of soil. In Panama, glyphosate dissipated quickly in soil and no residues were detected after 3 months. None of the herbicides were detected in streams near the application sites and replanting in Hawaii was dependent on the herbicide rate and the crop. New growth was observed a few months after treatment in Peru and Panama. Under field conditions, metalochlor had a half-life of 27 days though residues remained in the soil after the growing season [22]. In laboratory studies, the half-life of metalochlor under flooded conditions was 44.3 days as opposed to aerobic conditions when it was considerable longer being more than 130 days. The presence of residues even during the harvest showed the persistent nature of metalochlor with leaching to lower depths being one of the possible main dissipation routes.

the latter being more important under moist conditions. All the metabolites formed on hydrolysis are toxic with the exception of the amide. Fipronil also undergoes photolytic degradation, rapidly forming a sulfate metabolite [24]. Under tropical conditions of high temperature and low humidity as found in Northern Senegal, Fipronil degrades very fast (half life between 3.9 to 5.4 days). Following the initial rapid degradation, the process seemed to stabilize with the parent compound and the metabolites disappearing very slowly. Under the conditions mentioned, fipronil undergoes chemical transformation in soils. It is assumed that in moist soils with high organic matter content it will degrade more rapidly. On the above evidence, it appears that factors such as: degradation, volatilization, leaching and run-off control the fate of pesticides in tropical soils and contribute to the off site transfer of the residues into water bodies, e.g. fresh water streams, that discharge into coastal waters. These residues are then eventually taken up by aquatic organisms.

3 Pesticides in Tropical Waters

3.1 Pesticides in Surface Waters and Sediments

Extensive monitoring of pesticides has been carried out in Guatemalan water systems [25, 26]. Monitoring included sampling sites in rivers, municipal water systems of major towns, lakes and lagoons. Samples were analyzed for 13 organophosphates, trifluraline, propanil, chlorthalonil, 20 conventional organochlorine pesticides and the pyrethroid, cypermethrin. The rivers were in irrigated agricultural productive areas, where the main crops are coffee, sugar cane, tobacco, water melons, tomatoes and large banana plantations. The results from the Los Esclavos watershed showed that cypermethrin (0.95 $\mu g \cdot kg^{-1}$) was present in the highest amount, other pesticides found included endosulfan (0.06 $\mu g \cdot kg^{-1}$), γ chlordane (0.17 $\mu g \cdot kg^{-1}$) as well as some organophosphate residues. In the Motagua watershed, comprising of about 20 rivers of significant size, residues of the organophosphates; diazinon (0.1 $mg \cdot l^{-1}$), azinphos-methyl (1.3 $\mu g \cdot kg^{-1}$), dysiston (1.0 $\mu g \cdot kg^{-1}$) and malathion (0.25-0.35 $\mu g \cdot kg^{-1}$) were found, in addition to traces of cypermethrin, propanil and endosulfan. The concentration of residues were high in the Michatoya watershed and quite consistent over all the sampling periods, especially at the Villalobos and the Michatoya rivers sites. This was attributed to the intensive and diverse farming along their course, as well as from industries in Guatemala City. This was especially noticeable in the Villalobos river, which receives up to 60% of basically untreated discharges from the city, including both home and industrial sewage. Lake Amatitlán gets approximately 60% of its water from underground sources, and consequently had lower residues of pesticides during the year. The low levels were attributed to a dilution factor by the lake. This lake has a high microbial activity as it is undergoing eutrification. Cyclic levels of cypermethrin were found and attributable to agricultural use as well as illegal fishing in the lake. In general, it appears that rivers of the southwest and northeast regions showed higher pesticide residues levels (0.5 to 4.5 $\mu g \cdot l^{-1}$) than those in the southeast region. Lagoons contained only low levels of residues, the small rivers that feed the lagoons being mainly activated in the rainy season, although they have no natural drainage and are located in areas of warm climate. The levels of micro-organisms in these lagoons is probable very high so microbial degradation of pesticides residues may be an effective dissipation mechanism resulting in the very low pesticide residues levels found, even though these lagoons are surrounded by cattle farms or coffee plantations.

In following project, the watersheds of Motagua , northeast region and Amatitlán (Michatoya), southwest region were studied [27]. The sampling sites were rivers, lakes, and drinking water sources. The Amatitlán watershed (southwest region) had higher levels of pesticide residues than the Motagua watershed (northeast region). In both watersheds the highest level of pesticide residues were found at the beginning of the rainy season (May and June). In some cases total pesticide residues levels reached up to 8 $\mu g \cdot l^{-1}$, the maximum level for organophosphates were 2.1 $\mu g \cdot l^{-1}$, and for cypermethrin (1.47 $\mu g \cdot l^{-1}$) and organochlorine pesticides, 0.56 $\mu g \cdot l^{-1}$. Diazinon was the most frequently found organophosphate with an average level of 0.1 $\mu g \cdot l^{-1}$ and at maximum concentration of 0.6 $\mu g \cdot l^{-1}$. Diclorvos was the next with a higher average level of 0.3 $\mu g \cdot l^{-1}$and a maximum level of 2 $\mu g \cdot l^{-1}$. Azinphos-methyl was the least frequently found pesticide at an average level of 1.5 $\mu g \cdot l^{-1}$. Endosulfan and endosulfan sulfate, were found in many sampling points although their concentrations were low, a few samples reached levels up to 0.2 $\mu g \cdot l^{-1}$ .In the southwest sampling sites, organochlorine residues ranged from 0.1 to 0.5 $\mu g \cdot l^{-1}$. Organophosphate residues were predominant in the superficial water in the first months of the rainy season but as the season progressed, the organochlorine residues predominated. From the pesticide levels found, it appeared that due to the pattern of residues in the Villalobos river-lake, the Lake Amatitlán is becoming a reservoir for all pesticides carried down by rivers comprising its watershed. This finding differs from those reported in the first monitoring program [25, 26]. The traditional organochlorine compounds were an important part of the pesticides found, both for the frequency and number detected. The low levels were attributed to the ban on traditional organochlorine usages in Guatemala in 1988. Since organochlorine pesticides are persistent they will remain as background contamination in future monitoring studies.

In Guanacaste, northwest Costa Rica, pesticide residues were found in a monitoring program [14]. The percentage of pesticides present in water samples were propanil (20%) at levels between <0.1 to 0.8 $\mu g \cdot l^{-1}$, chlorpyrifos (32%) between <0.1 to 0.2 $\mu g \cdot l^{-1}$, lindane (18%) between 0.02 to 0.04 $\mu g \cdot l^{-1}$ and DDT's (71%) at levels between < 0.01 to 0.50 $\mu g \cdot l^{-1}$.

Another extensive monitoring project was carried out in Nicaragua from 1996 through to 1997 [28]. It involved areas of Sebaco valley and the Viejo and Matagalpa rivers, the area from the mouth of the Atoya river and Corinto port to El Viejo and Chichigalpa, areas around Real estuary from Somotillo to Buenos Aires towards Mailpaisillo. Pesticide residues found in river water samples were generally below the limits of detection, but toxaphene, methoxychlor and hepthachlor epoxide residues were detected in the suspended matter from filtered river water at levels up to 3$\mu g \cdot l^{-1}$. In Río Viejo, the highest level of toxaphene in sediments was 0.3 $mg \cdot kg^{-1}$, in the Atoya and Sasama rivers sediments contained up to 7.5 $mg \cdot kg^{-1}$ of toxaphene while in shrimp lagoons, heptachlor epoxide levels were found up to 18.5 mg·kg^{-1}. Similar results, were reported in an earlier study [29] for the Atoya river watershed, with residues of DDT, DDE and toxaphene being most frequently found in water at levels up to 0.046, 0.024 and 1.47 μg·l^{-1}, respectively. Some organophosphates were also present; e.g. ethion (0.020-0.074 μg·kg^{-1}), methyl parathion (0.015-0.242 μg·kg^{-1}), and ethyl parathion (0.015-0.355 μg·l$^{-1)}$. In sediments, residues of DDE (1.3-450.2μg·kg^{-1}) DDD (2.4-288.2μg·kg^{-1}) and DDT (2.0-95.5 μg·kg^{-1}) were found together with toxaphene at levels of 91-1020 μg·kg^{-1}.

Pesticide residues were found in the Suerte river basin, which drains into the coastal wetlands conservation area of Tortuguero, Costa Rica [30]. Water and sediment samples were taken from the river, streams and banana packing plant effluents. The compounds most frequently detected in surface water were thiabendazole, propiconazole, imazalil, terbufos, cadusafos, and chlorpyrifos. In effluents from banana packing plants, and streams draining into the Suerte river, the concentration and frequency of residues found were general high. In Suerte river, thiabendazole was found in one sample but imazalil was not detected in any sample. In the river and conservation areas, propiconazole and cadusafos were the pesticides most frequently found. Heavy rainfall (annual average, 4000 mm) could contribute to the runoff into surface water of pesticides applied in banana plantations. The mean concentrations found were reported to exceed the aquatic quality criteria values, indicating a chronic risk to aquatic organisms. No organochlorine insecticides were detected in samples from the study. In sediments from the Suerte river, propiconazole was found in one sample at 17 μg·kg^{-1} (dry weight) and in the conservation area, in 3 out of 10 samples up to 19 μg·kg^{-1} level. Sediments from streams contained residues of imazalil (mean 446 μg·kg^{-1} dry weight), propiconazole (mean 33 μg·kg^{-1} dry weight), thiabendazole (mean 435 μg·kg^{-1} dry weight). Sediment samples from the packing plants contained the same compounds but at higher concentrations, especially for thiabendazole.

Pollution of coastal zones due to agrochemical residues has become a serious problem, especially in tropical regions where the rainfall is usually high. The distribution and fate of DDT in a microcosm simulating the marine environmental conditions of north Viet Nam, found that DDT was rapidly deposited on sediment, with the residues concentrated in the 0-5 cm layer below the sediment surface. Only DDT, DDE and DDD were found in sediment samples at low concentrations [31]. Another marine microcosm experiment simulating the conditions in the Bay of Bengal coast, Saint Martin Island, South of Bangladesh determined the distribution and fate of DDT in water, sediment, algae and mussels. It showed that the disappearance of DDT from water was associated with an increase in DDT residues in sediment and algae, and a substantial DDT volatilization was observed [32]. A study of organochlorine residues in sediment and biota from the coastal region south of Pinar del Río Province, Cuba, was conducted to assess the levels in the coastal area near rice fields [33]. Only DDT, DDE and DDD were found at low concentrations, which tended to decrease when the rainy season started. No residues of other pesticides, e.g. endosulfan or pyrethroids, used in the rice fields were found. It was thought that the residues accumulated in the sediments of drainage canals and rivers before reaching the coastal region. At the mouth of the Zhujiang river, China, HCB and DDT were found in seawater, sediment and marine organisms, the concentration was highest for DDT and was found to increase in the dry season rather than in the rainy season [34]. The residues of both pesticides in sea water near the Zhunjiang river were from sediment suspended in seawater and not from runoff.

In Mexico, the Chantuto-Panzacola lagoon system in South Chiapas is conformed by five bodies of water. La Encrucijada Biosphere reserve was monitored for pesticides residues [35]. The importance of the area is its high diversity of aquatic species, many of which are of economical importance. The area is susceptible to pesticide contamination since the watershed irrigates a large agricultural area with a heavy pesticide usage. Samples of sediments and aquatic organisms were collected on five occasions in 1997. The sediments contained residues of HCB isomers, heptachlor, heptachlor epoxide, aldrin, α-endosulfan and sulfate, DDT's, dieldrin and endrin, with heptachlor epoxide having the highest concentration of 515 $\mu g \cdot kg^{-1}$ (dry weight).

The coastal lagoons of the Pacific Coast of Nicaragua were monitored in 1995 and 1996 for pesticides residues in sediments and water [13]. One system monitored was the Real estuary in the northwest, where sesame, sugar cane, bananas are grown as well as shrimp farmed. The Padre Ramos estuarine system, in the south of Nicaragua is an agricultural area but far from the lagoons. The estuarine system, Maderas Negras-Naranjo-Paso Caballos, which is the most extensive system of coastal communicated lagoons surrounded by mangroves, receives the discharge of the Atoya, El Realejo, Posoltega and Amalia rivers. They irrigate the main cotton producing area of Nicaragua. In southwest Nicaragua, the San Juan river connects the Nicaragua Lake with the Pacific Ocean and ends in a small estuary near San Juan del Sur City. In this region, agriculture is dedicated to the production of maize, sugar cane and tropical fruits.

In the Naranjo estuary near river mouths, dichlorvos at levels up to 0.41 $\mu g \cdot l^{-1}$ and chlorpyrifos at 0.083$\mu g \cdot l^{-1}$were found. These levels were from samples collected during the rainy season and were probably the result of runoff from agricultural areas. In the lagoons and canals of the estuarine system, the concentration of residues were lower possibly the result of dilution and rapid degradation. In the same estuarine systems, lagoon sediments contained a variety of organochlorine residues including α- and δ-HCB, lindane, DDT's, heptachlor, aldrin, dieldrin, endosulfan, chlordane and toxaphene as well as PCBs [36, 37]. Residue concentrations in the Real estuary, were low, with DDT's being the most prominent with the maximum of 8.8 $\mu g \cdot kg^{-1}$ (dry weight), toxaphene was also present in levels up to 9.3 $\mu g \cdot kg^{-1}$. Toxaphene was also found in the Padre Ramos estuary, up to 4.3 $\mu g \cdot kg^{-1}$ (dry weight). In the San Juan river estuary, DDT's concentrations of 11 $\mu g \cdot kg^{-1}$ (dry weight) were found. Other organochlorine compounds were detected but at concentrations below 1 $\mu g \cdot kg^{-1}$. The low levels of toxaphene and the DDT's in these estuaries suggest either minimal use in adjacent agricultural areas or an efficient retention of residues in wetlands and mangroves surrounding the lagoons. Based on the high residue levels found in the Maderas-Negras-Naranjo-Paso Caballos lagoons, an area where pesticides are extensively used implies that the former suggestion was correct. The highest concentration of toxaphene was up to 1420 $\mu g \cdot kg^{-1}$ (dry weight) and for DDT's, it was up to 270 $\mu g \cdot kg^{-1}$ (dry weight), measured in samples from the mouth of the Atoya river. Since these two organochlorines are persistent, their levels should decline slowly over the years. The levels of these compounds reported in soils from around the system will continue to be introduced into the lagoons by surface runoff and river discharge [13]. Residues of α- and β-Endosulfan and the sulfate were found at

levels up to 1.23 μg·kg^{-1}(dry weight), the highest concentration being found in these estuarine systems. Endosulfan residues in soils from the surrounding area has also been reported [13]. It is believed that in this environment, residues of the endosulfan isomers would have a short life but residues of the sulfate metabolite may be more persistent. Levels of cis, trans chlordane and trans nonachlor found were up to 7.8 μg·kg^{-1} (dry weight), most likely due to its use as a termiticide. Other organochlorines compounds were found at concentrations below 1 μg·kg^{-1} (dry weight) [37].
Organochlorine residues were also found in the sediments from Gold Button and New River lagoons, north Belize, with levels of aldrin (up to 48 μg·kg^{-1}), heptachlor epoxide (up to 620 μg·kg^{-1}), endrin (up to 110 μg·kg^{-1}), α-,β-endosulfan (up to 110 μg·kg^{-1}), p,p'DDT (up to 120 μg·kg^{-1}), p,p'DDE (up to 100 μg·kg^{-1}), α-HCB (up to 1900 μg·kg^{-1}), β-HCB (up to 430 μg·kg^{-1}) and γHCB (up to 270 μg·kg^{-1}) [38]. A direct relation between the pesticides found in soils, waters and sediments can be established in some of the studies discussed, and residues were eventually taken up by aquatic organisms. [13,14,25, 28, 29 35, 36, 38].

3.2 Pesticides in Groundwater in the Tropics

Reports on pesticide residues in groundwater in the tropics are particularly scarce. Carbaryl residues (up to 2.4 μg·l^{-1}) and 2,4-D residues of 0.32 μg·l^{-1} were reported in groundwater from Heredia, Costa Rica [39]. In the Philippines, residues of endosulfan (0.03 μg·l^{-1}), monocrotophos (1.84 μg·l^{-1}) and chlorpyrifos (0.032 μg·l^{-1}) were found in ground water from Calauan and Calamba [21]. In Nicaragua, residues of endrin and toxaphene were detected in wells [29] and NRI-PROMAP-MARENA reported residues of toxaphene up to 0.6 μg·l^{-1}, in groundwater [28]. Cypermethrin residues were present in groundwater samples from Guatemala used for human consumption [27], in most cases it was responsible for exceeding levels of 0.5 μg·l^{-1}, although some organochlorines and organophosphates were also present.

4 Conclusions

Few studies have been carried out on the fate of pesticides in tropical ecosystems. If reported, they often lack an appropriate assessment of the processes that affect fate and factors influencing those processes under specific conditions. Due to the complexity of the environmental problems, modeling is seen as an essential tool in resolving them.
In many cases, studies in tropical matrices have not been carried out using Quality Assurance (QA) systems, leading to the question of confidence in the data produced. For this reason it is important that laboratories reporting pesticide residue data have an adequate QA system.
Scientists in tropical countries carrying out research on pesticide fate are encouraged to develop strategic alliances between countries such as the ones that exist in some Latin America countries. The exchange of information, instrumentation and training have been an important aspect of the development of better expertise and facilities in this region.

Many of the new pesticides are active at low application rates, resulting in the need for sophisticated instrumentation for the traditional residue analysis. Such instrumentation is lacking in developing countries thus restricting analysis to older pesticides. The analytical capability of laboratories in developing countries needs to be improved in order that both research and monitoring can carried out on the environmental and health risks associated with the newer pesticides in tropical regions.

5 References

[1] A. Van Wambeke, Soil of the Tropics: Properties and Appraisal, McGraw-Hill, Inc. New York, 1992, 343p.

[2] P. Sánchez, Suelos del Trópico: Características y Manejo, IICA, San José, Costa Rica, 1981, 634p.

[3] N. Balakrishman, R. Borgström, S.W. Bie, Tropical Ecosystems: A Synthesis of Tropical Ecology and Conservation, International Science Publisher, Lebanon, NH, USA, 1994, 441p.

[4] UICN-UNEP-WWF, World Conservation Strategy: Living Resources Conservation for Sustainable Development (Gland:IUCN), 1980.

[5] MINAE/PNUMA, Informe Nacional: Costa Rica, Proyecto Reducción del Escurrimiento de Plaguicidas al Mar Caribe, 2001, GEF /100-99-04-2101, p. 5.

[6] A. Ambrus, Main Provisions of the International Code of Conduct on the Distribution and Use of Pesticides for Registration Requirements: in *Proceedings of a Symposium Environmental Behavior of Crop Protection Chemicals*, IAEA- SM-343/ 10 Vienna, Austria, 1997, pp.11-34.

[7] J.R. Cox, Requirements for the Safe and Effective Management of Pesticides in Less-Developed Countries, in *Monograph N°61, Crop Protection in the Developing World*, R. Black, A. Sweetmore (Eds.), BCPC, Surrey, UK, 1994, p. 21.

[8] K.S. Kookana, S.Baskaran, R. Naidu, *Aust. J. Soil Res.*, 1998, 36, p.716.

[9] M.R. Blumhorst, *Weed Tech.* 1996, 10:156.

[10]. I. Scheunert, R. Schroll, G. Cao, A Microcosm System to Evaluate the Fate of ^{14}C-labelled Pesticides in Soils from Different Climate Zones, in *Proceedings of a Symposium Environmental Behavior of Crop Protection Chemicals,* IAEA-SM-343/ 21, Vienna, Austria, 1997, pp. 181-186.

[11] K.D. Racke, M. W. Skidmore, D.J. Hamilton, J.B. Unsworth, J. Miyamoto, S.Z. Cohen, *Pure & Appl. Chem.*, 1997, 69(6), 1349-1371.

[12] I.M, Beck, Factores que Influyen en el Uso de Plaguicidas en Nicaragua, in *Memorias, Congreso Nacional: Impacto de Plaguicidas en Ambiente, Salud, Trabajo y Agricultura,* OPS-OMS, Managua, Nicaragua, 1997, pp. 20-30.

[13] F.P. Carvalho, S. Montenegro, I. Tolosa, J.P. Villenueve, C. Cattini, J. Bartocci, M. Lacayo, A. Cruz, Destino y Efectos de los Plaguicidas sobre el Ambiente Estuarino en Nicaragua, in *Memorias, Congreso Nacional: Impacto de Plaguicidas en Ambiente, Salud, Trabajo y Agricultura.* Managua, Nicaragua, 1997, pp. 429-440.

[14] O.M. Rodríguez, Informe Final, Proyecto SENARA-BID-UCR, Residuos de Plaguicidas en la Zona de Influencia del Proyecto de Riego Arenal-Tempisque, Centro de Investigación en Contaminación Ambiental, Universidad Costa Rica, 1997, 96p.

[15] A.C. Alvarenga, A.R. Cárdenas, Determinación de Residuos de Paraquat en Suelos Agrícolas de la Zona Occidental (Región I) de El Salvador, in *Recopilación de las Investigaciones de Plaguicidas Realizadas en El Salvador,* G.R. Calderón, L. Meléndez (Eds.),PLAGSALUD,2001,p.64.

[16] L.C. Luchini, T.B. Peres, M.M. de Andréa, *J. Environ. Sci. Health*. 2000, B 35 (1), 51-59.

[17] J. Espinoza, Fate and Effects of Pesticides under Tropical Field Conditions: Implications for and Research Needs in a Developing Country, in: *Proceedings of a Symposium Environmental Behavior of Crop Protection Chemicals,* IAEA-SM-343/32, Vienna, Austria, 1997, pp 93-110.

[18] J.E. González, E.B. de Saiz, L. González, O. Pérez, J. Ceballos, Dissipation, Degradation and Uptake of ^{14}C-Carbofuran in a Panamanian Alfisol Soil, in *IAEA-TECDOC-1116, Use of Nuclear and Related Techniques in Studies of Agroecological Effects Resulting from Use of Persistent Pesticides in Central America*, Final Report FAO/IAEA, 1998, pp. 71-79.

[19] J.E. González, J. Ceballos, A. Amor, E.B. de Saiz, 1999, A Study of Dissipation, Degradation and Binding of ^{14}C-Labeled Endosulfan to Soil in Model Lysimeter, in: *IAEA-TECDOC-1116, Use of Nuclear and Related Ttechniques in Studies of Agroecological Effects Resulting from Use of Persistent Pesticides in Central America*, FinalReport FAO/IAEA, 1998, pp.87-98.

[20] D.E. Robinson, A. Mansingh, T.P. Dasgupta, Fate of Endosulfan in Soil and in River and Coastal Waters of Jamaica, in *Proceedings of a Symposium Environmental Behavior of Crop Protection Chemicals,* IAEA-SM-343/28, Vienna, Austria, 1997, pp. 301-311.

[21] A.W. Tejada, L.M. Varca, S.M.F. Calumpang, C.M. Bajet, M.J.B. Medina, Fate of Pesticides in a Model Rice Paddy Ecosystem, in *Proceedings of a Symposium Environmental Behavior of Crop Protection Chemicals,* IAEA-SM-343/33, Viena, Austria, 1997, pp. 265-278.

[22] D. Sanyal, N.T. Yaduraju, G. Kulshrestha, J. *Environ. Sci. Health*, 2000, B35 (5), 571-583.

[23] C.S. Helling, Environmental Fate of Herbicides in Hawaii, Peru and Panama, in *Proceedings of a Symposium Environmental Behavior of Crop Protection Chemicals,* IAEA-SM-343/7, 1997, Vienna, Austria, pp. 389-406.

[24] H. Fenet, E. Beltran, B. Gadji, J.F. Cooper and C.M. Coste, *J. Agric. Food Chem.*, 2001, 49, 1293-1297.

[25] W. Knedel, J.C. Chiquín, J. Pérez, .S. Rosales, Determination of Pesticides in Surface and Ground Water Used for Human Consumption in Guatemala, in *IAEA. TECDOC-1116, Use of Nuclear and Related Techniques in Studies of Agroecological Effects Resulting from Use of Persistent Pesticides in Central America*, Final Report, 1998, pp.41-49.

[26] W. Knedel, J.C. Chiquín, J. Pérez, .S. Rosales, Determination of Organochlorine, Organophosphorous and Pyrethroid Pesticides in Surface and Ground Water used for Human Consumption in Guatemala, *FAO/IAEA/SIDA Coordinated Research Programme: "Agroecological Effects Resulting from the Use of Persistent Pesticides in Central America"*, 1998, 47p.

[27] W. Knedel, J.C. Chiquín, J. Pérez, S. Rosales, Estudio de los Niveles de Residuos de Plaguicidas en las Cuencas de Amatitlán y del Motagua, Proyecto Conjunto de Investigación entre Universidad del Valle de Guatemala y el Consejo Nacional de Ciencia y Tecnología de Guatemala, 1999, 95p.

[28] NRI Inglaterra-PROMAP-MARENA, Avances del Estudio de Contaminación por Plaguicidas de Cuencas Hidrográficas, in: *Memorias. Congreso Nacional: Impacto de Plaguicidas en Ambiente, Salud, Trabajo y Agricultura*, Managua, Nicaragua, 1997, 102-105pp.

[29] J.A. Alvarez, Niveles de Contaminación de las Aguas de la Cuenca del Río Atoya por Residuos de Plaguicidas Organoclorados y Organofosforados Aplicados en la Cultivación del Algodón, in: *Memorias. Congreso Nacional: Impacto de Plaguicidas en el Ambiente, Salud, Trabajo y Agricultura.* Managua, OPS-OMS, Nicaragua 27-31 de octubre, 1997, pp. 111-118.

[30] L.E. Castillo, C. Ruepert, E. Solís, *Environ. Toxicol. Chem.*, 2000, 19(8), 1942-1950.

[31] D.N. Dang, V.T. Vo, M.A. Nguyen, Distribution and Fate of ^{14}C-DDT in the Esturine Environment of the North of Viet Nam, in *Proceedings of a Symposium Environmental Behavior of Crop Protection Chemicals,* IAEA-SM-343/40, Vienna, Austria, 1997, pp. 313-319.

[32] M.A. Matin, E. Hoque, J. Khatoon, Y.S.A. Khan, M.M. Hossain, A.J. Mian. Distribution and Fate of ^{14}C-DDT in Microcosm experiments simulating the Tropical Marine Environment of the Bay of Bengal, in *Proceedings of a Symposium Environmental Behavior of Crop Protection Chemicals,* IAEA-SM-343/15, Vienna, Austria, 1997, pp. 279-287.

[33] G. Dierksmeier, R. Hernández, P. Moreno, K. Martínez, C. Ricardo, Organochlorine Pesticides in Sediment and Biota in the Coastal Region to the South of the Pinar del Río Province, Cuba, in *Proceedings of a Symposium Environmental Behavior of Crop Protection Chemicals,* IAEA-SM-343/42,Vienna, Austria, 1997, pp. 343-358.

[34] F. Cai, Z. Lin, Y. Chen, S. Chen, J. Yiang, F. Cai, L. Qian, Behaviour of BHC and DDT at the mouth of the Zhujiang River, China, in *Proceedings of a Symposium Environmental Behavior of Crop Protection Chemicals,* IAEA-SM-343/43, 1997, pp. 349-358.

[35] T. Alvarez Legorreta, Plaguicidas Organoclorados en el Sistema Lagunar Chantuto-Panzacola, Costa de Chiapas, México, in: *Resúmenes de la Primera Conferencia Latinoamericana sobre Aguas y Suelos Contaminados*, Quito, Ecuador, 1998, p. 59.

[36] S. Montenegro, M. Lacayo, F. Picado, A. López, Organochlorine Pesticidas in Sediment and Biological samples from the Coastal Lagoons of Nicaragua, in *IAEA-TECDOC-1116 Use of Nuclear and Related Techniques in Studies of Agroecological Effects Resulting from use of Persistent Pesticides in Central America*, Final Report FAO/IAEA, 1998, pp. 29-40.

[37] F.P. Carvalho, S. Montenegro-Guillén, J.P. Villeneuve, C. Cattini, J. Bartocci, M. Lacayo, A. Cruz, *Environ. Contam. & Toxicol.*, 1999, 36, 132-139.

[38] T.H. Wu, T.R. Rainwater, S.G. Platt, S.T. McMurry, T.A. Anderson, *Chemosphere*,2000,(40)671-678.

[39] A.G. Pérez, J. Reynolds. Incidencia de plaguicidas en las Aguas Subterráneas de la Cuenca del Río Virilla, in Preparation to be Submitted for Publication, 2002.

Residues and Consumer Safety

26 European Food Agency and Consumer Risk Assessment in the EU

Michael B Walsh

European Commission, Health & Consumer Protection Directorate General, Office: Genève 1.346, B-1049 Brussels, Belgium.

1 Introduction

The European Food Safety Authority was first mooted in the Commission's White Paper of January 2000 [1] which outlined the strategic objectives, priorities and work programme in relation to food safety. It also provided for the establishment of a European Food Safety Authority.

In the period prior to launch of the White Paper, the European Community had experienced consumer disquiet due to a number of food scares. These included contamination of animal feed, bovine spongiform encephalopathy (BSE), concern about the use of antimicrobial feed additive used to increase productivity in animal production, and genetically modified food. The White Paper identified the need to reestablish consumer confidence in the European food supply and to reassure trading partners of the Communities' ability to regulate and control the safety of its food supply.

The development of European food legislation over a period of forty years has been influenced by a number of different policy objectives, such as the Common Agricultural Policy and the creation of the Single Market. These different policy objectives has led to some divergence in the approach to food legislation and resulted in certain gaps and inconsistencies. It was also recognised that food legislation in the Member States varied in approach and detail which may cause disruption to the internal market. It was clear that it was necessary to further harmonise, at Community level, existing national requirements by placing them in a European context.

Having identified the problems and the work which need to be accomplished, it was clear that the present systems needed to be fundamentally overhauled. Given the increasing level of technological sophistication applied in the food sector, the need for rapid, reliable scientific advice was identified. The present systems lacked capacity and it was necessary to address this need by better using the considerable existing resources available in the Community. The White Paper proposed the creation of the European Food Safety Authority to address the scientific issues relevant to food safety and to communicate openly on such matters. Following a period of public consultation, it was agreed that this approach was appropriate as the most effective mechanism to address the growing need for scientific advice and to reestablish public confidence. This lead to the Commission's proposal [2] for a regulation of the European Parliament and of the Council to establish the European Food Safety Authority in November 2000.

2 Regulation on the European Food Safety Authority

The European Food Safety Authority was established on 28 January 2002 by Regulation (EC) No 178/2002 of the European Parliament and of the Council [3].
In addition to establishing the European Food Safety Authority, the regulation establishes fundamental principles, definitions and requirements of food law. These include recognition of the importance of function of the Community internal market, the precautionary principle, principles of transparency, and the role of international standards. Importantly from the point of view of the Food Safety Authority it establishes that food law must be based on the principles of risk analysis. It also establishes that in addition to direct factors influencing food safety, indirect factors also need to be taken into account when carrying out risk assessment on food safety. This would, for example, include materials in contact with food, animal feed and other agricultural inputs at the level of primary production, such as plant protection products. These principles will guide and influence the Food Safety Authority as it organises itself and carries out its scientific assessments.

2.1 Mission and Tasks of the Authority

The primary responsibility of the Authority will be to provide independent scientific advice on all matters with a direct or indirect impact on food safety. Previous experience with BSE demonstrates forcibly the dangers of not taking indirect effects adequately into account. BSE transpired not to be just an animal disease but also to have significant implications for human health and the environment. Similarly, indirect effects are also important in the assessment of plant protection products where in addition to the necessity for consumer risk assessment, attention also must be focused on worker safety and protection of the environment. Similar considerations will need to be taken into account in other areas of remit of the Authority, which would include for example, genetically modified organisms and feed additives. As an exception to the general rule, the Authority will provide scientific advice on non-food/feed genetically modified organisms as well as those used both in food/feed. The intention is to ensure the application of a coherent Community approach to the risk assessment of these products by adhering to the one-door one-key principle.

The Authority will provide scientific advice which will serve as the scientific basis for the preparation of Community legislation in the field falling under its mandate. In this respect there is a clear separation of the responsibility of the risk assessment function and that of risk management which will remain the responsibility of the Community Institutions, i.e. the European Commission and Parliament and the Council. The Authority will become the point of reference with respect to risk assessment. Its precise role will evolve over time but its role with respect to the authorisation process had become clearer following the publication of a number of recent Commission proposals. In such a proposal on smoke flavourings used or intended for use in or on foods [4], the Commission proposed that dossiers should be submitted directly to the Authority, and within a predetermined period of time the opinion of the Authority should be returned to the Commission which will remain responsible for the development of proposals on food law. The Commission following consultation of Member States in the newly created Standing Committee for the Food Chain and Animal Health will ultimately take decisions.

The Authority will also have an important role in the collection and analysis of food safety data and this will result for the first time in the existence of a centralised mechanism for the carrying out of this task within the Community. This would include, for example, the collection and analysis of pesticide residue dietary exposure data. This role will contribute to the Authorities capacity to contribute to the evaluation of dossiers submitted for Community approval. Similarly, the "intelligence gather" role which will extend beyond the Community will improve the capacity to ensure that the early warning signs of emerging risks are identified and acted on appropriately when they are more readily manageable. Normally such problems develop over a considerable period of time but are frequently not identified and acted on before they become major problems. The Authority will have both the remit and means in hand to advise risk managers of the early indications of such emerging risks. The resources available to the Authority will include its "in house staff", its scientific committee and panels and the expertise available to it through its network of expertise. In addition, the Authority will have the financial resources to commission studies in order to provide the necessary information to ensure that risk managers can be advised of the appropriate risk reduction measures which could be applied. In the event of a food safety crisis the Authority will provide scientific support to the Commission. Communication will also be an important task for the Authority. In addition to carrying our its duties in a transparent manner, the Authority will not only need to communicate with the Community Institutions but it will need to ensure that complex scientific advice is also communicated clearly and accurately to the public.

2.2 Bodies of the Authority

The Authority shall comprise: (1) the Management Board, (2) the Executive Director, (3) the Advisory Forum, (4) the Scientific Committee and Scientific Panels.

2.2.1 Management Board

The Management Board will have 15 members, four of whom will have a background in organisations representing consumers and other food chain interests, and a representative from the Commission. Members will be from different profession backgrounds and with a broad range of expertise and be geographically representative of the Community. The European Council following consultation of the European Parliament will appoint it: the Authority's first board on 17 July 2002. As the Authority is an independent agency with its own legal identity, it is not responsible to a Community Institution or to any other external organisation. It is the Management Board whose responsibility it is to ensure that the Authority carries out its mandate in accordance with the Regulation, i.e. that it functions efficiently with a culture of independence and transparency. The Management board is also responsible for the work programme, internal rules and the financial rules of the Authority and for the appointment of the Executive Director.

2.2.2 Executive Director

The Executive Director is to be appointed by the Management Board following a open selection process. He is responsible for the day-to-day administration of the Authority and for its staff. His other main task will include preparation of the draft work programme and its subsequent implementation, for organisation of the selection process or the Scientific Committee and the Panels.

2.2.3 Advisory Forum

The Advisory Forum will be comprised of one member from the national body with the same remit as the Authority. It will advise the Executive Directive on the Authority's work programme and work prioritisation. The Forum will allow input from the Member States and will accordingly contribute to the avoidance of duplication between the work of the Authority and that of the Member States. It will also promote networking between organisations in Member States.

2.2.4 Scientific Committee and Scientific Panels

Scientific advice is a core activity and central to this function is the Authority's Scientific Committee and the eight Scientific Panels. These will replace the six existing scientific committees which currently advise the Commission in relation to the safety of the food chain: the Scientific Steering Committee, which advises on BSE, TSEs (transmissible spongiform encephalopathies) and multi-disciplinary questions, the Scientific Committees on food, animal nutrition, veterinary measures relating to public health, plants, animal health and welfare. These existing scientific committees are composed of independent scientists appointed on the basis of an open call for interests as provided by in the two decisions [5, 6] of the Commission setting up the Committees. The Authority's Scientific Committee and Panels will take over the tasks of these scientific committees as soon as their members are nominated. The members of the Scientific Committee and Panels will continue to be independent scientists selected through an open process based on a call for expression of interests. The Authority's permanent Scientific Committee and Panels shall be responsible for providing the scientific opinions of the Authority, each within their own spheres of competence. The Scientific Committee will be responsible for general coordination and it will provide opinions on multidisciplinary issues falling within the competence of more than one Scientific Panel, and on any issue which may fall outside of the immediate remit of the Panels. It will be composed of the chairpersons of the Scientific Panels and six independent scientific experts who do not belong to any of the Scientific Panels. The Authority will establish the following eight Scientific Panels:

- Panel on food additives, flavorings, processing aids and materials in contact with food
- Panel on additives and products or substances used in animal feed
- Panel on plant protection products, their residues and plant health
- Panel on genetically modified organisms
- Panel on dietetic products, nutrition and allergies
- Panel on biological hazards
- Panel on contaminants in the food chain
- Panel on animal health and welfare

3 Conclusions

Following the creation of the European Food Safety Authority in January 2002 the Community had been given a significant additional resource in its activity on food safety. Due to its wide mandate to cover the entire food chain and all matters having a direct and indirect effect on the safety and health of consumers arising from the consumption of food, the possibility that gaps could arise will be minimized. Given the Authority's own resources and those that it will be able to mobilize through it networking role with Member States, it should be able to respond quickly to requests for scientific advice. This increased speed and thoroughness in the risk assessment exercise should be quickly reflected in the regulatory process. It should result in speeding up the authorization process and the timely identification of emerging problems. This, in association with the existence of an independent and transparent body with a capability to communicate with the public, should contribute to improved consumer confidence.

4 Summary

The European Food Safety Authority has been established to provide independent scientific advice and support for the Community legislation and policies that may have a direct or indirect effect on food and feed safety. It will provide independent information on all matters within its remit and communicate with the public and other interested parties. In order to discharge it responsibilities it will have its own staff resource, access to external independent scientific experts and will set-up networks to closely cooperate with similar bodies in Member States. In addition to its remit on food safety it will also provide scientific advice on non-food genetically modified organisms and nutrition in so far as Community legislation is concerned.

5 References

[1] Commission of the European Communities, White Paper on Food Safety, Brussels, 12 January 2000, COM (1999) 719 final. http://europa.eu.int/comm/dgs/health_consumer/library/pub/pub06_en.pdf.

[2] Commission of the European Communities, Proposal for a Regulation of the European Parliament and of the Council laying down the general principles and requirements of food law, establishing the European Food Authority, and laying down procedures in matters of food. COM (2000) 716 final. OJ C 96 E, 27.3.2001, p 247. http://europa.eu.int/comm/food/fs/efa/efa_prop_en.pdf.

[3] European Parliament and of the Council, Regulation (EC) No 178/2002 of the of 28 January 2002 laying down the principles and requirements of food law, establishing the European Food Safety Authority and laying down procedures in matters of food safety. OJ L31, 1.2.2002, p.1. http://europa.eu.int/eur-lex/pri/en/oj/dat/2002/l_031/l_03120020201en00010024.pdf.

[4] Commission of the European Communities, Proposal for a European Parliament and Council Regulation on smoke flavourings used or intended for use in or on foods, Brussels, 15 July 2000, COM (2002) 400 final. http://europa.eu.int/comm/food/fs/sfp/addit_flavor/flavourings/proposal_en.pdf.

[5] Commission of the European Communities, Decision of 10 June 1997, setting up a Scientific Steering Committee, Commission Decision No 97/404/EC of June 1997, OJ L169, 27.06.1997, p.86. http://europa.eu.int/comm/food/fs/sc/ojl169_en.html.

[6] Commission of the European Communities, Decision of 23 July 1997, setting up Scientific Committees, Commission Decision No 97/579/EC of 23 July 1997, OJ L 237, 28.08.1997, p.18. http://europa.eu.int/comm/food/fs/sc/ojl237_en.html.

27 A Comparison of Dietary Exposure and Risk Assessment Methods in US and EU

James R. Tomerlin

Exponent, 1730 Rhode Island Avenue N.W., Washington, DC 20036 USA, E-mail: btomerlin@exponent.com.

1 Introduction

Most countries around the world have passed legislation requiring that contaminant levels in the food supply, including pesticide residues, are at levels low enough to ensure safety. In the United States (US), the Federal Fungicide, Insecticide, and Rodenticide Act (FIFRA; [1]) and the Federal Food, Drug and Cosmetic Act (FFDCA; [2]) are the two major pieces of legislation that mandate this protection. These two acts were modified, and to some extent reconciled, by the Food Quality Protection Act (FQPA; [3]) in August of 1996. In the European Union (EU), the safety of the food supply with respect to plant protection products is mandated in Directive 91/414/EC [4] and is implemented within the various member states.

The purpose of this paper is to describe the policies and procedures of dietary exposure analysis and risk assessment in these two portions of the world, to highlight their differences, and to demonstrate how they both aim for the same goal, a safe food supply. Unless otherwise noted, all references to exposure in this paper pertain to the intake of pesticide residues in food. Nomenclature in other parts of the world refers to dietary intake, but the general concept is the same, regardless of the terminology. Likewise, all references to risk and risk assessment pertain to the risk assessment with regard to dietary exposure to pesticide residues.

Two types of dietary exposure are understood. Chronic exposure represents exposure to pesticide residues in food over a relatively extended time period, frequently assumed to be the course of a lifetime. In contrast, acute exposure represents exposure to pesticide residues in food over a relatively short time period, usually over the course of one day. Most of the significant differences between the EU and US approaches to dietary exposure analysis and risk assessment focus on acute dietary exposure. Some of the differences have their sources in the different types of data available in the two areas. Other differences have their sources in different regulatory policies or societal viewpoints.

2 Basic Dietary Model

2.1 The Exposure Algorithm

Regardless of whether chronic or acute dietary exposure is being estimated, and regardless of whether the model used is deterministic or probabilistic, dietary exposure is a simple function of the amount of food consumed and the residue concentration on the food:

$$\text{Dietary Exposure} = \text{Amount of Food Consumed} \times \text{Residue Concentration} \quad (1)$$

In a deterministic assessment, point estimates of the parameters are used, whereas in distributional analyses, distributions represent the amount of food consumed, and distributions of residue values may also be used. A significant advantage of the deterministic analysis is that few resources are required. A deterministic assessment typically is conducted using summary data and may be calculated relatively quickly. In contrast, a distributional analysis, in particular a probabilistic or Monte Carlo analysis requires relatively large datasets, experienced exposure assessors, and sophisticated computational algorithms. Additionally, science policies often are necessary to guide the use of data for probabilistic exposure analyses because the way in which data are used can have a significant impact upon the outcome of the model. Distributional analyses may be conducted based upon variation in either the food consumption data or the residue data. An assessment is not probabilistic unless model input variables are sampled in proportion to their probability of occurrence in nature. It is important to recognize that non-probabilistic distributional analyses can be used to estimate acute dietary exposure [5].

2.2 Risk Assessment Includes Toxicology

The equation presented above shows the basic model for dietary exposure. The dietary risk assessment, however, only occurs when the exposure estimated with the model is evaluated relative to an appropriate measure of toxicity. In the US, chronic exposures are evaluated relative to a chronic reference dose (cRfD). The US cRfD is analogous to the Acceptable Daily Intake used in the EU. In both cases, the toxicity measure is based upon a No Observable Effect Level (NOEL), adjusted as deemed appropriate by safety (or uncertainty) factors to account for variation between and within species. Additional factors may be applied as judged necessary by appropriate regulatory authorities. The NOEL is determined in long-term toxicity studies using experimental animals. In both the US and the EU, batteries of studies are required as part of the pesticide registration process. The cRfD/ADI is set using the lowest NOEL from the entire set of long-term toxicity studies.

In the US, quantitative risk assessments are conducted for carcinogenicity. The risk assessment is performed by multiplying the chronic dietary exposure estimate by the $q_1{}^*$, a coefficient that represents the carcinogenic potential of the chemical. The compound is judged acceptable if the result of the cancer risk assessment indicates that there is a probability of less than $1x10^{-6}$ that the compound produces cancer. This quantitative approach is not practiced in the EU.

For assessing acute dietary risk, an acute aRfD/ADI is determined from short-term toxicity studies. As with the chronic cRfD/ADI, the aRfD/ADI is established using the lowest NOEL from a range of short-term toxicity studies.

As the focus of this paper is dietary exposure, it will not discuss the toxicological aspects of dietary risk assessment any further. The significant point is that in both the US and the EU, chronic exposures have been evaluated for many years. In the US, acute dietary exposures have been evaluated for approximately 15 years, whereas acute dietary exposures have been considered in the EU for perhaps five years.

2.3 Point Estimates of Chronic Dietary Exposure

Deterministic models are used to estimate chronic, or long-term, dietary exposures. Although research is currently being conducted on the use of distributional techniques for chronic dietary risk assessment, the author is not aware of any working model at this time.

When calculating chronic dietary exposure, the deterministic models use point values for both food consumption and residue concentration, thereby yielding a point estimate of dietary exposure. In the US, the initial chronic dietary exposure estimate is the Theoretical Maximum Residue Contribution (TMRC) and is analogous to the Theoretical Maximum Daily Intake (TMDI) used to estimate chronic dietary exposure in the EU. Both the TMRC and the TMDI are relatively conservative estimates of dietary exposure. The TMRC is calculated as the product of the mean consumption value and the US pesticide tolerance [6]. In the EU, the TMDI is calculated as the product of the mean consumption value and the Maximum Residue Limit (MRL) [7]. The objective of both calculations is essentially identical: to calculate an estimate of the central tendency of the dietary exposure. Both calculated values use the central tendency dietary exposure estimate as the estimate of chronic (long-term) dietary exposure and calculate it using mean consumption data and the maximum residue permitted on the commodity.

Mathematically, the TMDI is calculated according to the following equation:

$$\mathrm{TMDI} = \frac{\sum(\mathrm{Consumption\ (kg/day)xMRL(mg/kg)}}{\mathrm{BodyWeight(kg)}} \qquad (2)$$

Note that the TMDI represents the total dietary exposure from all registered uses of the pesticide.

The methodology is different in the UK, where the TMDI is calculated as an estimate of high-end dietary intake, using the 97.5^{th} percentile consumption data for the two most highly consumed commodities and mean consumption estimates for the remaining commodities [8], as shown in the following equation:

$$\text{UK-TMDI} = \Sigma \left[\begin{array}{l}\text{two highest } 97.5^{th} \\ \text{percentile intakes}\end{array} + \begin{array}{l}\text{mean population} \\ \text{intakes from all} \\ \text{other foods}\end{array} \right] \quad (3)$$

The TMDI calculation is typically expressed as mg/kg bw/day, using standard body weights for the conversion.
The equation for the US-based TMRC is similar to that for the TMDI. However, exposure is expressed as mg/kg bw/day using body weight information included in the US food consumption database.

$$\text{TMRC} = \Sigma(\text{Consumption (kg/kg bw/day) x Tolerance (mg/kg)}) \quad (4)$$

The only real difference between the UK method of calculating the TMDI and the method used elsewhere in the EU is in the way the consumption data are used. The UK has set as their policy the calculation of a high-end chronic exposure value.

2.4 Point Estimates and Distributional Analyses of Acute Dietary Exposure

As discussed in the preceding paragraphs, a chronic dietary exposure analysis considers the potential impact of long-term exposures on health. In contrast, the acute dietary exposure analysis evaluates the potential health impact from current intakes, the foods that are being consumed "now." The question addressed by the acute dietary exposure analysis is "What is the likelihood that residues on the food consumed right now might have some effect upon health?"
In the US, acute dietary exposures have been assessed using distributional analyses since the mid-1980's. In the mid-1990's, probabilistic techniques started being used in the US to evaluate potential acute dietary risk [5]. The issue of acute dietary exposure and risk has arisen in the UK and elsewhere in the EU. Beginning with the "Geneva Consultation" a joint FAO/WHO expert consultation on food consumption and chemicals risk assessment held in 1997 [7], several workshops and conferences have been held to investigate both the scientific rationale of acute dietary exposure assessment and the correct methodology to use in the EU. The 1998 "York Conference" was another important meeting in the development of thinking within the EU regarding acute dietary exposure and how to estimate it [9]. The "York Conference" brought together scientists from the US and Europe to exchange ideas and discuss how the data that are available within Europe might be used for acute dietary risk assessments.

2.4.1 National Estimates of Short-Term Intakes – The NESTI Approach

At this time, acute dietary exposure in the EU is typically estimated by calculating National Estimates of Short Term Intakes (NESTIs) at the national level or International Estimates of Short Term Intakes (IESTIs) at the international level. Practically, however, the only readily available data for calculating estimates of short-term exposures are those from the UK. In practice, NESTIs calculated with the UK food consumption data are used to represent potential acute dietary exposures across the EU. As will be shown in the following paragraphs, the NESTI calculation provides a point estimate of acute dietary exposure.

Much of the development of the NESTI model has been conducted in the UK; most likely because the UK consumption data are suitable for this type of calculation [8]. Although the NESTI is a point estimate, the NESTI value may be calculated in three distinct ways, depending upon the characteristics of the food commodity being considered.

2.4.2 NESTI Case 1

Case 1: The available composite residue data reflect the residue levels in the food commodity as consumed. This is the case where either the commodity is well mixed during processing e.g. cereals, or where the normal portion size reflects the consumption of many units of the commodity e.g. cherries. Essentially, Case 1 is used when a "serving" of a food commodity is comprised of numerous units.

In this case the NESTI is calculated using the following equation.

$$\text{NESTI} = \frac{\text{F x (HR-P)}}{\text{mean bodyweight}} \qquad (5)$$

where F is the full portion consumption data for the commodity unit, and HR-P is the highest residue level detected incorporating processing or edible portion factors.

The residue value, HR, used to calculate the NESTI comes from residue field trials conducted according to Good Agricultural Practice (GAP) for the pesticide:commodity combination. Note that the NESTI calculation is strongly biased towards calculating a <u>high</u> exposure value. In addition to using the highest residue value, the consumption value used in the calculation is for a "large portion", usually a 97.5^{th} percentile consumption value. The rationale for this bias is that the acute dietary exposure estimate must account for the potential of high residues being consumed by individuals who consume large amounts of the commodity.

2.4.3 NESTI Case 2

Case 2: The residues in the food as consumed do not correspond to the available composite residue data. For example, a Case 2 commodity typically is consumed in discrete units, usually with no more than four units per eating occasion. Commodities of this type would be apples, bananas, oranges, and so on. Case 2 also applies for large commodities (such as melons) for which a significant portion of an individual commodity could be eaten during a single sitting. Finally, Case 2 also applies if the commodity is processed before consumption and the process leaves the commodity in the form of individual consumable units (e.g. baked potatoes).

There may be significant unit-to-unit variation in residues between units of Case 2 commodities, which has lead to a concern that using the composite residue value does not adequately represent the potential for high residues on some units. The NESTI calculation accounts for this variation through the use of default variability factors, *v*. If experimental data for individual units are available, it may be useful to calculate specific variability factors. These values are introduced into the equation shown below which represents the worst case.

$$\text{NESTI} = \frac{\{U \times \text{HR-P} \times v\} + \{(F-U) \times \text{HR-P}\}}{\text{mean bodyweight}} \qquad (6)$$

Where

U is the weight of the first commodity unit or if the full portion consumption data is less than one commodity unit then U is equal to the full portion consumption data and the second term of the equation drops out.

F is the full portion consumption value. When the amount of the full portion consumption is less than or equal to a single commodity unit then the second term of the equation drops out

v is the variability factor. Default values of *v* are:

- 5 for large commodities (unit weight over 250 g)
- 7[1] for medium commodities (unit weight 25 to 250 g)
- *v* does not apply for small commodities, such as berries, with unit weights less than 25 g

HR-P[2] is the highest residue level detected incorporating factors for processing or the amount of the edible portion.

2.4.4 NESTI Case 3

Case 3: Residue data are available for individual commodity units that reflect residue levels in the food commodity as consumed. The basic procedure is the same as for Case 2 except that the variability factors are not necessary.

---

[1] Originally, the *v* factor for medium commodities was 10. However, the residue workgroup of the “York Conference” recommended that the *v* factor be reduced to 7 for applications other than granular soil formulations.

[2] The 2000 Report of the Report of the Joint Meeting on Pesticide Residues recommended that the residue portion for the second and subsequent units be changed from the STMR to the HR.

The NESTI is calculated using the following equation:

$$\text{NESTI} = \frac{\{U \times HR\text{-}P_{ind}\} + \{(F\text{-}U) \times HR\}}{\text{Mean bodyweight}} \quad (7)$$

Where:

U is the weight of the first commodity unit or if the full portion consumption data is less than one commodity unit then U is equal to the full portion consumption data and the second term of the equation drops out.

F is the full portion consumption value. When the amount of the full portion consumption is less than or equal to a single commodity unit then the second term of the equation drops out

$HR\text{-}P_{ind}$ is the highest residue detected in individual commodity units incorporating processing or edible portion factors

As described by the UK Pesticide Safety Directorate, the NESTI method basically is a point estimate [8]. The NESTI methodology calculates the dietary exposure using high-end consumption and the highest residue value, adjusted by a variability factor, to account for potential variability between individual commodity units. Nonetheless, the calculated NESTI statistic is a point estimate.

2.5 The Distributional Approach

As implied by the name, distributional risk assessments use the entire data distribution to calculate exposure and risk. As described for the EPA Tier 1 assessment [5], a distributional analysis is not necessarily a probabilistic analysis. The EPA Tier 1 acute dietary assessment produces a dietary exposure distribution using the entire food consumption distribution and point estimates of residue concentration. There is no probability sampling in the Tier 1 assessment.

To some extent, the term "probabilistic risk assessment" is a misnomer. Even the most sophisticated probabilistic assessment uses a wide spectrum of data, from point estimates to full data distributions. Typically, a probabilistic assessment is a hybrid assessment, often using distributional data for some commodities and point estimates data for other commodities. The probabilistic assessment occurs when the data distributions are sampled on a probability basis, meaning the likelihood of a given value being sampled is according to the probability of that value occurring in "nature."

One difference between the US and the EU approach to probabilistic dietary risk assessment is in the method of sampling. In the US approach, probabilistic sampling is done for the residue distribution. However, Exponent's Dietary Exposure Evaluation Model (DEEM™; [10]) currently used by the EPA for dietary risk assessment uses <u>all</u> of the food consumption data. In the EU, it appears that probability sampling may be performed from both the residue and consumption distributions [11]. In the US, the CARES [12] and LifeLine [13] approaches apparently sample from both the residue and consumption distributions.

3 Food Consumption Data

Clearly, the data for food consumption is a very important component of the dietary exposure model. There are some differences in the approaches taken in the US and in the EU, although food consumption data are not collected or treated the same way across the EU. In addition to differences in survey design and collection methods, the organization of the data can also have an impact upon the dietary risk assessment.

3.1 Methods of Obtaining Food Consumption Data

This discussion is not intended to be an exhaustive treatment of food consumption survey techniques, bur rather an overview of the ways in which the food consumption component of the dietary risk calculation differs between the US and the EU.

Food consumption surveys are often conducted using the dietary recall or food diary methodologies. In the dietary recall technique, either direct face-to-face interviews or telephone interviews may be used to obtain food consumption data. Consumption amounts frequently are estimated using standard equipment, pictures, and diagrams to aid participants in estimating the amount of food consumed. Data from such recall surveys may be used for both chronic and acute dietary risk assessment. The food consumption surveys on which the US dietary risk assessment models are based are of the dietary recall type [14].

The food diary method of collecting food consumption data requires survey participants to maintain a journal of the foods they consume. The food diary methodology is much more rigorous than the dietary recall. One variation of the food diary method requires survey participants to measure (perhaps by weighing) the amount of food they consume. The UK food consumption surveys use the food diary technique, and participants are required to weigh the amount of food that is consumed [15,16,17,18].

This paper has treated the food consumption data from the UK almost as representative of EU food consumption data in general. Doing so, however, is mainly out of convenience at this time. The UK food consumption database is well documented, among the most readily available in electronic format, and suitable for detailed distributional analysis. The 97.5th percentile consumption estimates from the UK database have formed the basis of the NESTI acute dietary exposure estimates discussed previously [8]. In addition to the UK data, a Dutch dietary exposure model using Dutch data is under development, but is not readily available at this time [19]. The German data have also been used to estimate chronic dietary exposure [20].

In the UK, the Ministry of Agriculture, Fisheries, and Food and the Department of Health have conducted four sets of food consumption surveys between 1983 and 1997. These surveys make up the UK's National Diet and Nutrition Survey (NDNS) program. The UK has conducted surveys for infants [15], schoolchildren [16], adults [17], and most recently, young people [18]. The food consumption information from these surveys is used for the various UK dietary exposure models, including the TMDI and the NESTI calculations. Elsewhere in the EU, dietary risk assessments may be based upon national food consumption surveys, such as the German data [20], or upon the regional diets developed by the FAO/WHO GEMS/FOOD program [7]. At this time, the UK database provides the only food consumption data that is regularly used to estimate acute dietary exposure. However, the Dutch data also may be suitable for use in assessments of acute dietary exposure [19].

In the US, the food consumption component for dietary risk assessment is provided by data from the Continuing Survey of Food Intakes by Individuals [21-27]. Unlike the UK surveys, the US surveys are designed as an integrated whole that samples the entire population during the same sampling period. The data may be sorted into categories according to gender, age, region, ethnicity, and so on. The current design of the CSFII collects two days of data from each survey participant, with the days of collection being separated by three to ten days. In contrast, the UK surveys sample consumption for either four or seven consecutive days.

The UK and US surveys differ in their general approach to sampling. In the US, the household is the sampling unit, and data are collected for all members of the selected households. In the UK, however, census and other records are used to identify the types of individuals that are targeted, for example, 6-year old boys. Households are queried, and if a 6-year old boy resides in the household, consumption data are collected, but only for that individual. If a 6-year old boy does not reside in the household, data are not collected for anyone in the household.

The US food consumption database encompasses the entire country. Different consumption patterns may be distinguished for different population groups, defined according to gender, age, region of the country, ethnicity, and so on. There is not, however, a single food consumption survey that encompasses all of the national cuisines within the EU. It does seem, however, that if detailed recipe translation schemes could be developed, it should be possible to consider EU-wide consumption patterns on the basis of raw agricultural commodities.

4 Pesticide Residue Data

Dietary exposure is calculated as the very simple product of the amount of food consumed and the magnitude of the residue on the food. Unlike the food consumption value, which often is available from only one source (a government survey), the value of the residue portion of the equation may be provided from a wide variety of sources. These values range from the tolerance or MRL value, to data obtained from residue monitoring programs. The use to which various residue data may be put is often dictated by specific science policies within a given country.

4.1 Maximum Residue Limits and Tolerances

Almost always, the US and the EU want to consider potential risks resulting from worst-case dietary exposure assessments. These assessments typically use the tolerance or MRL value for the residue portion of the dietary exposure equation. In the NESTI calculation, the highest residue (HR) is used. The point of each of these calculations, however, simply is to provide the risk manager with an estimate of high-level exposures. The details may differ slightly between the US and the EU, but the general objective is the same.

4.2 Residue Data from Field Trials

When, as is often the case, dietary exposure estimates calculated with tolerances or MRLs exceed acceptable toxicity limits, revised dietary exposure estimates may be calculated using residues derived from field trials. For chronic dietary exposure, the mean field trial residue is used in the US, and the median residue (STMR – Supervised Trials Median Residue) is used in the EU. In addition, as shown in the NESTI calculation, the high residue from field trials can be substituted for the tolerance or MRL value.

In both the US and the EU, refined estimates of chronic dietary exposure are based upon estimates of the central tendency. For estimating acute dietary exposure, estimates of high-end residues are used. In the US, it is relatively common to conduct a probabilistic dietary risk assessment, whereas this procedure is less common in the EU.

4.3 Residue Monitoring

In most countries, residue monitoring programs only survey the general food supply to confirm that illegal residues are not detected. The potential utility of monitoring data for dietary risk assessment, for the most part, remains untapped in the EU [28]. In the US, monitoring data are used frequently in dietary risk assessment [29]. One US monitoring program, the Pesticide Data Program, was designed specifically for use in dietary risk assessment [30-33].

5 Additional Risk Assessment Factors

5.1 Processing Factors

Processing factors represent the influence of various processing steps, such as juicing, or cooking on the magnitude of the residue. A processing factor is the ratio of the residue concentration in the processed commodity to the concentration in the raw agricultural commodity. For example, if residues in apples are 1 ppm and 0.12 ppm in apple juice, the apple juice processing factor is 0.12. Processing factors are used in both the US and the EU, and in both chronic and acute dietary risk assessments. Once a processing factor has been calculated, it can be used to modify any residue value used in the assessment.

In order for processing factors to be used to best advantage, food consumption data for processed foods are needed. The US and UK food consumption databases both report consumption values for processed commodities such as juice, sauce, or purée. The risk assessment models used in the US are somewhat more detailed with respect to processed food forms than those in the UK. Some of the other food consumption data used in the EU do not include much information for processed foods.

5.2 Pesticide Use Data

In the US, surveys of percent crop treated are conducted by the federal government, by some state governments, and by pesticide manufacturers. The utility of information about the percent of the crop that is treated is probably one of the areas in which the EU and US differ most. In the US, estimates of the percent of the crop that is treated are used for both chronic and acute dietary risk assessments. In acute assessments, crop treated information is used in the probabilistic sampling to define the probability of consuming treated food. In the EU, information about the percent of the crop that is treated is used much less often than it is in the US. One reason for the hesitation to use this type of data is the opinion that the food distribution patterns within Europe cannot be determined or described easily.

It may be more reasonable to consider the percent of the commodity supply that is treated instead of the percent of the crop that is treated. By doing so, we can account separately for the amount of domestically produced commodity that is treated, as well as the amount of commodity imported into the country that is treated. Using percent of the commodity supply that is treated should allow EU dietary exposure assessments to avoid the practice of assuming that 100% of a commodity is treated. If evidence indicates that the entire food supply in a particular area is treated, additional assessments may be conducted with that assumption.

6 Comparative Assessments

Several assessments were conducted to illustrate the impact of different procedures on dietary risk assessment. In all cases, consumption data from the UK surveys were used. One of the differences between the US and the EU is the food consumption data. However, conducting assessments with both US and UK food consumption data will confound the comparisons, so the assessments will be run using only the UK food consumption data. All exposure estimates are presented as percent of the chronic Reference Dose (cRfD) of 0.005 mg/kg bw/day or the acute Reference Dose (aRfD) of 0.01 mg/kg bw/day (both toxicity values are hypothetical for illustrative purposes).

Hypothetical residue data for an exemplary chemical, CHEMX, were derived from actual residue data for four crops: apples, peppers, potatoes and tomatoes. The data are presented in Table 1.

Table 1. Summary Of Input Data For Dietary Risk Assessments For Chemx

	Apple	Pepper	Potato	Tomato
Number of Residue Values	16	16	16	16
Residue Distribution (ppm)	0.057	0.015	0.037	0.03
	0.089	0.024	0.04	0.037
	0.092	0.036	0.043	0.047
	0.32	0.077	0.072	0.047
	0.019	0.01	0.332	0.047
	0.02	0.01	0.031	0.05
	0.03	0.01	0.045	0.11
	0.045	0.01	0.065	0.18
	0.05	0.01	0.075	0.01
	0.06	0.012	0.24	0.01
	0.084	0.016	0.31	0.02
	0.09	0.03	0.71	0.023
	0.14	0.04	0.075	0.025
	0.022	0.049	0.24	0.03
	0.022	0.01	0.31	0.059
	0.026	0.13	0.71	0.06
Mean Residue (ppm)	0.084	0.03	0.21	0.05
Median Residue (ppm)	0.06	0.016	0.075	0.042
Percent Commodity Supply Treated	33	2	11	23

For chronic dietary risk assessment, the TMDI was calculated using the UK methodology – namely the 97.5th percentile consumption estimates for the two most highly consumed foods and median residue values. The UK TMDI does sum all the exposures for all four commodities. A second chronic risk assessment analogous to the TMDI was conducted using Exponent's DEEM™ model and the US procedures: mean consumption values and mean residue values. Finally, a TMDI calculation was conducted following US risk assessment procedures, incorporating the percent of crop treated data (from Table 1) into the assessment. None of the assessments made any adjustments for the effects of processing. The results of these calculations are presented in Table 2. None of the chronic risk assessments exceeded the cRfD, although the assessment that incorporated percent crop treated did have the lowest exposure estimate.

Table 2. Summary of Chronic Dietary Exposure Estimates For Children Expressed as Percent of The Chronic RFD Of 0.005 mg/kg bw/day

Type Of Assessment	Estimated Chronic Exposure
UK TMDI Methodology	21%
US Methodology	25%
US Methodology With Percent Crop Treated	4%

NESTI calculations were calculated using the procedures discussed in section 2.4.1 above. The results of these calculations are shown in Table 3. As seen in Table 3, all dietary exposure estimates calculated using the NESTI methodology, except for peppers, exceed the aRfD. Single commodity probabilitistic assessments assuming that 100% of the crop is treated result in acceptable exposure estimates for all commodities except potatoes. Single commodity probabilistic assessments that use the percent crop treated data from Table 1 resulted in acceptable risk levels for all four commodities.

Table 3. Summary Of Acute Dietary Exposure Estimates Expressed As Percent Of Acute RfD

COMMODITY	NESTI	100 PERCENT COMMODITY TREATED	USING PERCENT COMMODITY TREATED
Apple	443%	45%	25%
Pepper	5%	4%	< 1%
Potato	1078%	120%	73%
Tomato	112%	19%	7%
Total Diet	Not Applicable	122%	67%
Probabilistic assessments conducted for			

Two assessments were conducted using the US procedures with the UK food consumption database and the DEEM-UK™ model, in which a total dietary exposure estimate is calculated for all four foods at the same time. When 100% of the crop was assumed to be treated (so that probabilistic sampling was from the residue distributions), the resulting exposure estimates resulted in unacceptable estimates of risk. When percent crop treated was included in the assessment, the probabilistic assessment resulted in acceptable risk levels for all four commodities at the same time.

7 Conclusions

This brief review has shown that there are some differences in the way dietary risk assessment is practiced between the US and the EU. Many of the differences have to do in the types of input data that are available, or in some of the methods used to collect the data. Regarding the food consumption data, the US surveys use a dietary recall method, whereas the UK approach is a diary method that weighs and measures the amount of food consumed. In the US, data are collected for two non-consecutive days, whereas in the UK surveys, data are collected for either 4 or 7 consecutive days. The US survey is conducted as an integrated whole and includes all segments of the population. In contrast, the UK surveys are conducted for specific population groups. Finally, the US survey collects data at the household level, whereas the UK survey targets demographic characteristics, based upon census information.

With respect to residue data, the US procedures have used residue data from monitoring programs extensively. In fact, the Pesticide Data Program was designed specifically to provide data for dietary risk assessment. In contrast, monitoring data are not routinely used for dietary risk assessment in the EU.

Regarding the models used for dietary risk assessment, the most notable difference between the US and the EU is the routine use of probabilistic techniques in the US. In the EU, probabilistic techniques are currently being investigated. However, the probabilistic approach is not common in the EU at this time.

Linked to the differences between the US and the EU regarding the use of probabilistic techniques is the use of estimates of percent crop treated in risk assessment. Percent crop treated information is commonly used in dietary risk assessment in the US. This is generally not the case in the EU, although considering the percent of the total commodity supply that is treated may be a better conceptual model for the EU.

In spite of these differences in procedure and data utilization, the basic goals of dietary risk assessment are the same in the US and the EU – to protect the populace from unsafe pesticide levels in food, while permitting agriculturists to successfully protect their crops from harmful pests. It should be noted that this paper was written for the proceedings of the 10th Conference of the International Union of Pure and Applied Chemistry in Basle, Switzerland. Consequently, the paper was intended to provide a brief overview of the topic. A comprehensive review was recently published that provides much more detail about the subject [34].

8 References

[1] US Congress Federal Food, Drug, and Cosmetic Act (FFDCA) 21 U.S.C. 301 *et seq.* 1938.

[2] US Congress (1947) Federal Insecticide, Fungicide and Rodenticide Act (FIFRA); 7 U.S.C. s/s 135 *et seq* (1972).

[3] US Congress (1996) Food Quality Protection Act (FQPA) Public Law 104-170, Aug 3 1996.

[4] EC (1991) Council Directive 91/414/EEC of 15 July 1991 concerning the placing of plant protection products on the market. OJ L 230, 19 August, 1991. European Commission.

[5] US EPA (1996) Final office policy for performing acute dietary exposure assessment. Memorandum from D. Edwards to EPA OPP Staff 13 June 1996. US Environmental Protection Agency.

[6] US EPA (2000). Available EPA Information on Assessing Exposure to Pesticides in Food--A User's Guide (Revised) (PDF). http://www.epa.gov/fedrgstr/EPA-PEST/2000/July/Day-12/6061.pdf.

[7] WHO (1997) Food consumption and exposure assessment of chemicals. Report of a FAO/WHO consultation, Geneva Switzerland, 10-14 Feb 1997. WHO/FSF/FOS/97.5.

[8] PSD (1999) Guidance on the estimation of dietary intakes of pesticides residues. Pesticides Safety Directorate UK. 4 November 1999.

[9] C. A. Harris, J. R. Mascall, S. F. P. Warren, S. J. Crossley (2000). Summary report of the international conference on pesticide residues variability and acute dietary risk assessment. *Food Additives and Contaminants* 17(7):481-485.

[10] Novigen Sciences, Inc (now, Exponent, Inc.) (1996) Dietary exposure evaluation model (DEEM™).

[11] Development, validation and application of Stochastic Modelling of human exposure to food chemicals and nutrients. http://www.tchpc.tcd.ie/projects/montecarlo/.

[12] Cumulative & Aggregate Risk Evaluation System Conceptual Model (1999). http://alphacares.org/contents/CARES_Concept_Paper.pdf.

[13] The LifeLine Group (2001). Phase 1 Report: Findings from Literature Search and Review of Modeling Projects Currently Available or Under Development.

[14] USDA (1997a) Design and operation: continuing survey of food intakes by individuals 1994-96. Food Survey Research Group, Agricultural Research Service (ARS), United States Department of Agriculture.

[15] J. Gregory, D. Collins, P. Davis, J. Hughes, P. Clarke (1995) The national dietary and nutritional survey: children aged 1½ to 4½. HMSO.

[16] J. Gregory, K. Foster, H. Tyler, M. Wiseman M (1990). The dietary and nutritional survey of British adults. HMSO.

[17]A. Mills and H. Tyler (1992) Food and nutrient intakes of British infants aged 6-12 months. HMSO.

[18] Office for National Statistics Social Survey Division et al., National Diet and Nutrition Survey : Young People Aged 4 to 18 Years, 1997. [computer file]. Colchester, Essex: UK Data Archive [distributor], 25 January 2001. SN: 4243.

[19] Minutes of the EFSN/ERIS workshop held 28-29 January 2002, Brussels. http://www.efsn.net/efsnnieuw/meetings/Minutes%20of%20meeting%2028%20jan%202002.htm

[20] Biologische Bundesanstalt für Land-und Forstwirtschaft (BBA) (1993). Guideline Part IV, 3-7 (January, 1993).

[21] USDA (1992) Nationwide food consumption survey: continuing survey of food intakes by individuals 1989-90. Human Nutrition Information Service, United States Department of Agriculture. Dataset.

[22] USDA (1993). Nationwide food consumption survey: continuing survey of food intakes by individuals 1990-91. Human Nutrition Information Service, United States Department of Agriculture. Dataset.

[23] USDA (1994) Nationwide food consumption survey: continuing survey of food intakes by individuals 1991-92. Human Nutrition Information Service, United States Department of Agriculture. Dataset.

[24] USDA (1995) Nationwide food consumption survey: continuing survey of food intakes by individuals 1993-94. Food Survey Research Group, Agricultural Research Service (ARS), United States Department of Agriculture. Dataset.

[25] USDA (1996a) Nationwide food consumption survey: continuing survey of food intakes by individuals 1994-95. Food Survey Research Group, Agricultural Research Service (ARS), United States Department of Agriculture. Dataset.

[26] USDA (1997b) Nationwide food consumption surveys: continuing survey of food intakes by individuals 1995-96. Food Survey Research Group, Agricultural Research Service (ARS), United States Department of Agriculture. Dataset.

[27] USDA (2000) CD-ROM: CSFII 1994-96, 1998 Data Set, National Technical Information Service Accession number PB2000-500027. Food Survey Research Group, Agricultural Research Service (ARS), United States Department of Agriculture. Dataset.

[28] SCPH (1998) Monitoring for pesticide residues in the European Union and Norway. Report 1996. Standing Committee on Plant Health.

[29] US EPA (2000). Data for Refining Anticipated Residue Estimates Used in Acute Dietary Probabilistic Risk Assessments (Revised) (PDF). http://www.epa.gov/ pesticides/trac/science/residues.pdf

[30] USDA (1996b) Pesticide data program (USDA PDP). Annual summary calendar year 1994. Agricultural Marketing Service. Science and Technology Division, United States Department of Agriculture.

[31] USDA (1997c) Pesticide data program (USDA PDP). Annual summary calendar year 1995. Agricultural Marketing Service. Science and Technology Division, United States Department of Agriculture.
[32] USDA (1998) Pesticide data program (USDA PDP). Annual summary calendar year 1996. Agricultural Marketing Service. Science and Technology Division, United States Department of Agriculture.
[33] USDA (1999) Pesticide data program (USDA PDP). Annual summary calendar year 1997. Agricultural Marketing Service. Science and Technology Division, United States Department of Agriculture.
[34] R. Kroes, D. Müller, J. Lambe, M. R. H. Löwik, J. van Klaveren, J. Kleiner, R. Massey, S. Mayer, I. Urieta, P. Verger, A. Visconti (2002). Assessment of intake from the diet. Food and Chemical Toxicology 40:327-385.

28 MRLs and International Trade, a Developing Country Perspective

Cheah Uan Boh

Strategic Resources Research Centre, Malaysian Agricultural Research & Development Institute (MARDI), P. O. Box 12301, 50774 Kuala Lumpur, Malaysia.

1 Pesticide Residues in Food Moving in International Trade

Pesticide residues in food remain a trade problem for many developing countries, as is evident from the data on trade violations from the GEMS/Food Programme. Data were derived from importing countries from 1993 to 1998. The bulk of the data were from the USA, with additional information from Germany, Australia, Japan and Austria. The data were analysed according to pesticide-commodity combinations, frequency of violations, the total number of violating lots, exporting region and whether national or Codex MRLs are available (Table 1). Based on these criteria, commonly encountered insecticides include methamidophos, BHC, chlorpyrifos, permethrin, acephate, endosulfan, hexachlorobenzene, lindane and monocrotophos. Using the same criteria as for insecticides, residues of iprodione, thiabendazole, dithiocarbamates, diphenylamine, imazalil, procymidone and prochloraz also commonly result in trade problems.
It has been shown that pesticide-commodity combinations encountered in trade problems varied depending on the period sampling is conducted. Insecticides such as azinphos-methyl and aldicarb, which were identified as common insecticides facing frequent violations during the 1990–1994 period [1], were no longer posing a problem in the updated data (Table 1). Chlorothalonil, captan and vinclozolin identified as fungicides frequently involved in trade violations in the similar period, were not apparent in the 1993-1998 data. It can however, be concluded that pesticide residues remain a problem in agricultural commodities moving in trade.

2 Lack of National/Codex MRLs

It was apparent from the compilation of pesticide-commodity combinations of trade violations that national or Codex MRLs of these pesticides for many commodities, both minor and major, are not available (Table 1). Commodities of particular interest to developing countries where national or Codex MRLs are lacking include mango, papaya, banana, peppers, chillies, paprika, Chana dall (chickpea) and sesame seed.

MRL values are set according to good agricultural practices which differ in different countries as the type and incidence of pests and diseases are different and so are the treatment regimes. The MRL serves as a good guidepost for assessing the extent of pesticide contamination in food. It is also used to facilitate the movement of commodities in international trade.

Codex MRLs are often used as reference standards for the import and export of commodities by many countries, especially developing countries. Setting of MRLs demand vast financial and technical resources and these are often established by developed countries only. Residue trials are usually conducted for major crops in most countries by the agrochemical industry. Although this helps to overcome several problems pertaining to crops providing substantial economic returns, it does not address the problem of setting MRLs for minor food crops.

Codex MRLs are not available for many agricultural crops originating from developing countries. There are no MRLs of pesticides for starfruit and only one for passion fruit. For papaya, which is traded in substantial amounts in regional and global trade, Codex MRLs are available for chinomethionat, dithiocarbamates and prochloraz only. There are a limited number of Codex MRLs for several other crops such as pineapple, mango and watermelon.

Realising the importance of food safety and international trade, national governments worldwide are taking steps to address the issue. Efforts are also being made by the less developed economies to plan , initiate and implement development projects pertaining to food safety where pesticide residues are concerned.

3 Non-Tariff Barrier

As a result of a lack of international reference standards for many commodities, particularly those produced in developing countries, pesticide residues have increasingly become a non-tariff barrier in international trade. This has been apparent in several recent episodes involving the movement of minor crops in both regional and international trade. The export of many tropical fruits to Europe and the US is subject to the limit of determination for enforcement when national tolerances are not available. Though outright rejection usually does not take immediate effect, notice of the intention is always served to relevant national authorities. Many countries in the developing world are solely dependent on the Codex standards and in the event of their non-availability, the limit of determination is almost automatically applied. Such a practice is a common phenomenon in many developed countries where national MRLs for such commodities are not available. With the resources at their disposal, developed countries should however, conduct a dietary intake risk assessment. Very often rejection of a commodity is based solely on the limit of determination of a particular pesticide on the commodity. In most cases, the levels of pesticide residues found would probably not be a public health issue if subject to appropriate risk assessment; the current practice amounts to erecting a non-tariff barrier to trade.

4 Issue on Minor Crops

Minor crops such as tropical fruits, vegetables, spices, tea, etc. are traded in substantial amounts in international trade. The total amount traded is usually small in comparison with other major crops. These may however, contribute significantly to the national coffers of small developing countries. As illustrated in Table 1, Codex MRLs for many of these commodities are not available for reference as international food standards. This may be attributed to inadequate baseline data on residue trials and the low priority accorded by Codex to the elaboration of these MRLs.

The generation of science-based data for the elaboration of international standards should not be compromised. Flexibility however, should be allowed in some cases, for crops such as spices and tropical fruits where consumption is extremely low. In this regard, the recent agreement at the 34th Session of the WHO/FAO Codex Committee on Pesticide Residues to establish Codex MRLs for spices based on monitoring data is a step in the right direction. For some minor crops, 3 residue trials (sampling at recommended PHI only) may suffice if this is feasible without compromising on the scientific principles on the setting of Codex MRLs. An in-depth evaluation of this possibility should be seriously examined by the JMPR.

5 Portions of Commodities to which MRLs Apply

Many commodities traded internationally are not classified in the Codex Classification of Commodities. The portion to which MRL should be applied for international trade is not clearly defined. Since there is no international agreement on this, an arbitrary imposition of portions to which MRLs apply to reject or accept a crop is a common practice especially in developing countries. This has occurred on several occasions and has given rise to unnecessary trade dispute. These commodities should be elaborated at the international level and portions of the commodity should be clearly defined. The MRL for mango for example, is clearly defined as the whole fruit, but analysis is conducted without the seed for practical reasons. This however, is not applicable to other tropical fruits such as durian, jackfruit, lychee, duku, langsat, wax apple, rambutan, etc.

6 Regional WHO/FAO JMPR Evaluation

The possibility of regional evaluation of residue trial data for minor crops should be explored by the WHO/FAO and JMPR. This could ameliorate the problem of low priority accorded to the elaboration of minor crops moving in international trade.

Guidelines for data submission for minor crops such as spices, chickpea, sesame seed, etc. should be re-examined in view of the low consumption of these commodities. These could be re-established by the JMPR for regional evaluation. These guidelines could also be used by national governments to conduct supervised residue trials for the purpose, as the agrochemical industry would usually not support data submission for these crops due to obvious economic reasons. A senior member of the JMPR could be employed to guide the progress of these meetings where data on minor crops could be discussed, evaluated and MRLs elaborated. The CCPR should remain the risk management body. The regional effort could reinforce the efficiency of the Codex Alimentarius Commission by establishing more MRLs particularly for regionally traded commodities. The procedure could also be a capacity building process for developing a greater pool of expertise for regional evaluation and elaboration of Codex standards. The expected elevated cost for the process could be offset through cost-sharing by national governments, with the WHO and FAO playing a major coordinating role. International trade could be further facilitated through these efforts with the inclusion of additional Codex MRLs. MRLs for many of the crops in regional trade could be elaborated via this mechanism, and sound science should never be compromised in the process.

7 Continued Use of Deleted Codex MRLs

Codex MRLs which have been deleted arising from the lack of commitment for support by national governments, industry and other organizations could continue to serve as reference standards in international trade, particularly among the developing countries themselves, at least for a further 5 years after the official date of deletion, provided that the GAP has not changed and that the deletion was not the result of human health concerns. An understanding on this between the exporting and importing countries would further enhance the facilitation of international trade.

The number of Codex MRLs deleted each year far outnumbers the enactment of new ones, as the process of establishing international standards is a long and arduous one, and may result in unwarranted trade vulnerabilities particularly for developing countries.

8 Extrapolation

Extrapolation for major crops is almost impracticable. Developing countries should therefore, take steps to generate residue trial data for the elaboration of MRLs where Codex MRLs are not available for commodities of economic importance to their countries. Collaborative regional efforts could play a pivotal role in meeting this objective. Under the framework of the WTO's SPS Agreement, Codex recommendations relating to food safety should be based on sound science and risk assessment.

The Codex classification of crops could be reviewed to include commodities from developing countries. The classification may be revised in a major way to take into consideration the possibility of extrapolation for minor crops.

9 Process of Elaboration of Codex MRLs

The current Codex procedures of establishing MRLs consume a period of between 4-8 years and may result in potential trade disruption. This is evident from several incidents involving the export of agricultural produce, particularly those from developing countries, which have experienced trade problems in the past. Negotiations on a government to government basis have curtailed but not eliminated the problem entirely.

Current procedures should be improved to further facilitate the process of setting Codex MRLs. In order to avoid potential trade disruption, the process should be improved to between 2-4 years, and should not exceed 4 years to provide any practical value. The established Codex MRLs should stay longer than 10 years before a periodic review takes place. To avoid trade vulnerabilities, interim MRLs should be made available immediately after the JMPR evaluation. This should remain for 2-3 years, a period sufficient for further comments and evaluation by national governments and other organizations. Importing countries should justify with science-based evidence for the non-acceptance of these interim MRLs.

10 Level of Acceptance of Codex MRLs

Almost all developing countries are dependent on Codex MRLs for the import of foodstuff into their countries. While the level of acceptance is extremely high in these countries, which consider Codex MRLs as safe to human health as these have undergone elaborate risk assessment evaluation. Some developing countries however, for lack of understanding and information on the GAP for which these MRLs are established, did not consider the GAP in their countries when incorporating Codex values into their national legislation for food safety.

The USA and the EU, the two biggest trading blocs for developing countries however, often establish tolerance limits which differ from those of Codex. The level of acceptance of Codex MRLs is these countries is generally low. The reasons for this should be identified and rectified in the spirit of the WTO Agreement and for promoting and promulgating international fair trade practices. Tremendous amounts of resources have been channeled into the process of elaborating Codex MRLs and countries should harmonise with Codex MRLs wherever possible.

11 Harmonisation of MRLs

Harmonisation with Codex MRLs is a key process in the facilitation of global trade. Codex MRLs should remain the international reference standards for pesticide residues in food should a trade dispute arise, unless a clear and sound science-based assessment showed an intake likely to present a public health concern.

Many countries use differing MRLs in the enforcement of pesticide residues in food, which further aggravates the problem of unfair trade practices.

12 Regional Cooperation/Role of Regional Codex Committee

Regional cooperation remains a mechanism to overcome many of these problems for developing countries. Individual countries have limited resources to deal with the problem. An example is the regional cooperation on the harmonization of MRLs of pesticides among the Association of South East Asian Countries. Collaborative efforts are also being made to generate data on crops of common interest for the purpose of setting maximum residue limits of pesticides.

Regional Codex Committees could play an important role in the coordination of field trials to be conducted for the purpose of setting MRLs. These countries could embark on studies pertaining to their respective countries' interest and a coordinating effort be made on the pesticide-commodity combinations, timing of the trials, expected time for the results and the JMPR could be informed on the schedule of these for evaluation and elaboration. Technical assistance from the developed countries would be extremely beneficial in achieving the objective.

13 Long-term Remedial Measures

Developing countries should take efforts to initiate short and long-term programmes to overcome the problem of the lack of Codex MRLs for commodities frequently violating regional or international trade. A long-term objective of generating residue trial data for the elaboration of Codex MRLs of major crops should be set.

Developing countries should identify on a priority basis minor crops of economic importance to them. These could be put forward to the CCPR for inclusion in the Priority List or directly to the JMPR for consideration of extrapolation. The submission must also be accompanied by detailed information as outlined in the FAO Manual on *Submission and Evaluation of Pesticide Residues Data for the Estimation of Maximum Residue Levels in Food and Feed* for the estimation of group MRLs [2].

Table 1. Pesticide-commodity combinations frequently encountered in international trade.

Explanations: I = number of lots; II = National Status; III = Codex Status; (a) processed food in italics; (b) N = no MRL or permitted level exist; A = Residue exceeds permitted level for environmentally persistent pesticide that is no longer used; X = Residue exceeds MRL or permitted level; OK = Level under MRL; (c) Latin America includes Caribbean Region

Pesticide	Commodity (a)	I	II (b)	III (b)	Exporting Region (c)	Note (Codex MRL)
(002) Azinphos-methyl	Pepper, Serrano	1	X	OK	Latin Amer.	1 mg/kg: sweet pepper
(007) Captan	Grapes	1	N	N	Latin Amer.	25 mg/kg, Step 3
	Physalis	2	X	N	Latin Amer.	
(017) Chlorpyrifos	China celery	1	X	OK	Asia	0.05 mg/kg: celery
	Dill	1	X	N	Asia	
	Grapes	7	X	X	Asia	1 mg/kg
	Honeydew melon	1	N	N	Latin Amer.	
	Mushroom	1	X	X	Asia	0.05 mg/kg
	Strawberry	2	X	N	Africa	
	Spinach	8	N	N	Latin Amer.	
(021) DDT, p,p'-	Sesame seed	1	N	N	Asia	
(025) Dichlorvos	Cucumber	2	X	N	Asia	
(026) Dicofol	Chili, dried	1	N	X	Asia	1 mg/kg
	Spice	1	N	N	Asia	
(001) Dieldrin	Pumpkin seed	2	X	N	Europe	
(027) Dimethoate	Chana dall (chick-pea), Chinese	1	X	N	Latin Amer.	
	Peach	8	N	OK	Latin Amer.	2 mg/kg
(030) Diphenylamine	Grapes	3	N	N	Latin Amer.	
	Banana	1	N	N	Latin Amer.	
	Pear	16	N	N	Latin Amer.	
	Pear	2	N	N	SW-Pacific	
	Peach	9	N	N	Latin Amer.	
(032) Endosulfan sulfate	Papaya	1	N	N	Latin Amer.	
(034) Ethion	Chillies, dried chillies	1	N	N	Asia	
	Indian chillies	10	X	N	Asia	
(039) Fenthion	Kaki (Japanese persimmon)	3	X	N	Europe	

Pesticide	Commodity (a)	I	II (b)	III (b)	Exporting Region (c)	Note (Codex MRL)
	Kaki (Japanese persimmon)	3	X	N	Unknown	
	Litchi	2	X	N	Africa	
(041) Folpet	Raspberry	1	N	N	Latin Amer.	
(044) Hexachlorobenzene	Carrot	7	N	N	Latin Amer.	ADI withdrawn
	Grape	2	N	N	Latin Amer.	ADI withdrawn
	Winter squash	2	N	N	Latin Amer.	ADI withdrawn
(047) Bromide (inorganic)	Rice	1	X	X	Asia	50 mg/kg: cereal grains
(048) Lindane	Chana dall (chick-pea)	10	N	N	Asia	
	Sesame seed	10	N	N	Latin Amer.	
	Sesame seed	10	A	N	Asia	
	Spice	1	N	N	Asia	
	Split peas	10	A	N	Asia	
(049) Malathion	Dill	1	X	N	Asia	
(054) Monocrotophos	Longan	2	N	N	Asia	
	Pepper, Pasilla	2	N	OK	Latin Amer.	0.2 mg/kg: peppers, chilli
	Peppers, hot	1	N	X	Latin Amer.	0.2 mg/kg: peppers, chilli
(055) Omethoate	Peach	4	N	N	Latin Amer.	2 mg/kg, Step 6
	Chana dall (chick-pea), Chinese	1	X	N	Latin Amer.	0.2 mg/kg: beans at Step 3
(060) Phosalone	Grapes	2	X	OK	Asia	5 mg/kg
(064) Quintozene (PCNB)	Carrot	3	N	N	Latin Amer.	
(065) Thiabendazole	Banana	4	X	OK	Latin Amer.	3 mg/kg
	Green beans	12	N	N	Latin Amer.	
	Kumquat	2	X	OK	Europe	10 mg/kg: citrus fruits
	Mango	2	X	N	Europe	
	Mango	15	X	N	Latin Amer.	
	Papaya	6	X	N	Latin Amer.	
	Peach	8	N	N	Latin Amer.	
	Pitahaya	2	X	N	Latin Amer.	
	Winter squash	2	N	N	Latin Amer.	
(069) Benomyl	Chana dall (chick-peas), green	2	X	X	Latin Amer.	2 mg/kg: common beans (covered by MRL for carbendazim)
(072) Carbendazim	Mango	2	X	OK	Africa	2 mg/kg
	Mango	4	X	OK	Latin Amer.	2 mg/kg
	Papaya	4	X	X	Latin Amer.	0.5 mg/kg (0.1 mg/kg, Step 3)
(081) Chlorothalonil	Grapes	2	N	OK	Latin Amer.	0.5 mg/kg
	Spinach	6	N	N	Latin Amer.	
(083) Dicloran	Spinach	3	N	N	Latin Amer.	

Pesticide	Commodity (a)	I	II (b)	III (b)	Exporting Region (c)	Note (Codex MRL)
(085) Fenamiphos	Tomato	2	N	OK	Latin Amer.	0.2 mg/kg
(094) Methomyl	Asparagus	1	X	X	Latin Amer.	2 mg/kg
	Fresh lychee nuts	1	N	N	Asia	
(095) Acephate	Apple juice	2	N	N		
	Melon, Honeydew	3	N	N	Latin Amer.	
	Spinach	6	N	N	Latin Amer.	
	Winter squash	4	N	N	Latin Amer.	
(096) Carbofuran	Potato	2	X	X	Unknown	
(096) Carbofuran-3,3-OH	Grapes	8	X	N		
(100) Methamidophos	Apple juice	6	N	N	Latin Amer.	
	Basil	1	N	N	Latin Amer.	
	Green beans	3	N	N	Latin Amer.	
	Custard apple	2	N	N	Asia	
	Green beans	7	N	N	Latin Amer.	
	Longan	2	N	N	Asia	
	Papaya	1	N	N	Latin Amer.	
	Snow pea	3	N	N	Latin Amer.	
	Spinach	14	N	N	Latin Amer.	
	Winter squash	9	N	N	Latin Amer.	
(103) Phosmet	Pepper, Serrano	1	N	N	Latin Amer.	
(104) Daminozide	Peanut	10	X	N	Asia	
(105) Dithiocarbamate	Grapes	3	X	OK	Latin Amer.	5 mg/kg
	Mango	4	X	N	Latin Amer.	2 mg/kg, step 6
	Papaya	4	X	N	Latin Amer.	5 mg/kg, Step 6
	Pear	2	X	X	Unknown	3 mg/kg
(110) Imazalil	Banana	3	X	OK	Latin Amer.	
	Kiwi	2	X	N	SW-Pacific	
	Kumquat	2	X	OK	Europe	5 mg/kg: citrus fruits
	Mandarin	2	X	OK	Europe	5 mg/kg: citrus fruits
	Pear	5	N	N	SW-Pacific	
(111) Iprodione	Apple	2	N	OK	Africa	5 mg/kg: pome fruits
	Apple	4	N	OK	Latin Amer.	5 mg/kg: pome fruits
	Dried chrysanthe-mum	1	N	N	Asia	
	Kaki (Japanese persimmon)	3	X	N	Europe	
	Kale, Chinese (Choin-san)	2	X	N	Unknown	
	Kumquat	2	X	N	Europe	
	Papaya	2	X	N	Latin Amer.	
	Pear	35	N	N	Africa	
	Pear	14	N	N	Latin Amer.	
	Pear	13	N	N	SW-Pacific	
	Tomato	6	N	OK	Latin Amer.	5 mg/kg
(117) Aldicarb sulfoxide	Beans, French	1	N	N	Latin Amer.	

Pesticide	Commodity (a)	I	II (b)	III (b)	Exporting Region (c)	Note (Codex MRL)
(118) Cypermethrin	Chillies, dried	1	N	OK	Asia	0.5 mg/kg: peppers
	Sultanas	2	N	N	Asia	
(119) Fenvalerate	Grapes	1	N	N	Latin Amer.	
	Spinach	6	N	N	Latin Amer.	
	Tea, green	4	X	N	Asia	
(120) Permethrin	Tea, green	2	X	OK	Asia	20 mg/kg
(120) Permethrin, cis	Long beans, Chinese	2	N	OK	Latin Amer.	1 mg/kg: common beans
	Passion fruit	1	N	N	Latin Amer.	
(123) Etrimfos	Malt	2	N	N	Europe	
(136) Procymidone	Coconut cream	1	N	N	Asia	
	Grapes	2	X	X	Asia	
(138) Metalaxyl	Mandarin	2	X	OK	Europe	5 mg/kg: citrus fruits
(142) Prochloraz	Sweet potato	2	X	N	Latin Amer.	
	Mango	2	X	X	Africa	2 mg/kg
	Mango	2	X	X	Europe	2 mg/kg
	Mango	2	X	X	Latin Amer.	2 mg/kg
(144) Bitertanol	Banana	8	X	X	Latin Amer.	0.5 mg/kg
	Beans	2	X	N	Africa	(0.5 mg/kg: common bean (pods and/or immature seeds)
(159) Vinclozolin	Tomato	6	N	OK	Latin Amer.	3 mg/kg
(178) Bifenthrin	Pear	2	N	OK	Latin Amer.	0.5 mg/kg
(181) Myclobutanil	Tomato	1	N	N	Latin Amer.	0.3 mg/kg at Step 3
(189) Tebuconazole	Kale, Chinese (Choin-san)	2	X	N	Unknown	
(191) Tolclofos-methyl	Kohlrabi	6	X	N	Unknown	
2-Chloroethanol	Paprika, Powder	12	X	N	Unknown	
BHC, Alpha	Chana dall (chick-pea)	10	N	N	Asia	
	Sesame seed	10	A	N	Latin Amer.	
	Split dal	10	A	N	Asia	
	Split peas	10	A	N	Asia	
	Urad, washed	10	A	N	Asia	
BHC, Beta	Sesame seed	20	N	N	Asia	
	Split dal	10	A	N	Asia	
BHC, Delta	Sesame seed	10	N	N	Asia	
Chlorpropham	Grapes	2	N	N	Latin Amer.	
Chlorthal	Kale, Chinese (Choin-san)	2	X	N	Unknown	
Chlozolinate	Kiwi	2	N	N	Unknown	

Pesticide	Commodity (a)	I	II (b)	III (b)	Exporting Region (c)	Note (Codex MRL)
Copper	Vine leaf	4	X	N	Asia	
DCPA	Spinach	4	N	N	Latin Amer.	
Esfenvalerate	Spinach	6	N	N	Latin Amer.	
Pentachlorobenzene	Carrot	4	N	N	Latin Amer.	
Quinalphos	Chillies, Indian	10	N	N	Asia	
	Chilli, red	10	N	N	Asia	
Tetradifon	Papaya	2	X	N	Latin Amer.	

References

[1] CCPR, 1996. 28th Session of the WHO/FAO Codex Committee on Pesticide Residues. Problems relative to pesticide residues in food in developing countries. CX/PR 96/11.

[2] FAO/WHO 2002. Submission and evaluation of pesticide residue data for the estimation of maximum residue levels in food and feed and on the estimation of group maximum residue levels.

29 Consumer Perception of Food Risk and Consumer Trust in Europe

Beate Kettlitz

BEUC, the European Consumers' Organisation, Av. de Tervueren 36, B-1040 Brussels, Belgium, E-mail: bke@beuc.org

1 Introduction

Access to safe food supplies continues to be of key importance in public health. Recent issues such as BSE, the use of antibiotics for other purposes than treatment of disease in animals, the dispute regarding hormones, recent food scandals, such as MPA or nitrofen in feed or foodstuffs and, at another level, the introduction of genetically modified foods have led many consumers in Europe to question the safety, quality, or wholesomeness of their food.

Recent developments have significantly affected the relationship of consumers to and their perception of the food they eat. These include

- Internal market/enlargement/globalisation:
  As a result of the Single Market, progressive enlargement and globalisation, food and food ingredients may come from more countries than before and travel through many more different paths to the ultimate consumer. In a single market such as the EEA and the enlarged EU there are no internal border controls. Consumers need to feel confidence in the production, processing and regulatory regimes of other countries.
- Eating and living habits:
  There are many changes in eating habits including a move towards greater convenience. More often than before consumers are eating more processed, more handled or more packaged food.
- Changes in production:
  For many reasons, food production and processing are continually changing. In many cases production is becoming more intensive and more industrialised with higher inputs at the production level, such as pesticides or fertilisers.
- Changes in distribution:
  Nowadays, even very small enterprises may supply ingredients for a very wide range of food products from different countries.

- Changes in composition: People consume a wider range of food, food ingredients and food additives than ever before.
- Commercial promotion:
  Food, food products, "snack foods" and the like are advertised and promoted as never before. The current proliferation of claims is likely to increase even more with the prospective increase in functional foods, fortified foods, food supplements and the like.
- Consumer sentiment:
  In Europe, consumer confidence in food safety is increasingly volatile. Environmental and other factors are becoming more important in consumers' minds (although the question, how important the are, is a matter of argument).

2 Food Safety

Again and again consumers are concerned about the safety of their food and this concern is affected by a range of factors including (but not confined to) "objective" safety characteristics. Against this background, the following physical sources of danger can be identified:

- Microbiological contamination, e.g. *Salmonella, Listeria, Camphylobacter, E. coli* and others
- Residues of pesticides and veterinary drugs in food, e.g. organochlorines, antibiotics
- Environmental contaminants, e.g. mercury, lead, dioxins
- Migration from material in contact with food
- Food additives, particularly those to which significant numbers of people are hypersensitive, e.g. sulphur dioxide, tartrazine
- Others, e.g. mycotoxins, nitrosamins

It is recognised, that the final consumer can influence all of these factors to a greater or lesser degree by handling methods, including the handling of food. However, the ranking of food safety concerns has changed over time. While back in 1985 additives and contaminants were in the spotlight, in 1995, according to a study of the Dutch Agricultural University Wageningen [1], it was microbiological spoilage that was of most concern. Nowadays it is BSE, but two years ago the dioxin crisis in Belgium ranked top of the list, not to forget the GMO debate, which obviously also has been triggered by societal concerns.

3 Risk Perception of Consumers

Ongoing public demand for an adequate and safe food supply has led to major developments in the field of health science, toxicology, analytical chemistry, microbiology, food

hygiene, nutrition, regulation and control. Researchers in these disciplines have interacted to minimise the potential health hazards to consumers by establishing standards, that both guarantee the safety and nutritional value of food items consumed and permit regulatory control measures to be implemented.
It is recognised that good science, which is considered accountable, independent, responsible, and transparent in its approach, is imperative for effective and trusted regulation, and that adequate control must be based on these regulations. Also a very simple rule can be recorded: risk acceptance is very much depending on perceived benefits. If consumers are to have confidence in food safety and the quality of food in Europe, it is important that a coherent and consistent approach is taken to analysing and controlling food risks. Also there should be consistency in the way standards are set in Europe, including the efficiency of the rapid alert system in Europe to identify emerging risks.

Table 1. Perceived Risk in Different Product Categories [2]

Food product category	% consumers feeling unsafe
Bread and bakery products	9.7
Cheese	12.6
Fresh fruit	15.2
Fresh milk	14.8
Fresh vegetables	15.2
Eggs	20.9
Fresh seafood	22.8
Fresh meat	31.9
Frozen Food	33.7
Canned food	39.7
Other processed food	44.2
TV dinners	50.7

Table 2. Prioritisation of Food Safety Measures by EU consumers [2]

Food Safety Measure	%
Justified restrictions on pesticides or hormones	21
No preservatives	28
Appropriate preservation methods	29
Additives below recommended level	33
Preservatives below recommended level	35
No additives	36
Testing for microbiological safety	49
No hormone residues	54
No pesticide residues	56

It is obvious that societal concerns play a major role in how the consumer perceives risks and benefits (Tables 1 and 2). Furthermore, consumers do not distinguish between food safety risks and diet-related risks. Their risk perception depends on different traditions and perceived benefits. It also depends on demographic characteristics such as age, sex, or education, as well as psychological characteristics such as knowledge, experiences, or

beliefs. Effective communication about food risks may be influenced by the extent to which people perceive the information source to be reliable. Trust in information influences risk perception:
Confidence in government information, for instance, has been badly shaken by the BSE crisis in the UK. The Lord Phillips report published in 2000 revealed that the risks were underestimated by the government and had not been taken seriously enough.
In relation to new products or production methods it is, therefore, of paramount importance to prevent suspicion through early two-way dialogue between consumers, scientists and decision-makers.

4 Pesticides in Agriculture

In May 2002 the Food and Veterinary Office of the European Commission published the report of the year 2000 pesticide residue monitoring program. The report highlighted the results of the analysis of pesticide residues in 45 000 samples of fruits, vegetables and cereals and revealed that 96% of the food samples did not exceed Maximum Residue Limits (MRLs). It is known in the consumer movement, that the coordinated program exists since 1996 and that it aims at checking the most important pesticide-crop-combinations. Although the results might not seem to be alarming, there is still a series of remaining questions, which express the need for

- harmonised monitoring and surveillance systems in Europe,
- comparable consumption data,
- harmonisation of substances allowed for use in Europe,
- the establishment of particularly designed control groups if residues are ubiquitous.

On top of it, it is important for consumers to have reassurance that special attention is being paid to

- vulnerable groups of populations,
- bound residues of pesticides and pharmaceuticals,
- the combined effect of multiple pesticide residues, including possible synergistic effects of multiple residues,
- multiple sources of exposure to pesticides.

Given the challenges that are involved in the use of pesticides in general and the risk perception by consumers in particular, it is all-important that pesticides must be used at levels "as low as reasonably achievable". Consequently, the use of chemicals for crop protection should be part of an integrated crop protection system:

- Specific targets should be designed at national levels to progressively achieve more stringent qualitative and quantitative reductions in the use of pesticides on specific crops and overall.

- Principles for Good Agricultural Practices (GAP) should be developed which specifically aim at reducing the dependency of agriculture on plant protection chemicals. This may be facilitated by establishing an EU-supported 'GAP research facility'.
- Least-harmful plant protection practices should be developed and promoted wherever feasible and practicable.
- Best environmental practices should be applied.
- The use of pesticides should ultimately give reason to appropriate control measures of the residue levels.
- To achieve this aim all players in the chain should have the appropriate knowledge about the responsible use of pesticides, and ongoing training should take place.

5 The Process of Risk Analysis

How best can consumers be given good reasons to have confidence in risk assessment and risk management, and how best can this be communicated to consumers?
The process of risk analysis consists of three individual circles: risk assessment, risk management and risk communication. They are all linked to each other and even overlap in tasks. Proper functioning of this process is of key importance.

5.1 Risk Management

Risk management is the process of weighing policy alternatives in the light of risk assessment results and other factors and, if required, of selecting and implementing appropriate prevention and control options including regulatory measures. The decision-making process should begin with the risk managers. They, in consultation with all stakeholders, should define and frame the food safety problem and develop a risk assessment policy. This policy should guide the risk assessors, for example by setting out their remit, who should participate, questions that need to be addressed, how uncertainties should be dealt with, factors that the assessors need to consider, and how they should express their findings. Based on the results of the risk assessment and other relevant factors, the risk managers should decide what actions need to be taken in consultation with all relevant stakeholders including consumers. It is essential that the reasons for policy decisions are transparent and clearly communicated. Furthermore, the decision making process must distinguish itself through transparency and accountability.

5.2 Risk Assessment

Risk assessment is the scientific evaluation of known or potential adverse health effects resulting from human exposure to food-borne hazards. The remit of any scientific committees should be regularly reviewed to ensure that the risk assessors are able to address the relevant issues raised by the risk managers, and that no issues are being missed.

- The appointment of members must be open and transparent, based on scientific merit and relevant experience.
- There should be full declaration of interests, both of the experts and the source of research upon which they base their decisions.
- Advisory committees should consist of a broad range of qualified experts to ensure that an issue can be considered by all disciplines that may have an interest.
- Public interest representatives, at least with an observer status, should be appointed to all scientific advisory committees based on merit and expertise in the field of consumer movement and preferably with a technical (scientific) background.
- Agendas and minutes of meetings should be published, and papers considered by the assessors should be publicly available. Also a review of what is considered to be 'commercially confidential' is needed with the onus on disclosure.
- Scientific committees should encourage 'minority' scientific evidence to aid their deliberations. Wherever possible meetings should be held in public.
- The scientific advice provided by the committees should be communicated to the public in an understandable form. If there are uncertainties, they should be explained.

As a reaction of the various food scares in Europe the European Commission successfully launched the establishment of a European Food Safety Authority (EFSA). Although its final seat has not yet been agreed upon, it is understood that the Authority will soon start operating from its temporary location in Brussels. The task of EFSA will be a complex one, potentially of immense value, but it must not be burdened with unrealistic expectations. It is obvious that the success of the Authority will strongly depend on its recognition by the public: to provide a unified, excellent and independent scientific advice based on an accountable and transparent operation. Thus, it would be unrealistic to expect, that the majority of European consumers would accept EFSA as the ultimate authority on food safety already in the near future.

5.3 Risk Communication

Risk communication is the exchange of information and opinions concerning risk and risk-related factors among risk assessors, risk managers, consumers and other interested parties. The goals of this communication are to describe risks and benefits, to conduct a fair and factually oriented debate about different types of actions, and to find solutions for conflicts between stakeholders.

It would be a mistake, however, to believe that all these goals could be achieved just by supplying information about risks. In fact, success depends on each of the following building blocks:

- The quality of the relationship between those involved.
- The design of the information.
- The shaping of the dialogue.

Without this foundation, risk communication is, at best, just a patchwork and, at worst, ineffective.

How difficult risk communication might be, depends on the complexity of the risk issues to be handled [3]:

- The first level is to provide factual, scientific information to the public. It is the duty of the risk communicator to explain the assessment tools and results, and to clear the uncertainties in a way that the public may develop a better understanding of the risks. In terms of substance, the aim is to communicate facts and knowledge. This requires both, a high scientific expertise and the ability to translate technical information into a common sense language.
- The second level of the communication deals with participation and conflict. Risk communicators should create a culture of fairness and mutual understanding. All stakeholders should be empowered to express their views. In social terms, the aim must be to enhance the quality of the relationship between all those participating in the communication, i.e. to develop the social climate for talking to each other.
- The third level is dealing with different worldviews, social values and concepts of "good life". This is the case when "quality of life" is a part of the risk assessment and the search for appropriate risk management options. Here, neither scientific expertise nor discursive competence and openness are sufficient. The main issues are trust and social responsibility.

Everyone agrees that sound science is an essential base for food safety decisions. There is also general recognition that some "other factors," such as economic costs and benefits, good agricultural, manufacturing and veterinary practices, practicalities of measurement and enforcement, and national trade interests should also be legitimately considered when establishing food safety policies. Factors that affect consumer preference, including ethical and religious concerns, the desire to promote more sustainable food production practices, and many other things that consumers value, may affect consumer health only indirectly, if at all. But they are definitely related to "promoting a fair practices in food trade," and as such they are clearly legitimate "other factors", that the decision-making process must recognise. Risk communication, therefore, needs to change from the traditional top-down approach to communicating information, to an inter-active two-way exchange of information that enables effective public participation throughout the risk decision process. This will require a commitment from risk managers to consult with consumers from the outset.

New methods of inter-active ways of debating issues, including perceived costs and benefits of specific measures must also be developed and fed into policy decisions. Discussions and decisions in the context of risk issues are interdisciplinary and must be tackled across boundaries of expertise for an intensive exchange of ideas. Risk communicators should be aware of and remember the different risk perspectives and take them into account in their risk communication efforts.

An equally intensive reconciliation of opinions is essential in order to understand the complex correlations involved, and to be able to make "acceptable" decisions and take "appropriate" actions.

Risk communication aiming at consumer trust depends on a positive appraisal of the competence, fairness, and social responsibility of the partners in the communication process. Without such a positive appraisal, risk communication is bound to fail.

As the *Eurobarometer* from 2001 publications clearly demonstrates, it is not only facts related competence that drives consumer trust, but its inclusion in ethical/moral challenges. Therefore, it is important that scientists, consumers and risk managers talk to each other. Scientific advice is invaluable for well-founded consumer opinions, and dialogue with scientists is important to enable evaluation of scientific questions relating to food safety concerns. European consumer organisations, through research and contacts among their members, are able to identify major issues of consumer concern. Consumer needs, attitudes and responses to new food production, regulatory developments and reactions in emergencies are central. Consumer inputs can shape scientific advice with the aim of ensuring that consumer issues are addressed in a constructive dialogue for positive outcomes, while identifying matters of consensus, disagreement, or areas for further research. It is important to use trusted sources to provide consumers with answers to their questions and refute inaccurate scare stories in an understandable language. Risk communicators should make use of mass media more effectively, and the use of Internet and interactions between science, culture and society should become an additional tool. The successful and efficient risk communicator should help people to appreciate that there is no risk-free solution in food production.

6 Conclusion

- Food legislation should be precautionary and comprehensive in scope, secure and efficient in enforcement, and should respect consumer expectations.
- Feeding stuff legislation should be redesigned to meet food safety standards and precautions.
- Transparency at all stages of the legislative process is essential.
- Public representation, which cannot be replaced by policy makers, should be assured. This would include proper consumer consultation.
- As a consequence early involvement of consumers in decision making is of key importance for both increased transparency and acceptance of future food policy.
- The interactive-two way dialogue with scientists, risk managers, and consumers, as well as the openness and transparency of scientific advice and the decision-making process are paramount.
- An effective system of communication with and toward consumers on risks and policy measures needs to be established.
- It is necessary that consumers feel that full and comprehensive actions are taken to tackle major food safety concerns.
- Sound monitoring systems need to be in place to identify sources of infection, to guide the risk assessors, and to measure the effectiveness of food safety measures.
- It is crucial that consumers feel that efficient controls are in place.
- Without fail: Suspicion must be prevented!

7 Summary

Recent food scares in Europe like BSE and the dioxin crisis have led to a dramatic decrease in the trust of the general public in food safety, the quality of its assessment and management, as well as in the ability of public authorities to regulate and control the safety of food supply. The increasing complexity of food chain production systems with changes in processing and distribution, the increased globalisation of trade, the internal market in Europe, changes in the composition of food, changing eating habits, and consumer sentiments influence discussions in society on food safety and quality. Early involvement of consumers in decision making is, therefore, of key importance for both increased transparency and acceptance of future food policy. Sound monitoring systems to identify the level of relative risk should underpin food policy decisions.
There is a great need for a coherent European approach to design adequate early identification systems for any approaching risks. An effective system of interactive communication with consumers, risk assessors, and risk managers should be established. Food policy must aim at creating the conditions, in which consumers have good reason to feel confident about the food they eat.

8 References

[1] C.S. Schlotjes, Voeding 1996, 57, 15-17.
[2] INRA Europe, Eurobarometer 49, Prioritization of Food Safety Measures by EU Consumers, 1989.
[3] European Commission, Directorate C - Scientific Opinions, Preliminary report on scientific quality of life criteria in risk benefit assessment, meeting 16 May 2002.

Keyword Index

Numbers behind keywords indicate chapter numbers